ENVIRONMENT AND MAN

ENVIRONMENT AND MAN

RICHARD H. WAGNER
THE PENNSYLVANIA STATE UNIVERSITY

NEW YORK
W · W · NORTON & COMPANY · INC ·

This book is printed on recycled paper.

"... To the wide world and all her fading sweets ..."
Shakespeare, *Sonnet 19*

CONTENTS

Part III MAN MAKES NEW TRAUMAS   222

Part IV THE BIOTIC WORLD AND MAN   276

Part V MAN'S URBAN ENVIRONMENT   340

PREFACE

Traditionally, many colleges and universities have advised their students to take at least one course in general biology, on the premise that their graduates ought to have some idea of the fundamental structure and physiological processes that govern their lives. Such courses on biological form and function can be lively, vital, and relevant if the instructor relates his material to the needs and interests of his students. All too frequently, however, they give short shrift to, or beg altogether, the critical question of the interaction of man with the lives and activities of other organisms and with the environment itself.

Because of our sudden awareness of the magnitude of man-environment problems, efforts are being made on many campuses to fill this curricular gap with courses that recognize man's tremendous impact on the natural world and analyze it from various points of view. Although there are scores of books covering the many individual aspects of the man-environment field, as well as symposia, anthologies, and exposés, there is no balanced text that pulls all the threads together. This book was written to fill that need.

My intent was not to write an exhaustive or learned treatise; any one chapter in this book could easily have been expanded to fill a volume. What has been needed for some time is a general introduction to man-environment problems that presupposes no background in the sciences, one that any college student or other interested reader could read with understanding and profit.

Dealing with such an extraordinarily complex subject, I have been highly selective lest the book, quite literally, get out of hand. Realizing that many relevant aspects of man's effect on his environment could not be thoroughly discussed in the limited space available and perhaps should not be in a general survey, I have supplied annotated further reading lists at the end of every chapter to aid the interested reader in his search for more detailed information. Since the whole field is changing so rapidly, I have emphasized the understanding of *basic* problems. To help the reader keep up to date once he has established this basic orientation, I have provided a list of periodicals in the field (Appendix 1), most of which are readily available and comprehensible to the interested reader. Since I feel that scientific knowledge is only part of the solution to man-environment problems, a list of organizations actively concerned with preserving environmental quality and encouraging ecological sensitivity is also included (Appendix 2), so that the reader, if he wishes, may take an active role in what is increasingly being recognized as a struggle of heroic proportions.

The central theme of this book is that man's relationship to his environment has passed through several phases. In the beginning, man was shaped by an environment which acted as a selecting agent and controlled the evolution of his present features. Throughout this period man, like the other animals, remained in equilibrium with his environment. But then something happened that changed the face of the world, completely and irreversibly: man developed culture and thereby shattered this equilibrium. From that time on man has exerted an ever-increasing influence upon his environment.

Initially, he simply increased the amplitude of already existing environmental traumas. Smoke from natural fires, or polluted water from animal carcasses or mineral seeps, for example, was magnified in proportion to man's numbers and cultural diversity. My discussion of these natural traumas has been arranged in a sequence reflecting man's increasing technological involvement and seriousness of effect, from fire through radiation.

More recently, new traumas have been introduced by technology, which has synthesized compounds totally new to the environment, such as DDT, and has produced a great many new forms of known substances like beryllium and asbestos.

Man has also profoundly affected the organisms that share the earth with him, both by introducing exotic species in new locations and by driving other species to extinction. His biotic effects have not, of course, been limited to the lower species, for biological and chemical warfare have long been applied by man to man.

Finally, man's rampant technology has placed him, once again, under the direct selective influence of an environment of his own making: the city.

The book concludes with a discussion of man's ultimate problem—finding the balance of resources and population that will enable all of earth's organisms to coexist in an environment prejudicial to none.

While the order of the chapters reflects my own approach to man and environment, worked out over several years' experience teaching a course in this area, others may prefer a different sequence. The comprehensive scope of the book should allow the sections or even individual chapters to be read and used in any order. To provide this flexibility, a full index is included.

In preparing this book I had the gracious assistance and advice of Dr. William Niering, Dr. Diane Merner, Rollin and Merry-Jo Bauer, and Michael Capizzi, who read all of the manuscript. Special thanks are also due to the editorial staff of W. W. Norton.

Richard H. Wagner

State College, Pennsylvania
October 1970

ENVIRONMENT AND MAN

Part I

Photo by J. Paul Kirouac

CHAPTER 1

ENVIRONMENT AND THE EVOLUTION OF MAN

Liquids have no shape of their own; they assume the configuration of their container. Although the time scale is quite different, the form of a plant or animal is also shaped by its environment, the "container" in which it exists. This fitting into an environment is called evolution. When a species conforms to its environment perfectly, evolution is complete. Completed evolution is rare in nature, however.

Man also evolved to conform with his environment, and for the greater part of his existence on earth has constantly been molded by that environment. While man has existed on earth for only a few million years, at most, the thread of his evolution goes far back in geological time.

The earth's crust has undergone extreme changes during its 4.5 billion-year-long history. Surface temperatures have varied from that of molten rock to a hundred or more degrees below zero; seas have formed, receded, and reformed over what is now dry land; mountains have arisen from ancient sea beds, have eroded away, and risen again. In spite of these extreme changes, complex and delicately balanced forms of organized living matter have developed. These organisms are so fragile that they may be quickly disorganized into nonliving matter by relatively slight internal changes in chemical composition, temperature, or pressure. How, then, do organisms survive the widely varying environmental conditions on earth?

The key to survival is not resistance to change, but meeting change with change. Even though a single species must inevitably be modified or become extinct, continued life is assured through evolution.

If you were to view the evolution of life on earth as a thirty-minute film, you would see wave after wave of new species evolving, filling the environment with a diversity of life forms, and then receding—sometimes

FIGURE 1.1

Periods and epochs	Duration (millions of years)	Time prior to present day	Distinctive life
PHANEROZOIC			
CENOZOIC ERA			
Quaternary	1 or 2	1 or 2	Modern man
Recent			
Pleistocene	1 or 2	1 or 2	Stone-age man
Tertiary			
Pliocene	9 or 10	12	Great variety of mammals Elephants widespread
Miocene	13	25	Flowering plants in full development Ancestral dogs and bears
Oligocene	15	40	Ancestral pigs and apes
Eocene	20	60 }	Ancestral horses, cattle, elephants
Paleocene	10	70 }	and primates
MESOZOIC ERA			
Cretaceous	65	135	Extinction of dinosaurs. Mammals and flowering plants slowly appear
Jurassic	45	180	Dinosaurs abundant Birds and mammals appear
Triassic	45	225	Flying reptiles and dinosaurs appear First corals of modern types
PALEOZOIC ERA			
Permian	45	270	Rise of reptiles and amphibians Conifers and beetles appear
Carboniferous			
Pennsylvanian	35	305 }	Coal forests
Mississippian	45	350 }	First reptiles and winged insects
Devonian	50	400	First amphibians Earliest trees and spiders Rise of fishes
Silurian	40	440	First spore-bearing land plants Earliest known coral reefs
Ordovician	60	500	First fish-like vertebrates
Cambrian	100	600	Abundant fossils first appear
PRECAMBRIAN			
Late Precambrian	1000	2000	Scanty remains of primitive invertebrates: sponges, worms, algae, bacteria
Earlier Precambrian	1500	3500	Rare algae and bacteria back to at least 3000 million years for oldest known traces of life

Figure 1.1 Geologic time is divided into eras and periods of varying duration and remoteness from the present. The Tertiary period is particularly important to our study of man-environment interactions because of the development at that time of the broad-leaved forests. (After Holmes)

totally, but occasionally leaving a few of the best-adapted species behind. It is humbling to note that man's existence on earth would flash by in the last 3.5 seconds of the film!

THE NICHE CONCEPT

Every animal except man has a well-defined habitat, a place where it is generally found and to which it has become adapted: trout "like" cold water streams; a wood thrush, heavy woods; a woodchuck, weedy pastures. Within this habitat each species has a unique life style not shared by any other species. This complex of activity within a given habitat is called a niche. A species in a niche undergoes adaptation that is specialized; no two species occupy the same niche. Under experimental conditions, the best-adapted of two organisms occupying a niche outcompetes the other and causes its disappearance. As the complexity of the habitat increases, the number of niches increases, and with this, the number of species filling those niches. In an undisturbed habitat in a reasonable state of equilibrium, most, if not all, niches are occupied. As habitats change, the niches also change. If an animal cannot adapt to the changes and so evolve along with its niche, it becomes extinct; its place is usurped by a species which has evolved to fit that particular niche.

By the beginning of the Tertiary period, 65 million years ago (Figure 1.1), conifers had been replaced by broad-leaved trees, in just this way. A rich forest covered large areas that today are too rough, too dry, or too cold to support forests. This early Tertiary forest, by giving ample opportunities for new forms of species to evolve and survive, was responsible to a great extent for the rapid evolution of the most recent wave of animals—the mammals.

THE RISE OF THE PRIMATES

One group of mammals, the primates (forerunners of man), was quick to exploit the new environment of broad-leaved trees. Primates are difficult to characterize because they do not share obvious primary characteristics, such as the hooves of the grazers or the specialized teeth of the carnivores (meat-eaters). Perhaps the simplest way to describe primates is to say that compared with other animal groups, they are unspecialized.

Once an animal becomes specialized, its specialization either leads to success, if the environment is stable, or to extinction, if the environment is rapidly changing, for evolution tends to be irreversible. The early primates were able to adapt to a changing environment because they retained their generalized features. If the primates had lost toes, as did the horse, they could not have regained them for use in tree climbing. This lack of specialization allows man, the most advanced primate, to occupy a tremendous range of ecological niches.

The earliest primate species for which we have evidence is a squirrel-like animal who lived on the ground. A combination of competition and predation on the ground, together with increasing evolutionary opportunity in the expanding Tertiary forest, probably led such primates to shift their habitat from the ground into the trees.

Living in a tree, as any tree climber knows, requires two basic abilities: an ability to grasp branches and an ability to perceive three dimensions, so that movement can be rapid and sure. The danger of falling is a constant hazard for tree dwellers (squirrels *do* fall out of trees occasionally). It is not surprising then, considering the special qualities of an arboreal environment, that animals invading this new environment began to adapt to it.

The Opposable Thumb Leads to New Occupations

Primates characteristically have five digits terminating the fore- and hindlimbs. These digits are the prime contact the animal has with its environment; consequently, they are usually the first part of the animal to undergo adaptation in response to changes in its environment. When a five-toed animal grasps an object, the digits tend to separate into two groups. If the big toe or thumb is able to move so that it opposes the other digits, the hand or foot can lock firmly onto the object being grasped, making movement from branch to branch easy (Figure 1.2).

With a set of four appendages that were able to grasp effectively,

Figure 1.2 As animals adapted to the changing environment and moved to a niche in the trees, an opposable thumb developed on the hands and feet.

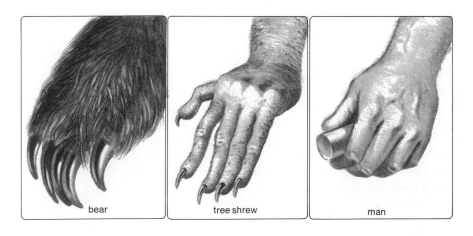

bear   tree shrew   man

the position of primates in the treetops was secure—literally and figuratively. A walking animal like the horse needs all four limbs for support, therefore, it has little opportunity for the evolution of other, independent functions for the limbs. The opposable toe made it possible for a primate to hold on to a branch with its hindlimbs while its forelimbs were freed for other activities. This freedom of one set of appendages to explore, feel, and touch opened a new avenue of evolutionary possibilities.

Snout Length as a Function of Survival

Yet the tactile sense in itself was incomplete without a sensor to evaluate the impressions collected. Here the second major aspect of the arboreal environment—its three-dimensional quality—placed great emphasis on a visual system that could perceive depth. The ground-dwelling primate had a long snout with an eye positioned on each side (Figure 1.3), precluding stereoscopic vision. After several million years of tree life a reduction in snout size permitted stereoscopic vision, giving an accurate representation of the three-dimensional environment to which these primates were becoming adapted.

A ground-dwelling mammal's snout serves two important functions: the long, powerful jaws secure and deliver food to the mouth, and the nose serves as an important environmental sensor. Because four legs are needed and adapted for support and locomotion, objects of interest must be picked up with the jaws and are rarely manipulated with the paws. The first two or three feet of air above the ground are usually rather moist, and carry scents and odors; the closer to the ground the animal, the more important is his nose as a sense organ.

Figure 1.3 The length of an animal's snout determines his capacity for three-dimensional sight as well as his ability to use his olfactory sensor.

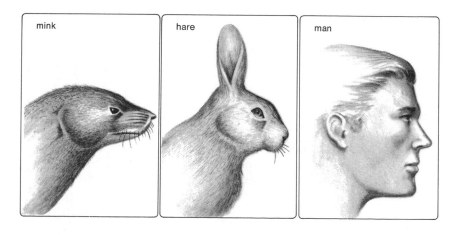

mink   hare   man

A primate abandoning the ground for the trees would certainly have found the dry, upper air much freer of odors and his nose correspondingly less useful. Procurement of food by the snout also became less important as the forelimbs no longer performed a strictly supportive function, but began to supplement and finally to supplant the jaws as a means of selecting food and placing it in the mouth.

These two factors allowed the snout to diminish in size and become less important in function, without adversely affecting survival.

NATURAL SELECTION

Adaptive change has, so far, been rather casually considered—forelimbs grasped, vision became stereoscopic, snouts shortened. How did these changes occur? Certainly animals weren't able to make these changes consciously; even man can manipulate the environment to his ends only at times, and cannot—at least now—make intentional structural improvements in his physical adaptation to the environment.

Every living cell of an organism contains information in its nucleus that governs the activities of that cell either singly, or in concert with other cells. But the reproductive cells are especially important because they produce the egg and sperm which, in higher organisms, are the only link between generations. The potential for change or evolution is greatly enhanced when the information-containing material (DNA) of the egg or sperm is altered. Such alterations of DNA are called mutations and the environmental factors which cause them, mutagens. Some of the more important mutagens are radiation (see Chapter 11), chemicals (see Chapter 14), temperature extremes, and ultraviolet light.

In man, a mutation causing the malfunction of one lung or kidney is not necessarily disastrous, since most humans are born with a pair of each of these organs. However, were the mutation to involve a change in the heart, or perhaps a block in an important biochemical pathway, the effect of the mutation would be more critical.

Occasionally a mutation permits an organism to interact with its environment in a more efficient manner, hence the mutant organism proliferates. Far more often, however, mutations result in an organism less able to cope with its environment.

Suppose you had a color television, but felt it could stand some improvement. One approach would be to expose the complicated circuitry and splash solder around at random. There is a remote possibility that you could somehow improve the quality of the picture, but the chances are overwhelmingly greater that you would make such a mess of things that the set would not work at all. If you had an infinite number of television sets and millions of years to experiment, sooner or later you would probably make an improvement. Such are the odds against a successful mutation.

Selection Enhances Favorable Mutations

Mutations move from an individual to future generations only by way of the egg and the sperm. If a mutation impairs the offspring's ability to cope with its environment, the animal may fail to mate or even die before it can reproduce. As a result, unfavorable mutations are lost. Changes that are neither favorable nor unfavorable can be carried indefinitely, simply because there is no selection pressure to get rid of them. Finally, any mutation that even slightly increases the ability of an organism to survive or to produce more offspring will be transmitted throughout the species population.

To see how mutation is related to evolution, consider snout length again. Snout length is determined by a complex interaction of genetic and environmental controls. Because of slight variations in the environment during development or perhaps minor mutations, no two individuals of a species are born with exactly the same snout length.

Since having a long snout means less effective stereoscopic vision, a primate with a long snout would be at a greater disadvantage in his arboreal habitat than a primate with a short snout and less likely to survive long enough to reproduce; hence, short-snouted primates would tend to replace longer-snouted primates.

Over a period of evolutionary time all organisms are subject to random mutational change. While the environment does the selecting, it selects differently for each species; a mutation that is useful to one animal could be a disaster for another. A spider monkey without a tail, for example, would be as inconvenienced as a man with one. While mutations producing monkeys without tails rarely occur, when they do, the aberrant individual does not proliferate; hence, monkeys have tails and men do not.

THE EMERGENCE OF MAN

Had the Tertiary period and its forest continued up to the present, man might never have evolved. Instead, there probably would have been a diversity of primates, well-adjusted to their arboreal world. This, however, was not to be. The luxuriant early Tertiary forest that extended across the continents from coast to coast resulted from the free movement of moist air from the oceans over the interior of the continents, providing enough rain to support the forest. About the middle of the Tertiary period, old mountain chains were rejuvenated, and new mountains like the Sierra-Nevada-Cascade complex, the Andes, and the Himalayas, were uplifted. The rising mountains deflected the moisture-laden air and the interior of the continents became drier.

As the humid ocean air was forced up and over the mountains, it was cooled, losing much of its water on the slopes facing the ocean. The air that flowed down the inland slopes became hot and dry. The higher the mountains, the drier the land in the lee of the mountains, until that land could no longer sustain the Tertiary forest. This effect, known as rain shadow, resulted in the desert-like conditions in the Great Basin of the United States, the Gobi desert north of the Himalayas, and the dry plains east of the Andes in Argentina. Climates were changing, and the plants and animals dependent on certain climatic conditions were obliged to change also.

The rich and diverse forest that had always required a generous supply of moisture could not adapt, and as the forest died out, a savannah— a mixture of grasses and small trees—gradually took its place. Today the remnants of the Tertiary forest are to be found in scattered patches throughout the tropical and temperate zones of the earth; in between are deserts, grasslands, and savannahs (Figure 1.4).

Figure 1.4 The remnants of the Tertiary forest are now scattered around the world. This forest once covered a large part of the earth's surface. (Dymaxion projection after Buckminster Fuller.)

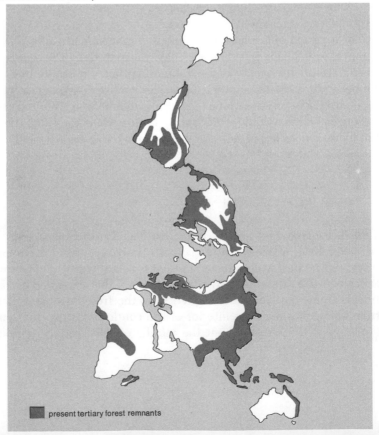

present tertiary forest remnants

Descent from the Trees as a Selection Pressure

What happened to the tree-adapted primates as their habitat changed? Some managed to stay with the retreating forest and are found today in its remnants—the orangutan in southeast Asia, the chimpanzee and gorilla in central Africa, lemurs in Madagascar, and various monkeys in the tropics of Central and South America. But some were able to survive the loss of trees and return to the ground their ancestors had abandoned millions of years before. It is within this group of primates that we again pick up the thread leading to man.

The descent from an arboreal to a grassland-savannah niche was not sudden: no primate could have adapted to an abrupt disappearance of the sheltering forest. But as the distance between trees increased, more and more time, of necessity, was spent on the ground. This opened a new area of selective pressure, which had an effect on appearance and activities of certain primates. For example, once on the ground it became important for primates to be able to see over the grass to avoid predators and to communicate with other primates. This encouraged standing upright and further reinforced the functional separation of forelimbs and hindlimbs. Whatever tail was present in the early primates became less important on the ground, and hence could be lost without adversely affecting survival.

Toward the end of the Tertiary period, when the environment began to stabilize again, there were several stocks of advanced primates—tailless, bipedal, possessing stereoscopic vision, with shortened snouts and smaller, less powerful jaws, and opposable thumbs giving great manipulative power to the hands.

The Hand-Eye-Brain Interdependence Leads to Man's Emergence

The primitive brain was rapidly changing at the same time that the more obvious adaptations, first to an arboreal, then to a grassland environment, were taking place. As the eyes were emphasized and the nose de-emphasized, the brain responded with greater development in the portion associated with seeing. The interplay between hands and eyes was of paramount importance in the further development of the brain. Species with manipulatable hands found new environments to explore, new food to eat, while their eyes, seeing the world in three dimensions, guided their hands with a deftness impossible in other animals. The brain responded to these new stimuli by further growth in size and complexity, which, in turn, stimulated manual dexterity. This unique interdependence of hand, eye, and brain enabled at least one primate stock to make that quantum leap from instinctive repetition to the creative innovation that separates man from beast.

In qualitative terms, brain development, stimulated by hand-eye interaction, seems to have moved through three stages: seeing and doing, remembering what was seen and done, and imagining what might be seen and done. Seeing and doing is common to most higher animals; remembering what was seen and done is characteristic of many species and indicates an ability to learn and be trained. But imagining what might be seen and done is limited to man and a few species of apes. Once this final stage was reached, self-sustaining intellectual evolution became possible. After an unusually short evolutionary period, man emerged. But why did only man evolve to this degree, leaving his fellow primates so far behind?

The answer lies in the comparative specialization of the higher primates. Many highly specialized species have survived through long periods of time, but they have not usually evolved into a variety of new species. Generalized species have the greatest evolutionary potential because they can respond to the greatest number of environmental options. As specialization removes these options, a species inevitably approaches an evolutionary dead end, becoming extinct in periods of environmental change. Because they are relatively unspecialized, sparrows and rodents are rapidly evolving; whereas ostriches and elephants, which are quite specialized, have reached this evolutionary dead end.

In the late Tertiary period the primate stock with the most generalized features had the best evolutionary opportunities. The ancestors of monkeys and gibbons had already begun specialized adaptation to the arboreal environment, developing long prehensile tails that functioned like another leg, and long arms that allowed swinging in an easy motion. The stock which gave rise to chimpanzees and gorillas never became fully independent of trees nor achieved a true bipedal stance; they walk knuckles-down on the ground during the day and return to the trees at night for shelter.

Only one primate stock developed the hand-eye-brain correlation before excessive specialization limited future evolutionary potential—from that stock, man emerged.

The fossil record indicates that there were at least two man-like primates in Africa about two million years ago: *Paranthropus*, a heavily built vegetarian, and *Australopithecus*, somewhat smaller, more graceful, and an eater of both plants and animals. Perhaps the disappearance of *Paranthropus* was just another example of specialization leading to an evolutionary dead end; or, in view of man's tendency for aggression towards other men, possibly *Australopithecus* speeded the extinction of *Paranthropus*. *Australopithecus'* extinction may then have been brought about by competition from the more advanced *Homo erectus*, who in turn fell before *Homo sapiens*.

CULTURATION: A TRANSITION

Up to the appearance of *Australopithecus,* evolutionary changes came rather quickly and markedly affected the features of the protomen of the day. After *Australopithecus,* outward changes were more subtle. Neanderthal man—or even the more primitive *Homo erectus*—might pass unnoticed on a city street today if dressed and shaven. Why the apparent turning point in man's evolutionary change? The answer, in a word, is culture.

Development of the Use of Tools

There are suggestions that *Australopithecus* had tools—crude pebbles, perhaps, but purposefully used. Many animals use tools in one way or another, but only man, and in rare instances, monkeys and apes, use them to defend themselves against predators or to kill their prey.

The consistent use of tools is of great significance in the effect of environment on man's evolution. For the first time an animal could use its brain to meet the challenges of its environment, instead of having to make gross structural adaptations. That is why further changes in the outward appearance of man have been small. When man first started living in the Arctic, he did not have to grow fur, he wore clothes; when he first flew, he did not have to develop wings by mutation, he used a tool. With culturation, that is, the achievement of culture, man was freed from most of the direct selective effects of his environment.

Culturation, however, had its own effects which further refined and modified the crude australopithecine stock. Monkeys and apes have rather impressive canine teeth (Figure 1.5). Neither early nor contemporary man has prominent canines. Why? Monkeys and apes use their canines for defense, exposing them in threatening grimaces, biting or slashing, if necessary. This action, however, requires a dangerous proximity to be effective. By using tools, sticks, or stones, early man could fend off at-

Figure 1.5 As man evolved physically and culturally, he no longer used his teeth as a weapon, and hence the jaw receded and the size of the canines was drastically reduced.

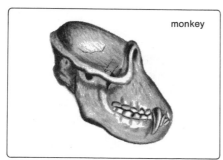
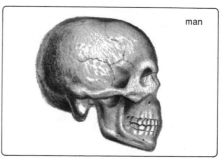

tackers without having to risk injury to himself in close combat. The large canine teeth became expendable, and if by random mutation they were reduced in size, survival was not threatened.

As the defense function of teeth and a powerful jaw was superseded by a brain directing hands that used tools, the massive musculature supporting the jaw was no longer necessary for survival. The ridges to which these muscles were attached, and which gave the skulls of early men such a formidable appearance, also became unnecessary. Mutations producing more delicate skulls, no longer selected against, spread throughout the population.

Later, man learned to control fire and use it to his advantage, just as he learned to use natural objects as tools. Cooked food is softer and more easily chewed than raw food, again lessening the importance of large, powerful teeth and jaws. In addition, new varieties of food became available through the use of tools and fire, broadening still further the tendency of early man to dietary generalization. The development of tools and the control of fire did more to make man truly human than any other cultural or biological advance.

HUMAN ADAPTATIONS

Although culturation seemed to remove man from some specific selective forces, environmental selection continued. If man had remained in the place where he evolved, the task of relating adaptation to environment would be relatively simple. But man has been on the move since the early days of his evolution; a primitive hunting and gathering culture must cover enormous areas of territory to survive. Climatic and cultural changes also stimulated people to migrate far from the site of their evolution and the environment responsible for their particular adaptations.

Despite the difficulty of unraveling the connection between environment and adaptation, some explanations can be pieced together from circumstantial evidence. Two adaptations are discussed in the next section: one—pigmentation—is very obvious; the other—sickle cell anemia—became apparent only through recent scientific study.

Pigmentation and Vitamin D Synthesis

Human skin, as well as the skin of most vertebrates and many invertebrates, contains a pigment known as melanin, which is formed from the amino acid, tyrosin. Skin color is determined primarily by the amount of melanin it contains; albinos, for example, have no melanin, and their skin looks pink because of blood-filled capillaries near the skin's surface. Among the human population, skin tones range from pink to very dark

Figure 1.6 The people of the world have widely varied concentrations of the basic skin pigment, melanin. (United Nations)

brown (Figure 1.6). If we look at the distribution of pigment types we see that the darkest pigmentation is in the tropics, generally, and the lightest skin is found in the high latitude temperate areas. The two extremes of the gradient might be represented by a Nigerian on the one end and a Laplander on the other, with a North African representing an intermediate degree of pigmentation.

How can we explain the lighter pigmentation of people living nearer the poles? One hypothesis involves pigmentation and vitamin D synthesis. The skin contains a precursor substance called 7-dehydrocholesterol, which under the influence of ultraviolet radiation from sunlight is converted into vitamin D. Since egg yolks and fish liver oil are the only common foods with much vitamin D, the sunlight–skin reaction, until quite recently, was the most important source of this important vitamin. Vitamin D controls the absorption of calcium from the intestine and the deposition of this mineral in the bones. If there is less than 0.01 milligram per day of vitamin D available from both sunlight and the diet, not enough calcium is absorbed, and the result is brittle and deformed bones. In children this condition is known as rickets; in women it may result in pelvic deformities leading to difficulties in childbirth. Too much vitamin D also causes problems: greater than 2.5 milligrams a day may lead to kidney stones, calcification in the joints, and deposits in the aorta and other large blood vessels. The body cannot regulate the amount of vitamin

D absorbed from food, or eliminate a possible toxic dose. However, control of vitamin D from sunlight is achieved by the rate of photochemical synthesis in the skin. The amount of melanin present in the skin controls the amount of ultraviolet light reaching the photochemical zone, and hence the quantity of vitamin D produced from the precursor.

Because of the high light intensity in the tropical and subtropical regions where man probably evolved, early man may well have had a darkly pigmented skin, allowing just enough ultraviolet penetration to produce vitamin D in the safe 0.01–2.5 milligrams a day range. But as early man moved out of the tropics into the more cloudy and stormy temperate and arctic regions, heavy pigmentation prevented adequate transmission of ultraviolet light through the skin, and too little vitamin D was produced. When this resulted in rickets, the most heavily pigmented individuals never reached reproductive age. With the environment selecting in favor of less heavily pigmented skin, it was just a matter of time before most of the inhabitants of northern Europe became quite light complexioned.

When the sun's rays fall at an angle of less than 35°, much of the ultraviolet is filtered out by the atmosphere. When, in addition, smoke and airborne dirt are present in the atmosphere, as is characteristic of most industrial cities, there is little ultraviolet light that can penetrate to the level of the inhabitants. In European cities during the eighteenth and nineteenth centuries there was a high incidence of rickets, probably because of the combination of factors. In Jamaica, where the sun's rays never fall at an angle of less than 50° and there is little industrialization, rickets is unknown.

What about Eskimos, who live farther north than most Europeans, and yet are darkly pigmented? Eskimos eat large quantities of fish, seal, and whale liver, all of which contain generous amounts of vitamin D. Thus, isolated from dependence upon sunlight for vitamin D production, there was no selection for lighter pigmentation.

Discovery of the relationship between vitamin D and rickets led to the supplemental use of cod liver oil and, more recently, the addition of irradiated ergosterol to milk. Today no child, regardless of his geographical location, need develop rickets from the lack of vitamin D. If, that is, he can get enriched milk in his diet.

Assuming that vitamin D synthesis was the main cause of pigment variability in man, the present level of culturation has usually removed this characteristic from active environmental control. There are, however, other aspects of skin pigmentation besides its role in vitamin D synthesis. Pigments also protect the skin from sunburn. Even pale skin becomes more heavily pigmented, or tanned, upon exposure to sunlight. This reversible mechanism allows maximum absorption of ultraviolet in the winter when especially needed, yet protects the skin in the summer, when people are less heavily clothed and vitamin D is abundantly produced. Apparently, in the past, the selective advantage of freedom from rickets has outweighed greater susceptibility to sunburn or, more rarely, skin cancer in lightly pigmented people.

Sickle Cell Anemia

The normal red blood cell in man has the shape of a biconcave disc. Occasionally hemoglobin, a protein that readily combines with oxygen, is imperfectly formed in certain red blood cells. Such cells have a caved-in or sickle-like appearance. These sickle cells are less efficient carriers of oxygen, resulting in a general oxygen deficiency called sickle cell anemia. This condition is found in a broad belt stretching from tropical Africa, Sicily, Greece, and southern Turkey to India. In some places up to 20 percent of the population carries this trait. If the inheritance of the condition is complete (or homozygous), the individual usually dies long before the reproductive age; but if the inheritance is incomplete (or heterozygous), survival is not only possible, but ironically, the individual has a selective advantage. Recently scientists noticed that although the homozygous individuals didn't live to reproduce, the frequency of sickle cell carriers remained stable in the population. Further investigation found that sickle cell anemia somehow gave the sufferer a resistance to a common form of malaria, hence the sickle cell mutant was selected for in tropical areas.

With improved sanitation and mosquito control programs in sickle cell areas, leading to a lower incidence of malaria, the selective advantage would be removed; in time one would expect the sickle cell carrier to be selected against and the incidence of sickle cell anemia reduced in the population. In environments where malaria is not common, sickle cell anemia may be debilitating enough to be selected against by the environment. Occasionally a sickle cell carrier dies or is severely incapacitated when traveling in a jet plane. Although jets have pressurized cabins, they maintain a pressure equivalent not of sea level but 5000 feet; a sickle cell carrier is not able to maintain adequate oxygenation of his body at this pressure, while a normal person experiences no problem at all.

Many biological adaptations have taken place in man since his culturation began. More recently, the use of technology in place of physical adaptation has made it possible for man to insulate himself from the effects of the natural world by creating artificial environments. Today, man can occupy an incredible variety of ecological niches without the necessity of any structural change in his body—the upper atmosphere without growing wings, the bottom of the sea without gills or fins, and most recently, outer space. We might conclude from this that man's technology has been superbly successful and that the environment is conquered.

But the environment has not been conquered. To the contrary, our independence of it is an illusion, engendered by our remoteness from a world we see through safety-glass windshields and thermopaned picture windows. We cannot elude the environment, for even our insulations have had their effects. Consider, for example, the effect on our health of mechanical locomotion and the resulting lack of exercise; consider the incidence of respiratory disease caused by overheated and underhumidified houses. The very technology we invoke to insulate ourselves from the natural world creates massive perturbations in that world. We respond by increasing our isolation from the natural environment to protect ourselves from the hazards we have loosed, when we should be directing our technological resources to the amelioration of these man-made disturbances. With each turn of this vicious cycle we become more remote from the natural world outside our artificial environments, and less willing to make the massive economic commitment that the restoration of environmental quality will require.

In the following chapters we shall try to understand the paradox of a technology that seems to protect man from his surroundings but has produced an environment more hostile than the ancestral cave.

FURTHER READING

Ardrey, R., 1966. *The territorial imperative.* Atheneum, New York. A lively book based on Lorenz's notion that aggression in man is instinctive.

Blum, H. F., 1961. "Does the melanin pigment of human skin have adaptive value?" *Quart. Rev. Biol.,* **36,** pp. 50–63. Thorough review of the pro and con arguments about the role of pigmentation in man's evolution.

Bresler, J. B. (ed.), 1966. *Human ecology.* Addison-Wesley, Reading, Massachusetts. A collection of papers skillfully culled from the extensive literature dealing mostly with man's physiological adaptation to environment.

Dubos, R., 1965. *Man adapting.* Yale University Press, New Haven, Connecticut. A bacteriologist-humanist looks at man's evolution. The perspective is broadly medical.

Harrison, R. J. and W. Montagna, 1969. *Man.* Appleton-Century-Crofts, New York. Honestly illustrated, no-nonsense approach to man's evolution from the anatomical point of view.

Howells, W. W., 1967. *Mankind in the making*. Rev. ed. Doubleday, New York. Beautifully written account of the evolution of man, intended for the layman, but fitting the pieces of the story together with great care.

Loomis, W. F., 1967. "Skin-pigment regulation of vitamin-D biosynthesis in man." *Science*, **157**, pp. 501–506. Logically presented case for the relationship between skin pigmentation and vitamin D production.

Lorenz, K., 1966. *On aggression*. Harcourt, Brace & World, New York. The classic work by a pioneer animal behaviorist, documenting the evidence for man's innate aggression.

Montagu, A. (ed.), 1968. *Man and aggression*. Oxford University Press, New York. A reasonable defense of the view (disagreeing with Lorenz, Ardrey, and Morris) that aggression is learned, not instinctive, in man.

Morris, D., 1967. *The naked ape*. McGraw-Hill, New York. Imaginative explanation of many of man's idiosyncrasies, not all of which is acceptable to anthropologists, psychologists, and animal behaviorists.

Pfeiffer, J. E., 1969. *The emergence of man*. Harper & Row, New York. The broad scope of this book offers something for everyone.

CHAPTER 2

MAN "MASTERS" THE ENVIRONMENT

Each advance in man's cultural evolution seems to have had a greater impact on his environment than the previous one. As hunting and gathering gave way to herding, agriculture, industrialization, and increased technological complexity, man seemed to lose touch with the magnitude of his effect on his surroundings.

MAN, THE SUPERPREDATOR

Examining the evolutionary record, we see the process of species gradually becoming extinct and being replaced by newly evolved species. The overall numbers of species probably remained constant, for there is only a finite number of available niches. It seems odd that in a relatively short period of geological time, the last twenty thousand years, dozens of species of mammals and birds have disappeared from their niches without replacement. Something unusual must have happened; the spacing of the extinctions over several thousand years makes it difficult to ascribe them to climatological or geological events alone. Furthermore, most of the extinct animals were large—the mammals over one hundred pounds, and the birds certainly large enough to be conspicuous and edible.

The large North American mammals were hardest hit—70 percent became extinct during the last 10–15 thousand years of the Pleistocene epoch. Horses and camels, which had evolved in the New World, became extinct on this continent, as did mammoths and mastodons, which had migrated into the New World over the once dry land of the Bering Strait. The ground sloth, the saber-toothed tiger, the tapir, the dire wolf, the giant buffalo, an antelope, and the giant beaver also disappeared, and yet there was no concomitant loss of small mammals, plants, or aquatic organisms.

A surprising number of the fossilized bones of these animals and birds were discovered associated with charcoal and stone tools, such as arrowheads or spear points, some still embedded in the bones (Figure 2.1). Was it possible that primitive man had hunted these animals to extinction? At first this idea was denounced as ridiculous. Who could imagine a Paleo-Indian with his crude weapons exterminating the mighty mastodon or the swift horse? We are not certain what really happened, but in a time of severe environmental stress caused by glacially-induced climatological and vegetational changes, certain animal populations may have been pushed beyond the point of survival by man, the superpredator. To envision man's possible role as exterminator, we have to look beyond the spear-carrying savage killing one mastodon or horse. As the Plains Indian of historic times demonstrated again and again, herds of animals could be driven by fire off cliffs or bluffs to their destruction, and, if their numbers had already been reduced by unusual natural stresses, this practice could have led to their extinction.

If man was instrumental in the extermination of so many large mammals, how did any escape? Possibly some, like mountain sheep and goats, were saved by occupying rough country; others, like moose, bear, and deer, found refuge in wooded country. Those escaping extinction on the open prairie may have been both abundant and fecund enough to keep pace with man's effort to kill them for food.

Figure 2.1 Many buffalo bones have been found with spear points embedded in them, suggesting this reconstruction of a prehistoric buffalo hunt. (McGraw-Hill Films)

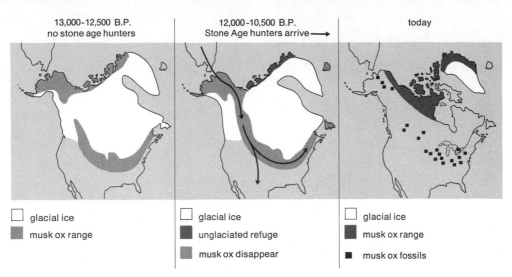

| 13,000-12,500 B.P. | 12,000-10,500 B.P. | today |
| no stone age hunters | Stone Age hunters arrive ⟶ | |

glacial ice	glacial ice	glacial ice
musk ox range	unglaciated refuge	musk ox range
	musk ox disappear	musk ox fossils

Figure 2.2   Musk ox survive to the present day in the arctic, but to the south only deposits of bones indicate that musk ox were ever present. One possible explanation for their extinction in the south is the migration of Stone-Age hunters into the areas south of the glacier. (Modified from P. Martin, *Nat. Hist.*, **76**, 1967.)

The musk ox might well have been saved by an accident of distribution (Figure 2.2). The range of musk ox was divided into two populations by the continental glacier.[1] Like the woolly mammoth which shared its range, the southern group was probably exterminated by the Paleo-Indians moving into North America from Asia. The northern population along the unglaciated Arctic Ocean coast seems to have survived because of less contact with these hunters. By the time the ice had melted, the Paleo-Indians had evolved into the present Eskimo stock of the area, with whom the musk oxen were apparently able to coexist.

But what of Africa, whose broad plains teemed with the greatest variety of grazing animals in the world? At least 30 percent of this fauna was also lost during the late Pleistocene. The fact that more African animals survived to the present day is explained by the concurrent evolution of man in Africa, giving some of the animals time to develop defenses against man's special predatory activities. In North America however, man suddenly appeared upon the scene from Asia, occupied the continent and began hunting animals that had never had a chance to adapt to this special type of predator.

Man's role as exterminator can be inferred from the wave of extinctions that followed his occupation of new continents or islands around the world: the disappearance of the giant kangaroo and other large animals when the Bushmen came to Australia; the loss of twenty-seven species of moa (a large ostrich-like bird) when the Maori reached New Zealand; and the elimination of the dodo from Mauritius (an island in the Indian Ocean) by the Dutch and Portuguese sailors in the sixteenth and

[1] Paul Martin, 1967. "Pleistocene overkill." *Nat. Hist.*, 76 (10), pp. 32–38.

seventeenth centuries. Most of the niches left by these extinctions remain unfilled; an interesting exception is the niche occupied by the horse in North America. Ten thousand years after its extinction in North America, the horse was reintroduced by the Spaniards and successfully reoccupied its old niche. In a relatively short time there were hundreds of thousands of wild horses ranging over western North America. Even today some mustangs remain free in the more remote areas of the Rocky Mountains and Great Basin.

DOMESTICATION OF THE LAND

Somewhere in the tangles of his cultural evolution, man learned that by taming certain animals he could not only provide himself with meat, hides, and wool, but also milk, butter, and cheese. It was much less work to follow a herd of domesticated cattle, sheep, or goats, than to chase down elusive wild game.

In contrast with their wild predecessors, most domesticated animals are not adapted to strenuous wandering and are dependent upon readily available water and food supplies. This means that unless the grazing range has particularly good soil and abundant rain, the stock may graze beyond the ability of the grass to recover, ultimately forcing the herd to move to another pasture. The repercussions of overgrazed range are seen in the Middle East and along the Mediterranean. This land originally supported thin, open forests with a good growth of grass on the mountain slopes and shrubs and grass on the plain. Homer, writing around 900 B.C., mentioned wooded Samothrace and the tall pines and oaks of Sicily. However, incessant grazing by growing herds of sheep and goats gradu-

Figure 2.3 Overgrazing by sheep in Egypt has left this ground unprotected by vegetation. Erosion has already begun. (Photo by Alice Mairs)

Figure 2.4 This blackjack oak in Tennessee, perched on a soil pedestal while denuded ground all around has been eroded away, demonstrates the importance of vegetation in soil stabilization. (U.S. Forest Service)

ally eliminated such forests by destroying the seedlings that replaced old trees. Because the trees were not replaced, a park-like forest developed with widely spaced trees and grass covering the ground (the origin, perhaps, of the western concept of park?). Finally there were no trees left at all. As grazing pressure increased, even the grass disappeared (Figure 2.3). Without trees or grass to keep the soil in place on steep slopes, extensive erosion occurred (Figure 2.4). Today much of the soil that once covered the mountain slopes, making possible the growth of trees and grass, lies in deep layers on the plains or in the heavily silted rivers and their deltas. Many of the seaports of the ancient Mediterranean world are now miles from the sea because of this greatly accelerated silting process. The steep mountain slopes are stripped to bedrock in many parts of Turkey, Greece, Yugoslavia, Italy, Spain, and North Africa. The lower slopes are often densely covered with a growth of evergreen shrubs, which is called chaparral, garrigue, or maquis. This vegetation is of little economic value compared with the grass and forest that it replaced.

From Forest to Field

The forests of central Europe were far more resilient than the Mediterranean forests. With more plentiful rainfall distributed evenly over the year, central Europe was covered with a luxuriant forest of beech and oak, so dense and forbidding that permanent settlement was delayed until culturation enabled man to clear trees and plant crops. This was apparently done by chopping down or girdling trees, burning the brush, and then planting grain in the ashes—a practice that produced a good crop. As the forest began to grow back, stock could be pastured on the grasses and weeds that persisted under the thickening cover of trees and shrubs. Then the farmers moved on and the forest returned.

These intermittent agricultural disturbances, while relatively minor, did create conditions that stimulated the evolution of weeds. Plants found new environments in the campsites adjacent to the clearings. Hard-packed soil eliminated competition from plants indigenous to the forest, while trash heaps, enriched with organic remains, gave those plants able to tolerate the bright sun an excellent place to grow and reproduce. To this day most garden plants and weeds depend on man to provide direct sunlight for their best growth, and quickly disappear when man allows the regrowth of forest trees or shrubs, which shade them. Moreover, man has encouraged the development of weed species by intentionally moving plants or seeds from one place to another, thus bringing together plants from widely separated areas as crop seed contaminants, and allowing them to hybridize and evolve into weeds. Then, by regular agricultural practices, the life cycles of weeds became closely synchronized with the crop plants, assuring the contamination of man's crops with certain weed seeds.

Clearing the forest had still another effect on man. When the forest vegetation and its animal population were disturbed or reduced, insects, such as the Colorado potato beetle, that lived on natural vegetation transferred their attention to man's crops. Also, some of the parasites of the vanishing animals were able to make the transference to man, for instance, the malaria-carrying mosquito. As farming replaced hunting as the principal means of getting food, human populations grew much larger and more stable, creating centers for maintaining and spreading parasites and disease. Indeed, the larger the population of a village, town, or city, the greater became the danger of epidemic or plague. Therefore, many diseases, pests, and weeds date only from the beginning of agriculture, which allowed permanent settlements.

Forest clearance in Europe reached an early peak during the Pax Romana, but after the Roman Empire fell to pieces and during the period of civil and religious wars, plagues, and other misfortunes which followed, the forest once again advanced, bringing about what was literally a "dark age." With the Renaissance and the rise of nation-states, clearing for farmland was resumed and the great European forest was again reduced to isolated patches. Because of the tradition of long-term ownership of land by one family, often for hundreds of years, the farmland was generally well cared for and today remains reasonably fertile and productive.

The European forests made this transition from forest to field over a period of at least a thousand years, but the North American deciduous forest fell in a far shorter time. To describe how dense and extensive this forest once was, it has been suggested that in 1620 a squirrel with an inclination for travel could have journeyed tree to tree from the Hudson to the Mississippi without once touching ground. Forgotten in the romance of this claim are the woodland Indians, who, unlike the buffalo hunters of the plains, burned clearings in the forest and raised pumpkins, squash, and corn. Indeed, the crops grown in those Indian clearings kept the Pilgrims alive that first winter in Massachusetts. Soon, however, the new colonists were clearing their own fields, and by 1820 much of New England was crop or pasture land. It took but a short time to discover that upland New England farms often produced more stones than pumpkins, and so farmers began moving west to New York State, then to Ohio, Indiana, and Illinois, abandoning the poor farmlands back East to the forest which covers much of New England today. Over two-thirds of Connecticut—a very small state, approximately 50 by 100 miles, with a population of 2.5 million—is covered by forest, much of it second or third growth. Worthless from a forester's point of view, this suburban forest is of incalculable value as an aesthetic and recreational resource. Some idea of the extent of former clearance can be seen from the number of stone walls running through forests over a hundred years old (Figure 2.5). Hardly a woodlot in New England lacks its crumbling stone wall

Figure 2.5 At one time stone walls separated fields or pastures in the rolling countryside of Westchester County, New York.

covered with poison ivy and inhabited by blacksnakes, skunks, and chipmunks.

Abandonment of poor or worn-out soil became a common pattern in the settlement of North America, especially in the south, as the boyhood wanderings of Abraham Lincoln attest. But today people no longer move from poor farms to better farms; the movement is usually from the farm, poor or rich, to the city. Hence, a second wave of land abandonment is now returning land that should never have been farmed back to forest.

INDUSTRIALIZATION

The effects on the environment of clearing the forests have been evidenced in many ways, not all of them obvious. To begin with, goats and sheep, insatiable as they have been for the past two thousand years, were not the only consumers of trees and vegetation in the Mediterranean world. The only source of fuel throughout the classical and medieval periods in this region was wood. Forests were also cut to build ships for use in wars as well as for trading, because the settlements around the periphery of the Mediterranean depended on ships for communication and trade. Responding to the stimulus of this overseas trade, industries such as iron-working, glassmaking, and pottery began using large quantities of charcoal. Initially, their needs were met near the cities, but as the forests were cleared and overgrazing by sheep and goats prevented their regrowth, charcoal had to be hauled from farther and farther away, opening even more areas to grazing. Finally, the supply of charcoal, a relatively clean-burning fuel, could not keep up with the growth of industry. The subsequent discovery of coal led to increased water and air pollution problems (Chapters 7, 10). But what concerns us in this chapter are the environmental repercussions of removing coal and ore from the ground.

Stripping the Earth

Digging large holes to extract ores causes little disturbance in the ecology of an area when the surrounding countryside is already barren, dry, and starkly eroded, as it is in the ore-producing states of the Rocky Mountains and Southwest (Figure 2.6). But in the humid, forested East, the disturbance is noticeably destructive. With the exception of the Mesabi Range in Minnesota, ore mines are rather small-scale and infrequent in the midwestern and eastern United States. Of far greater impact is the removal of coal in stripping operations. The desirable coal seam often lies below layers of soil and rock, which must first be removed.

Figure 2.6 An open pit copper mine in Utah does not seem nearly as out of place in the dry desert of the Great Basin as it might in the lushly vegetated East. The rows of black dots are railroad cars which contrast with the enormous size of the mine crater.

Not too many years ago, strip-mined land was abandoned; when the coal was gone, there remained the desolation of steep piles of discarded earth or overburden alternating with the trenches from which the coal had been removed. Because the soil was often acidic from minerals leached from the piles of overburden, very few plants could grow, and the area became a wasteland.

Today most states have strict laws requiring backfilling of the trenches, recontouring the ground surface to some semblance of its original state, and the planting of trees. Red pine or hybrid poplars can usually survive and even do well if the acidity of the soil is not too great. Such reclaimed areas, called spoil banks, may be used for recreation, particularly when they have collected water to form ponds for swimming or boating. Two problem areas remain, however: those areas strip-mined before reclamation laws came into effect, and certain parts of Kentucky where coal seams occur in horizontal strata near the ridge tops. When roads are built to mine these veins or mountain tops are removed to expose them, the refuse is dumped down the sides of the mountain to collect in the valleys. When wet, these unstable slopes constantly slip and cannot be planted with vegetation. Once the coal is gone, mountainsides, entire valleys, and streams are so disrupted that any economic or recreational use is impossible for many years (Figure 2.7).

The problem in Kentucky continues to worsen. Huge power plants, built to keep pace with the surging demand for electricity, have revitalized a dying strip mining industry by providing an outlet for cheap coal. But coal obtained in this fashion is cheap only if soil stabilization, reclamation of spoil banks, and purification of polluted streams are not added to the mining cost. It might make more sense to eliminate or control the environmental misuse at the beginning and pay the increased prices this would entail, than to live with spoil banks for years, then use public money for their rehabilitation. The cost of environmental rehabilitation is expensive now and is likely to become more expensive in the future.

Figure 2.7 Contour strip mining for coal in West Virginia has destroyed the natural ecology of the area. The ridge tops have been denuded and tailings slide into the river valleys below. (USDA—Soil Conservation Service)

TECHNOLOGY

Technology is the application of the laws of science to practical problems, frequently involving extremely clever and prodigious engineering skills. Unfortunately these skills were rather narrowly applied as man attempted to harness nature more economically. We will look at two examples of the way man has thereby altered the face of the earth: both the construction of canals and dams were dependent on the growth of technology, and both have had ecological implications.

Connecting the Waters

More than one hundred years ago, the Welland Canal was opened, allowing ships moving from Lake Ontario to Lake Erie to bypass Niagara Falls. Then in 1932 the canal was deepened to thirty-six feet. With this deepening, lampreys began to move from Lake Ontario, where they were part of the natural fauna, into the western Great Lakes, where they had never been before (Figure 2.8). Within a few years the valuable catch of whitefish and trout in the western Great Lakes was sharply reduced, with severe economic repercussions in the fishing towns around the Great Lakes.

The lamprey feeds on the blood of large fish by secreting a chemical to prevent the blood from coagulating. In this fashion the lamprey slowly kills the fish. The lamprey spawns in small streams that empty into the lakes. The larvae which develop live in the streams for a few years, then move back into the lakes to complete their development, feeding on the valuable lake fish.

At first it seemed that the lamprey was here to stay. But in the late 1950s a chemical was found that would selectively kill the lamprey larvae without affecting other organisms. By the immensely laborious and expensive expedient of treating all tributary streams, some measure of lamprey control has been attained. Without question, it was the deepening of the Welland Canal that allowed the lamprey to enter new territory with such disastrous results. Fuller knowledge of the lamprey's ecological adaptability might have forewarned us of the possible introduction of the lamprey into the western Great Lakes, and precautions could then have been taken in the construction of the Welland Canal to avoid this.

Two other canals, the Suez Canal and the Panama Canal, both connect widely differing bodies of water, but have until recently had features that effectively prevented faunal exchange between the connected water masses. The Suez Canal connects the Mediterranean Sea with the Red Sea. Before the last glaciation in Europe, the two seas were connected naturally and shared a rich tropical flora and fauna. Then, a fall in sea level, through the removal of water to form glacial ice, separated them, and the subsequent chilling of the Mediterranean eliminated many of its tropical species. Today there are many open niches in the Mediterranean, while the Red Sea has most of its biological niches filled. When the canal was dug across the Isthmus of Suez, it intersected a small lake that was highly saline because its bottom was covered with a thick layer of salt. Any organisms passing through the canal had to be able to tolerate a change of salinity from 45 parts per thousand in the Red Sea to 80 to 100 parts per thousand in the Great Bitter Lake section of the canal (Figure 2.9). Since few organisms could tolerate such a change, this built-in salt barrier prevented most Red Sea organisms from passing into the Medi-

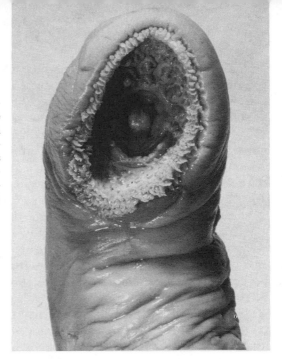

Figure 2.8 The conical structures of the oral disk of the lamprey are rasps which rub a hole in the side of the fish causing bleeding. The lamprey feeds on the blood and the fish eventually dies. (U.S. Dept. of Interior, Fish and Wildlife Service)

Figure 2.9 A view of the Nile Delta and the Suez Canal from Gemini 4 shows the Mediterranean (left) connected to the Red Sea (right). The Great Bitter Lake is the small body of water between them. (NASA)

terranean. After the Anglo-French debacle of 1956, the canal was deepened to thirty-six feet. The greater flow so diluted the hypersaline water of the Great Bitter Lake that today the salinity of the entire canal is approximately that of the Red Sea.

In the years since dredging, over one hundred species of invertebrates have entered the Mediterranean from the Red Sea. Significantly, one of the twenty-four species of fish that have made the transit, the lizard-fish, has become an important element in the fishing industry of Israel, a rather ironic sidelight to Nasser's deepening of the canal.

Because of available niches in the Mediterranean, introduction of new fauna from the Red Sea may very well increase the productivity of the eastern Mediterranean, without grossly upsetting the ecological balance. Some native Mediterranean species may be lost, but in this instance, the genetic and economic potential of the new species may be worth the risk.

Such is not the case with the Panama Canal. At the time of its construction, the spine of mountains running through the isthmus made the digging of a sea-level canal like the Suez impractical. The center of the canal is by contrast a fresh water lake to which ships passing between the Atlantic and the Pacific ascend and descend by means of locks. This lake serves as an effective barrier to the passage of most forms of marine life, but there is a possibility that because of the outmoded size of the Panama Canal locks, a new sea-level canal may be constructed or the present canal may be extensively rebuilt. A sea-level canal seems economically feasible only with the use of nuclear explosives to dig the ditch and expel the debris. When the costs of liability for property damage, relocation of nearby populations, and secondary excavation of rubble slipping back into the canal are added, the total expense may preclude the building of a sea-level canal.

But if, despite the exorbitant cost, the canal is constructed, we must make use of previous experience from the Welland and Suez canals to project what might happen if animals from the Atlantic and Pacific can move freely from ocean to ocean. The oceans are quite different in many respects. The Pacific is colder, water temperature fluctuates more during the year, and the tidal range is greater—almost twenty feet, compared with one and a half feet in the Atlantic.

The two oceans have been separated by Central America for only three to four million years, so the fauna are certainly similar. Yet in this time there has been some divergence. According to Ira Rubinoff, a marine biologist at the Smithsonian Institution, were the species to mix in one or the other ocean several possibilities exist:

1. If two populations were able to interbreed, the hybrids might eliminate either or both parent forms.

2. If the populations were farther apart genetically, the imperfect hybrids might eliminate the parents, but be incapable of long-term survival themselves.
3. If the populations did not interbreed but occupied similar niches, one might outcompete and replace the other.

Before construction of a sea-level canal, we should investigate these possibilities in a series of controlled experiments, looking at those species most likely to make the move from one ocean to another. Few would argue that a sea-level canal should not be built simply because a few species of fish might become extinct, but the implications are broader than this. The fisheries and sea life of both coasts of Central America are of great importance to the economy of the region. Until we can predict with some assurance which species will dramatically increase their populations, or conversely, which will become extinct, we cannot afford still another economically engineered, ecologically blind experiment.

Damming the Rivers and Flooding the Land

Dams are the opposite of canals in that they impede rather than facilitate access of organisms to an area. The most obvious feature of a dam is the water which it impounds, but of equal importance is the environment flooded and the land which now surrounds the new body of water. Many dams have caused problems in one or more of these environments, but one stupendous project, the Aswan High Dam, has had some unique effects extending far beyond the dam site or its lake.

Constructed on the Nile in the United Arab Republic, the Aswan High Dam contains seventeen times the volume of the pyramid of Cheops at Giza and creates behind it a lake (Lake Nasser) six miles wide, ninety-three miles long, and 250 feet deep. One unusual aspect of this project is its effect on the Nile River. Traditionally, the Nile has flooded every year, spreading fertile silt, derived from the nutrient-rich basaltic lavas of Ethiopia, over the Nile valley and delta. Over thousands of years this has created a fertile flood plain in the Nile Valley that has nurtured civilization almost from its beginning (Chapter 3). As the fertile Nile flood water spreads out into the Mediterranean, it dramatically increases the productivity of the coastal water from the Nile delta northeastward to Lebanon. This fertilizing effect was measured by a count of the primary producers, tiny algae cells which form the basic unit of the food chain in the sea. Before the Nile reached flood stage, there were 35,000 cells per liter; after flood stage, 2,400,000 cells per liter. One of the major harvesters of this primary food source has been the sardine; over 18,000 tons of sardines per year were caught in the UAR before the Aswan Dam was built.

Since the dam, the Nile silt has been sedimented out on the bottom of Lake Nasser. Without the annual floods and silt deposition, the Nile valley and delta will have to be fertilized artificially. Fortunately, the UAR is in a good position to do this: there are abundant deposits of phosphate available, and power from the Aswan Dam is already being used to produce nitrate.

The effect on the sardine industry has been more serious. After closure, the catch fell from 18,000 tons to 500 tons, cutting the overall production of Egyptian fisheries by almost 50 percent. However, much of this loss may be recouped by fish harvested from Lake Nasser. The present catch of fresh water fish from the lake is about 2,000 tons, but by the time the lake is completely filled in the 1970s, the catch is projected to reach 12,000 tons. One salutary effect of the Aswan Dam has been to bring about a long overdue reorganization and modernization of the fishing industry. At least a few of the unemployed sardine fishermen can be relocated to handle the growing production of fish from Lake Nasser. But Lake Nasser, like all man-made lakes, is a temporary phenomenon, accumulating the silt which was once deposited in the Nile Valley. Where will the fishermen be moved when the lake has filled?

The creation of Lake Nasser has had another far-reaching effect. Like Lake Kariba in Rhodesia, Lake Nasser has greatly increased the breeding sites of the snail which carries a parasitic blood fluke causing schistosomiasis, malaria-carrying mosquitoes, and the black fly that carries trachoma (an eye disease). Schistosomiasis, which has a debilitating effect on the population, can be expected to increase explosively as land surrounding Lake Nasser is brought under cultivation. Although this disease may be controlled somewhat with new drugs such as hycanthone, the only sure way to eliminate the infection is by increasing sanitation facilities and by education, neither of which is widely available to the fellahin of Egypt.

The economic and material needs of people will continue to demand the development and completion of projects such as the construction of canals and dams. We can continue to build on a purely economic and technological basis, picking up the ecological pieces as we go along; or we can become emotional and bemoan every lost earthworm or crushed ant, while turning a deaf ear to the crying of a hungry child. The only rational approach is to anticipate the massive technology-oriented development projects of the future and very carefully weigh the prospects. In this way we may preserve some semblance of the natural world. On the basis of what we have learned from past projects, we should have some idea of the potential ecological problems previously unthought of by engineers or economists. In the time interval between a project's conception and realization—often several decades—much descriptive and experimental environmental information can be obtained. Mathematical models might be developed to enable prediction of potential problems years in advance; at the very least, this would give sufficient research and development time to meet these problems.

There is a need to broaden the current approach to cost analysis, now largely based on what is economically and technologically feasible, to include other values, especially the recognition that projects exist in an environment that will be affected. Our first obligation is to other people, but we cannot ignore the plants, animals, and soil of the world as we scheme and dream, earthmovers and nuclear explosives ready. Our continued existence depends on the continued well-being of *all* plants and animals, not just corn and cows. Our disregard for the other organisms on earth is not just a manifestation of myopia; it is based on our abysmal ignorance of the interrelatedness of plants, animals, and their environments.

FURTHER READING

Caudill, H. M., 1963. *Night comes to the Cumberlands: a biography of a depressed area.* Little, Brown and Co., Boston, Massachusetts. A graphic, moving account of man's rape of the Cumberland Plateau.

Lowe-McConnell, R. H. (ed.), 1966. *Man made lakes; proceedings.* Academic Press, New York. A wide-ranging collection of papers discussing the environmental impact of large dams and their impoundments around the world.

Marsh, G. P., 1885. *The earth as modified by human action.* Scribner's, New York. The first synoptic recognition of man's impact on his environment. Contains a fascinating wealth of detail.

Martin, P. S., 1967. "Pleistocene overkill." *Natural History,* **76** (10), pp. 32–38. An absorbing and plausible rationale for man's first major environmental effect—extinction.

Rubinoff, I., 1968. "Central American sea-level canal: possible biological effects." *Science,* **161,** pp. 857–861. Some background on what might happen if a sea-level canal comes about.

Thomas, W. L. (ed.), 1956. *Man's role in changing the face of the earth.* University of Chicago Press, Chicago, Illinois. Excellent collection of papers dealing with a whole spectrum of man-environment interactions.

U.S. Department of Interior, 1967. *Surface mining and our environment.* Washington, D.C.

Veen, J. Van, 1955. *Dredge, drain, reclaim.* 4th ed. Nijhoff, The Hague, Netherlands. A thorough, well-illustrated treatment of the saga of dikes, polders, and land reclamation in the Netherlands over the last thousand years.

CHAPTER 3

SOIL, SAND, AND CYCLES

Despite tremendous technological advances since the agricultural revolution, man continues to be dependent on the soil for food. A complex mixture of compounds and elements necessary to man's existence on earth, soil plays a major role in the ecology of any particular area. Both plants and animals depend on it either directly or indirectly for nutrition. For this reason, it is important to understand something about its formation and nature.

SOIL FORMATION

Soil has a developmental history and is in some respects as alive as the plants that depend on it. It contains not only organic components, but so many interdependent organisms that the resulting complex behaves as though it were one large organism. Long ago, farmers noticed the association between parent rock and soil type and assumed that a simple relationship existed everywhere. Chalky soils were found on chalk, sandy soils on sandstone, and clay soils on limestone. Later, however, when large areas of Asia, Africa, South and North America were explored, it was discovered that rather uniform soils were found in regions with varying types of parent rock and consequently that a subtler relationship obtained between rock type and soil.

Weathering

Solid rock is continuously being broken down into small particles, a process referred to as weathering, which may involve either physical or chemical factors. Alternate heating and cooling of the rock itself or freezing and thawing of water in its cracks can, in time, fracture the hardest rock, as evidenced by the pile of rock fragments or talus found at the foot of most cliffs (Figure 3.1). This is the action of physical weathering. Other agents of physical weathering, running water and ice, can grind stones, pebbles, and sand into silt and clay, the finest components of soil.

The more the rock fractures by physical weathering, the faster chemical weathering takes place. A piece of granite, for example, may be broken down into its component minerals up to ten times more rapidly by the combined action of physical and chemical forces than it would be by chemical weathering alone. Since rocks are made of mineral crystals or older rock fragments cemented together, the minerals and the cementing substances are all soluble to a degree, and thus can be dissolved by rainwater. Pure rainwater would not make very great headway on an outcropping of rather dense and hard granite or basalt, but rainwater is rarely pure. Carbon dioxide dissolves in rainwater as it falls through the atmosphere, forming carbonic acid which, although it is a weaker acid than vinegar, still attacks rocks more vigorously than pure water.

Any parent rock can contribute only those minerals that it contains. Once released, however, the proportions of these minerals in the soil may be greatly affected by climate and the type of vegetational cover. A simple granite, for example, might contain three minerals: feldspar, mica, and quartz. Feldspar, a complex of potassium, aluminum, and silicon dioxide, breaks down in the weathering process into soluble potassium

Figure 3.1 Physical and chemical weathering on the cliff face formed this talus slope in Colorado. (U.S. Forest Service)

carbonate and kaolinite, a clay. Mica, containing magnesium as well as the oxides found in feldspar, decomposes into soluble carbonates, the sesquioxides of iron and aluminum (sesqui means 1.5 times, referring to the 3:2 ratio of oxygen to metal), and more clay. Quartz, that is, silicon dioxide simply weathers from large pieces to smaller pieces.

A soil weathered from granite, then, would be expected to contain carbonates, the source of many plant nutrients, sand, clays, and sesquioxides. But the climate strongly affects the rate at which these soil components are either retained or washed away, leached, by water percolating through the soil. In a cool moist climate, the sesquioxides and clays tend to be leached away by rainfall, leaving behind porous, sandy soil called a podzol (from the Russian word meaning ash). Many of the light, greyish or brownish sandy soils of the temperate zones of the world are podzols. In the tropics and areas with a warm moist climate, the silica or sand is leached out, leaving behind the sesquioxides and clays. Because of the quantities of iron and aluminum in sesquioxide form, the soil is brightly colored in shades of red, orange, and yellow, and because of the clay content it is heavy and plastic in consistency. These laterite soils (from the Latin for brick, an apt description of their dry state) or soils resembling them are found in the piedmont of the southern United States from Virginia to Texas, and in many other regions with warm moist climates.

SOIL STRUCTURE: A POROUS TRAP

If rock were to weather into tiny cubes like salt crystals, they would pack tightly together, leaving little room for either water or air. However, weathering produces rounded particles, and when these are packed together the interstices are filled with either air or water. In addition, the activities of plants and animals make large channels and holes that allow even more air and water to enter.

Immediately after rain most of the spaces between the soil particles are filled with water, much of which is lost to the water table by the action of gravity, or to the atmosphere by evaporation. The remaining moisture is gradually absorbed by plant roots and returned to the air via the plant's leaves. This transpiration—loss of water from the plant—is an integral part of the cycling of water in the environment (see Chapter 6).

Obviously, plants cannot grow without water, but what is less obvious is that the air content of soil is important too. Plant roots contain living tissues that are as vital to the plant as its leaves, stems, and flowers. While water can move into the plant root by simple diffusion, minerals dissolved in that water may require an output of energy by the root to be absorbed. The source of this energy is photosynthesis. Plant leaves are able to use sunlight to split water molecules into hydrogen and oxygen ions (charged particles). The oxygen ions combine to form gaseous oxygen, but the hydrogen ions are fed into a complex series of reactions along with carbon dioxide, and ultimately yield a sugar.

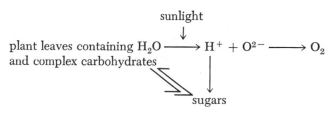

$$\text{plant leaves containing } H_2O \xrightarrow{\quad \text{sunlight} \quad} H^+ + O^{2-} \longrightarrow O_2$$
and complex carbohydrates → sugars

Sugars can be burned at once in an energy-yielding process called respiration, or transported to the roots where the sugars supply the energy needed to absorb certain minerals from the soil. But oxygen must also be available in the soil. In waterlogged soil all the spaces between particles are filled with water and oxygen is in very low concentration. Plant roots therefore die from being unable to respire properly, and because of their critical function in supplying water and minerals, the plant also dies.

The feeder roots of trees, those roots active in absorbing water and minerals, are quite close to the soil surface, and are killed if the soil in which they are growing is covered with an additional six inches or more of new soil or fill. Such filling prevents air from reaching the level where the roots are growing. The reason many trees die when a house is built on a wooded lot is that the feeder roots of the trees are killed when soil removed from the cellar hole is spread over the roots of adjacent trees to a depth of several feet. Within a year the leaves yellow, and the tree dies. This need not happen if care is taken to build dry wells around the trees where soil will be placed. By backfilling up to the dry well, a zone of aerated roots is left adjacent to the tree, which permits the tree to survive until its roots can occupy the new soil (Figure 3.2).

Figure 3.2 By constructing a dry well around the base of a tree sufficient feeder roots survive to keep the tree alive until it can develop new roots in the new soil backfilled against the wall.

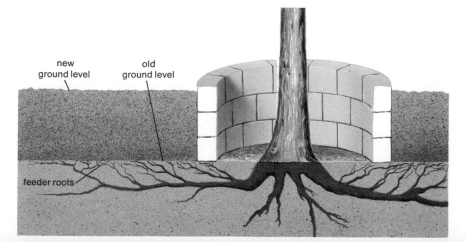

new ground level
old ground level
feeder roots

SOIL: THE ORGANIC COMPONENT

Soil with its inorganic components, sand, clay, air, and water, is still incomplete. The organic material which must also be present in fertile soil is called *humus* and is made up of partially decomposed vegetable material, leaves, rotting wood, and animal remains—their excrement as they live, their bodies when they die. Humus is the rich brownish-black crumbly top layer or leaf mold in temperate zone forests which you see when you brush away the dry leaves on the forest floor. Large animals, such as elephants, deer, or rabbits, make some contribution to humus; mice, insects, and worms contribute even more. But, because of their numbers, bacteria, algae, fungi, and other microorganisms contribute more humus to the soil than all the big trees and animals combined. These organisms are also of critical importance in the decomposition of organic material and the cycling of nutrients.

Humus, which gets mixed with the mineral components of the soil by the burrowing activities of animals and the dying and rotting of plant roots, forms a fairly stable association with the clay in the soil, a clay-humus complex (Figure 3.3). This complex then forms aggregations, or clumps, of various sizes that determine the soil texture, which in turn helps to determine the kinds of plants that will grow in a particular kind of soil.

As organic compounds and minerals in the soil gradually decompose, they dissociate into simpler and simpler compounds, finally forming ions that percolate through the soil and are attracted and held by the clay-

Figure 3.3 A schematic drawing of a clay-humus particle shows the relationship between mineral nutrients and soil fertility. Carbonic acid combines chemically with the mineral nutrient ions in the particle, leaving hydrogen ions behind.

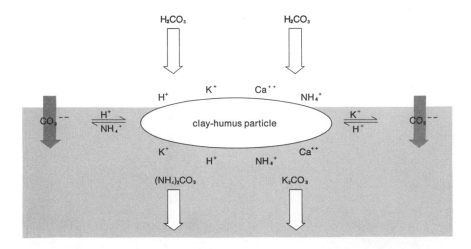

humus particles. These ions comprise a reservoir that determines soil fertility. If there is an abundant supply of clay-humus particles well saturated with minerals, the soil is said to be quite fertile.

Carbonate ions from rainfall can extract minerals from the pool, leaving hydrogen ions in their place. If positively charged ions continue to be added to the soil from leaf fall, their loss to the carbonate ions that pass through is balanced, and the soil remains fertile. But if the replenishment of ions is interrupted or diminished, they will be leached away as carbonates and the soil will be less fertile.

This is what has happened since the settlement of Virginia in the seventeenth century. When rich forest was cleared and the land plowed and planted in tobacco, minerals were not replaced; in fact they were removed both by the leaching action of rainwater and the harvesting of the tobacco leaves. Without replenishment, the fertility dropped until the land was worn out. Large areas of the South were farmed in this fashion; much of this land was subsequently abandoned because of its low fertility. The economic disruption of the Civil War caused further areas to be abandoned. After a few years pines invaded the land and the extensive pine forests covering so much of the present-day South were formed. If you enter one of these forests in the late afternoon when the sun is at a low angle you can still see furrow marks from the last plowing, over eighty years ago.

Soil Acidity

As the clay-humus particles lose their mineral nutrients the soil becomes less fertile. When these minerals are replaced, however, with hydrogen ions, the soil becomes more acidic. Acidity is defined as the concentration of hydrogen ions, which is usually expressed by the symbol pH. A pH of 1 means that one in 10,000 parts by weight of the soil is composed of hydrogen ions. A pH of 2 means 1 in 100,000 parts; 3, 1 in 1,000,000 parts, and so forth. The lower the pH, the greater the acidity. The pH scale runs from 0 to 14. A pH that decreases from 7 to 0 is becoming increasingly acid; a pH that increases from 7 to 14 is becoming increasingly alkaline. A pH of 7 represents neutrality.

If most of the minerals normally held by the clay-humus particles are replaced by hydrogen ions, the pH drops to about 4. On the other hand, if the clay-humus particles are saturated with minerals, the pH rises to about 7, close to neutral. If the soil is supplied with organic acids or lime, the pH may be lower or higher than these two values. For most soils the pH is between 5 and 7. A low soil pH may affect the solubility of various minerals and, hence, adversely affect the activities of vital microorganisms.

We cannot, however, infer the fertility of the soil from its pH alone. If we had a soil with a large quantity of clay-humus particles but they were only half saturated with minerals, the pH would probably be below 7. Another soil with a very much smaller quantity of clay-humus particles that were fully saturated with minerals would probably have a pH close to 7. An assumption that the latter was more fertile than the former on the basis of pH might be quite mistaken. A more accurate way to measure the fertility of soils is to find the number of clay-humus particles present, determine from that the nutrient exchange capacity of the soil, and from that its fertility.

Man has had his greatest impact upon the water, air, and organic components of soil. This impact is especially obvious in the beaches and dunes characteristic of many coastlines. At first glance, sand hardly seems a soil at all compared with that in our gardens or under our lawns. But if moisture is available, coastal sand qualifies as a soil by supporting vegetation, mostly simple but vigorous growth of a few grasses and shrubs. Beaches and dunes when thus viewed as a soil-vegetation complex demonstrate graphically how man may interfere with normal cycles that produce and maintain these fragile features of our landscape.

CYCLING OF BEACHES AND DUNES

Westhampton Beach is one of the northernmost of a chain of barrier islands stretching virtually without break along the east coast of North America from Long Island south to Yucatan, Mexico. Isolated from each other by inlets and from the mainland by lagoons or bays, until recently these barrier islands have retained a wild beauty that is characteristic of the interface between land and sea, just as the Great Plains embody the beauty of the land-sky interface. Because of this charm these island beaches have been popular as recreation areas for many years. Early visitors first made day trips by excursion steamers, then built crude summer cottages, and finally plush, year-round homes.

Barrier islands and their beaches, formed largely of sand, are quite unstable, despite their characteristically heavy cover of grasses, shrubs, or even low trees. Both the barrier island and its beaches have an erosion cycle: that of the island is hundreds or perhaps thousands of years long, that of the beach is annual. These cycles cause barrier islands to move about constantly, unlike rocky islands, which usually just get smaller as they are eroded away. Barrier islands form when breaking waves build sand bars just off the coast (Figure 3.4). Once these sand bars are built above the high water line, they begin to increase rapidly in size as sand is washed up by longshore currents and by waves. The islands get longer and wider as more sand is added. The movement of longshore currents

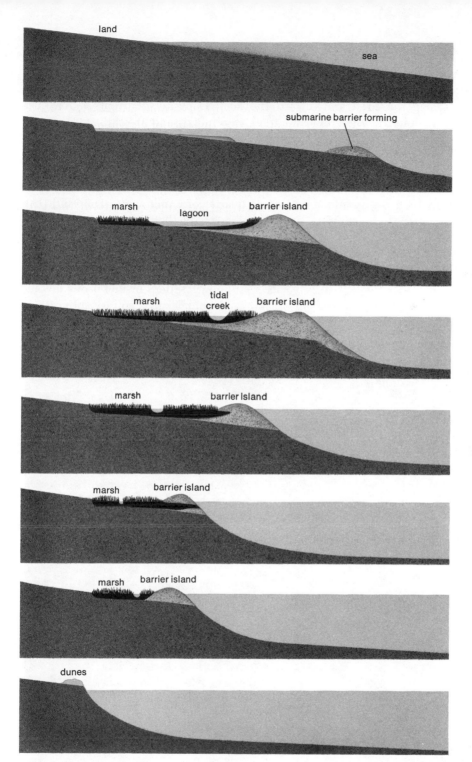

Figure 3.4 Formation, growth, and destruction of a barrier island is shown in this series of drawings. This erosion cycle is caused by the movement of water. (Modified from D. W. Johnson, *Shore Processes and Shoreline Development,* John Wiley & Sons, 1919.)

parallel to the mainland, and the force of storm waves coming in at various angles to the beach line, keep the island moving in two directions —along the coast and toward the land. Longshore currents move in a fairly constant direction and tend to remove sand from the up-current end of the island and to deposit it on the down-current end. Structures built on the eroding end eventually topple into the water; those on the growing end will soon be hundreds of yards from the water's edge.

Violent fall and winter storms often breach these barrier islands, at times scouring an inlet a tenth to a quarter of a mile wide. The sand that occupied these inlets is spread out over the bottom of the adjoining lagoon in the form of a delta or a series of sand bars (Figure 3.5). Over a period of years, as these inlets open and close, large quantities of sand are deposited in the lagoon. The lagoon then becomes shallower, and is gradually covered with water plants that convert the lagoon into a marsh.

Figure 3.5   The Moriches Inlet on Fire Island, New York, just after it was opened by a storm. Sand washed into the bay by the storm waves is piled up along the shore of the inlet. (U.S. Air Force)

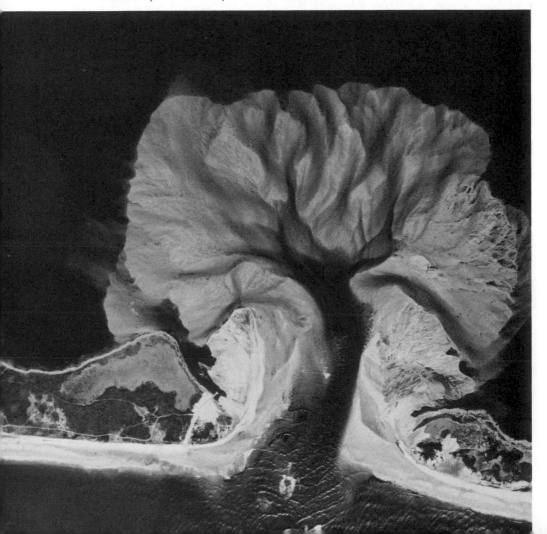

In time, then, the island slowly spreads landward, filling in its lagoon, until it is a part of the mainland.

As old barrier islands are absorbed into the coastline, new ones form so that at any given time there are barrier islands in various stages and positions along coastlines which characteristically support these features. Naturally, beaches follow the movement of the islands of which they form a part. But they have their own annual cycle too.

In the winter large storm waves stir up the sand on the beach face and pull it offshore, eroding the beach by as much as two or three yards. Anyone visiting the beach at this season might well wonder what became of last summer's broad expanse of sand. During the succeeding spring and summer, gentle wave action pushes the sand back onto the beach from the offshore bars where it was deposited during the winter storms. By Memorial Day or the Fourth of July weekend, the beach on the eastern coast of the United States is restored. This annual cycle is even more obvious along the California beaches. Although these beaches are not usually associated with barrier islands but are found at the foot of cliffs, they have the same rhythm of winter removal and summer replacement of sand.

The Disappearing Coastline

Barrier island beaches, then, are constantly on the move, and part of their charm as natural phenomena lies in this movement. But how many times has an eager summer resident built a house in the dunes or on the beach only to find to his dismay that his house and property are moving into the sea.

The fact is, beaches, barrier islands, and sand bars are ephemeral features of the environment. They will persist if sea level, climate, sand supply, and offshore currents remain constant, but not in exactly the same position from year to year. Problems arise when man wants to stabilize the unstable. A simple but revolutionary solution would be to put all such beaches and barrier islands under public ownership, and allow only the most temporary and replaceable recreational facilities to be built. Then we could let these ephemeral features come and go as they please. For those who fear that all beaches would soon slip away into the sea, it can be argued that there will always be sand beaches for recreation somewhere because sand eroded from one beach is not usually dumped into the middle of the ocean but is piled up on another beach somewhere. If man were the cause of the erosion and beach-forming actions, we might properly be concerned, but since these processes are natural we should have the good sense to let them happen as they will, rather than spending millions on remedial action.

If such a prescribed dose of public ownership is unpalatable, then some form of land use control might be established. This could be ad-

Figure 3.6 This groin was constructed to retain sand on a beach originally composed of large cobblestones. The current which moves from right to left in this picture has already begun to remove sand to the left of the groin.

ministered as a zoning regulation which would prohibit the construction of houses on the most unstable sites. The ultimate alternative to such measures is, of course, complete freedom to build a house wherever one wishes, on the top of a sand dune, the high tide line, or even a quarter mile out to sea. But in exchange for this right the builder should not expect public subsidy when his investment is threatened by shifting sand or storm-whipped waves.

Man also attempts to "improve" beaches. Because of strong longshore currents and lack of sand, many stretches of coastline are covered with cobbles and boulders. By pumping sand from offshore deposits, an artificial beach can be constructed over the rocks, but no one should be surprised when after a few years the sand has been returned by the waves to its former location. Such a man-made beach can be maintained only through constant effort and expense.

Artificial beaches can be stabilized to some extent by constructing groins or stone piers out into the sea at right angles to the shore. These intercept the longshore currents that sweep the sand away and allow some sand to be retained (Figure 3.6). Even so, sand must be pumped in every few years to maintain a continuous beach, and the groins must be kept in good repair.

On exposed, sandy coasts the strong onshore winds pile the sand up into dunes often 60–100 feet high, and since dunes form slowly, there is time for the growth of stabilizing vegetation. But a dune is no more

Figure 3.7 Sea-oats, a dune grass, help to hold these sand dunes together near Wrightsville Beach, North Carolina. (Photo by Jim Page)

permanent than the barrier islands on which it develops. Along the Outer Banks off the North Carolina coast the development of dunes follows a pattern that is probably true of all coastal dunes. The dominant plant growing on such established dunes is a grass called sea-oats (Figure 3.7). The growth of the sea-oats and other grasses keeps pace with the development of the dune, stabilizing it while it increases in size by intercepting windblown sand (Figure 3.8). Because this process takes place all along a beach, a long row of dunes is formed parallel with the shoreline. If the coast is building up and sand is accumulating, the process of dune formation is repeated again and again, each new row of dunes being formed on the upper beach in front of the older dunes. From the air they appear as a striking pattern of parallel ridges.

As long as the dune is covered with grass, all is well. The grass slows the wind velocity near the dune surface, protecting the sand already in place, and encouraging deposition of fresh sand. But when this thin, living skin is broken by fishermen, bathers, or livestock wearing a path through it, the wind is able to reach unprotected sand and a cycle of erosion begins. Channeled by such paths, the wind velocity increases, lifting still more sand from the dune surface. The dune grasses are not able to keep pace with sand erosion and are gradually undercut and

Figure 3.8 Young sea-oats plants intercept sand to form a dunelet. This seems to stimulate the sea-oats to grow vigorously, producing more shoots to intercept more sand. Thus the dune gets larger and larger.

Figure 3.9 From a small opening in the sea-oats cover, this dune was attacked by wind and is in the process of being blown away.

killed. From a relatively small opening worn in its protective surface, an entire dune often may be completely blown away by the wind in only a few years (Figure 3.9).

Because of this precarious stability, towns in the vicinity of large dune systems, like Provincetown on Cape Cod or the towns along the Bay of Biscay in southwestern France, have taken elaborate precautions to keep nearby dunes stabilized. Should the wind begin to move them, whole towns could easily be buried. Yet in some areas, dunes are still leveled in callous disregard of their function or aesthetic appeal; houses are built in tract style, and lawns and shrubs are planted as though the sea were a thousand miles away. Under such circumstances, no one should be surprised when the sea reclaims its own.

Beaches and dunes represent one type of cycling. But whatever the climate or type of soil, the vegetation found on most soils plays a critical role in another type of cycling. As we saw in our discussion of the weathering process, nutrients are captured by the roots of plants and ultimately made available to other elements of the biological community. Without this cycling mechanism, most soils would be so lacking in nutrient content that their cultivation would be impossible.

SOIL CYCLES

An ecosystem is any group of organisms interacting with each other and their environment; thus a drop of pond water, the entire pond itself, or the region in which the pond is located can all be considered ecosystems. The ultimate source of the energy which powers the ecosystem is sunlight. But this energy is dissipated as it flows through the system from the photosynthesizing primary producers, algae or grass, through a series of consumers, hunters, and harvesters. Without continued recharging from sunlight ecosystems would soon run down and their component organisms die. Water, foods, and minerals, however, recycle indefinitely in a regular, often subtle way, and it is precisely this cycling of materials through the ecosystem that can so easily be disrupted by man. The continued well-being of ecosystems—and of man, who is a part of nature—depends upon what man does about keeping these natural recycling mechanisms intact.

Almost all minerals, nutrients, and gases have distinct cycles and cycle periods. The most obvious cycles are those of oxygen, carbon dioxide, and water, because photosynthesis, respiration, and transpiration are interwoven and complementary (Figure 3.10). In the production of sugar by photosynthesis, green plants use carbon dioxide and water, releasing oxygen as waste. When these same plants respire, they use oxygen and release carbon dioxide and water, some of which is reused. Animals also need oxygen to produce energy from food, releasing carbon dioxide and water as waste. Much of the oxygen needed by animals comes from plants, and much of the carbon dioxide required by plants

Figure 3.10 Three cycles, oxygen, carbon dioxide, and water, are intimately interlocked through the action of biological, chemical, and geological processes.

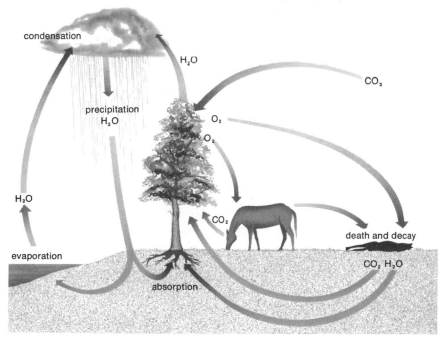

comes from animals. Thus oxygen, carbon dioxide, and water, interrelated in various physiological processes, cycle in the environment. Because cycles such as these involve living organisms as well as geological and chemical processes, they are called biogeochemical cycles.

Mineral Cycling in Temperate Forests

Plants, along with the animals that consume them, perform roles important to the maintenance of the environment as a whole, roles that are not always obvious to man. Gases are by no means the only materials cycled in ecosystems. The roots of many hardwood forest plants readily absorb minerals from deep in the soil. Some species are selective, absorbing greater quantities of some mineral ions than others. Dogwood, for example, selectively absorbs calcium, much of which goes into the developing dogwood leaves in the spring. When the leaves fall at the end of the summer, calcium is added to the nutrient pool held by the clay-humus particles, making it available to plants with shallower root systems. Dogwood thus acts as a pump, bringing calcium from deep in the soil to the surface.

When the leaves of mineral-rich deciduous plants decay after their fall, they decompose and release minerals to the clay-humus particles, resulting in a rich soil with a pH around 7. Let us suppose that because white pine is considered to be more valuable than most hardwoods, we decide to cut down a grove of hardwoods and plant pine. Pine roots are less efficient than hardwood roots at absorbing minerals, so pine needles have a lower mineral content. In a few years the supply of rich hardwood leaves would be replaced with pine needles. More minerals would be lost from the clay-humus particles than would be replaced from the decaying pine needles. As a result, the fertility would decline, the hydrogen ions would accumulate, and the soil would become acidic, that is, the pH would drop. The fall of both fertility and pH would be accelerated by the resin content of the pine needles, since resin forms organic acids in the soil which lower pH and retard decomposition, further decreasing the availability of minerals. Thus, man by his manipulation of vegetation affects the nature of the soil.

Mineral Cycling in Tropical Forests

In a temperate rain forest, most of the minerals made available by decomposition of organic litter or disintegration of the parent rock are quickly absorbed by plant roots and incorporated into the vegetation. If you were to stand in an oak-hickory forest in midsummer you would find several inches of slowly rotting leaves covering the rich topsoil, itself black with incorporated humus. Conversely, a tropical rain forest has such a continuing high rate of organic litter decomposition that no mineral pool has time to accumulate. Directly beneath the most recently fallen debris is a heavy, clay-containing, mineral soil. As a result of this tie-up of all available minerals in the standing vegetation, the cycling of minerals is rapid and direct. As soon as a leaf falls, it is decomposed and its minerals are absorbed by plant roots and channeled into the growth of another leaf. So tight is this cycling process that those few ions not absorbed by plant roots but leached through the soil into the water table, and then out of the system, are replaced by ions picked up by the tree roots from the slowly disintegrating bedrock below.

When tropical forest is cut, minerals are suddenly released faster than crop plants or the remaining trees are able to use them. They leach out of the system and fertility drops sharply. If the disturbance covers only a few acres, weeds and short-lived successional species quickly invade the area, shield the soil, and begin to restore the balanced mineral cycle. But when very large areas are cleared, this kind of recovery may be impossible. The lateritic nature of the soil also becomes part of the problem. When the forest is cleared, the heavily leached sesquioxides are exposed to high temperatures, and they bake into pavement-hard laterite. Once formed, laterite is almost impossible to break up, and areas that once supported lush forest quickly become scrubland at best, supporting only shrubs or stunted trees.

Ignorance of this aspect of forest ecology has had disastrous consequences in some tropical forests. A few years ago a large tract of rain forest in the Amazon Basin was cleared and cultivated. In five years the fields were virtually paved with hard-packed laterite. The ancient Mayan and Khmer civilizations of Mexico and Cambodia probably fell into decay in part because of the destruction of their soil by laterite formation following rain forest clearance.

In view of the abundant evidence that clearing tropical regions of their protective vegetation can often lead to ecological disaster, it is disturbing to read of short-sighted proposals to reclaim and develop large areas of "unproductive" forest. Schemes like the Mekong River Project, which would convert the Mekong River in Southeast Asia into a series of pools like the Tennessee River in the United States, are long on engineering skills and short on ecological knowledge. Much of the fertility of the Mekong Basin comes from the silt left by annual flooding. Deposition of this silt behind a series of dams would benefit no one. The large-scale forest clearance that would have to accompany such an extensive environmental change would probably accelerate the laterization of the soil and result in double-edged destruction under the guise of development: the resources of the forest would be destroyed, and the soil would be ruined for any future growth of vegetation.

THE NITROGEN CYCLE

Approximately 78 percent of the air we breathe is nitrogen. In combination with carbon, hydrogen, and oxygen, nitrogen is an important part of the proteins vital to life. Because of its role in protein synthesis, nitrogen is also one of the most important elements to plants. Absorbed as nitrate or ammonium ions, nitrogen contributes to rapid and luxuriant growth. We know that we can easily supply nitrate and ammonium ions to plants artificially, but how do plants get their supply naturally? Since over 90 percent of the world's supply of nitrogen exists as a gas in the atmosphere, let us examine a series of conversions within the ecosystem that began in the atmosphere. Certain bacteria are able to "fix" nitrogen into ammonia (NH_3), using as their energy source the foods produced by photosynthesis of a host plant. These organisms often live in swellings or nodules on the roots of plants such as beans and clover. These microorganisms are capable of fixing 200 pounds of nitrogen per acre per year into ammonia.

But many plants are unable to use ammonia directly and some are poisoned if ammonia accumulates. Fortunately another group of soil bacteria is able to oxidize ammonia into nitrite (NO_2), and still another oxidizes nitrites to nitrate (NO_3). From these sources, nitrogen is made

Figure 3.11   The nitrogen cycle is of critical importance to the growth of plants that supply the amino acids necessary to build animal proteins.

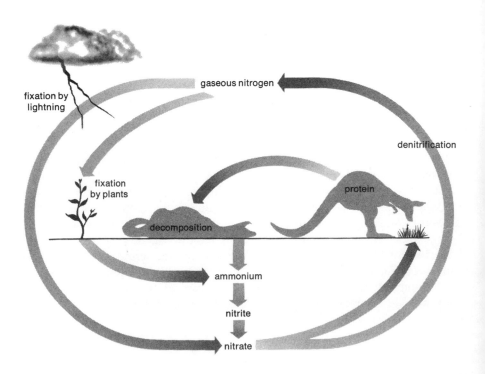

available for the synthesis of protein. But what is the fate of nitrogen contained in proteins when plants or animals die? These proteins are broken down by bacteria into their component amino acids, and the amino acids are in turn broken down into ammonium (NH_4) ions. Bacteria convert the ammonium into nitrite, then either degrade this to nitrogen gas or oxidize it back to nitrate, where it is absorbed by plant roots and the cycle begins again (Figure 3.11).

Perturbations and their Repercussions

In late 1966 a water company in California's San Joaquin Valley notified its customers that its water was considered unsafe for infants, for it contained more than forty-five parts per million (ppm) of nitrate, close to the level that may cause the occasionally fatal disease methemoglobinemia. Methemoglobinemia is the term used to describe a depletion of oxygen in the blood of infants. Because of the almost exclusive milk diet of babies, their stomach and small intestines are less acid than those of adults. This sometimes allows bacteria, which convert nitrate into nitrite and which are native to the alkaline large intestine, to survive in the upper portions of the digestive tract. If water with high levels of nitrate is fed to such infants, these bacteria convert the nitrate to nitrite. Although nitrate is relatively innocuous, nitrite is able to bind itself to hemoglobin in the red blood cells. Normally hemoglobin carries oxygen as oxyhemoglobin all over the body, supplying the respiring cells with oxygen. Nitrite, however, unites more readily than oxygen with hemoglobin-forming methemoglobin. The more nitrite produced, the more methemoglobin formed and the less oxygen available to the body's tissues, resulting in an anemia. This condition has been observed when the nitrate ranges from 66–1100 ppm in the water supply of infants.

The problem occurred in the San Joaquin Valley in California because of certain irrigation practices. Early settlers found a fertile valley but only six inches of rain a year. Wells soon provided irrigation water, but after a number of years of intensive pumping, the well water level fell. At the same time, leaching of both natural and artificially added sources of nitrates produced an enriched zone of nitrates deep in the soil. Then in 1951 the Friant-Kern Canal bringing water from the Sierra Nevada was opened. The water table began to rise again because of the plentiful surface source of water, the reduced demands from wells, and drainage from the irrigated fields. When the water table reached the enriched nitrate zone, the nitrates dissolved and began to contaminate the wells used for drinking water.

Water is not the only source of excess nitrate. Leaf crops that are heavily fertilized with nitrate to encourage vigorous growth may contain greater amounts of nitrate than are desirable. An infant given strained spinach that contains excess nitrate might risk conversion of enough nitrate to nitrite to cause methemoglobinemia. As a result of this danger, the Food and Drug Administration has established as maximum 40 ppm of nitrate in canned food. Thus the consumer is now better protected, but producers are faced with a new challenge. Because of leaching, only about 20 percent of the fertilizer applied to a crop is actually taken up by that crop. To counteract this, the tendency has been to fertilize more heavily, leading occasionally to dangerous accumulations of various minerals in the harvested crop. Yet if the land is not fertilized with nitrates the yield of leaf crops constantly falls. This presents the farmer with a rather serious problem. He must somehow balance meeting the FDA standards with maximum fertilization. If productivity of farms can be maintained only by overfertilization, then less soluble fertilizers that prorate their nutrient effect over an entire growing season must be developed. Other solutions have not as yet been sought because of the relative ease of fertilization. The alternative is pollution of underground water tables that recharge very slowly, some requiring thousands of years.

Logging and Nitrate Runoff

Overfertilization of cropland is not the only source of excess nutrients in the soil. A fascinating experiment set up by F. H. Bormann and G. E. Likens on a series of watersheds in a New Hampshire forest carefully monitored the minerals being added to the systems by rain and snow, and those being removed via stream flow, allowing a balance sheet to be drawn up. After a few years of undisturbed operation one watershed was logged with as little disturbance as possible. The trees were cut and delimbed but left on the forest floor and a herbicide was used to prevent sprouts and weeds from growing. After two years the nitrate ions increased dramatically from 0.9 to 53 ppm. An unexpected result of this experiment was that the overfertilization of a sparkling brook produced a dense growth of algae (see Chapter 7).

Normally nitrate is highest in winter when plants are dormant, lowest in summer when plants are actively growing and using nitrate. As was apparent from the experiment just described, clear cutting, or felling all the trees in a given area, increases runoff by as much as 40 percent over the year (418 percent in normally low water months when the vegetation loses much water from its surface by transpiration). Increased runoff increases the rate of leaching, but the reduced root absorption decreases the removal of minerals from the material leached out, and results in the loss of nitrate from the system. If watersheds are to be properly managed for high-quality water, logging must be recognized as a potential source of excess minerals as well as of silt from erosion.

Obviously we still need to know more about the details of the nitrogen cycle, in particular how it is affected by land use patterns.

NUTRIENT CYCLING

When we examine the nutritional needs of plants, we find that there are two groups of nutrients, those that are required in large quantities, macronutrients, such as nitrate, phosphate, calcium, potassium, and others; and those needed in very small quantities, micronutrients or trace elements, such as cobalt, boron, copper, vanadium, and so on. The need for macronutrients was recognized long ago and was supplied first by natural manures, then by synthetic fertilizers. The need for the micronutrients was a more recent discovery. In 1906 Sir Frederick Hopkins first broached the idea of accessory food factors, which were later recognized as micronutrients. Humans as well as plants need these micronutrients as components of vitamins, enzyme systems, hormones, and their precursors.

The process of harvesting, taking crop plants from the field for processing or consumption, removes many minerals incorporated in plant tissues. Unless these minerals are replaced, the fertility of the soil declines, gradually in temperate areas, precipitously in the tropics. Routine fertilization replaces lost macronutrients—sometimes too efficiently, as we have seen. But the micronutrients may be depleted without symptoms of conspicuous deficiency developing in the plants, and they are much more difficult to replace. Trace element deficiency is not confined to plants, however; it can occur in animals and in humans as well. In man, its subtle, debilitating effects have sometimes been mistaken for mental deficiency. The difficulty of diagnosing trace element deficiencies can best be appreciated by an analogous situation. Tropical natives, once thought by healthy Europeans to be simply lazy and shiftless, were subsequently found to be actually suffering from massive infestations of various kinds of tropical parasites whose cumulative effect was to drain them of that energy and drive so highly prized by "superior" cultures.

In Australia a certain chronic disease of cattle was found to result not from bacteria or viruses, but from a deficiency of cobalt in the soil. When this trace element was added to the soil in the proper amount and passed on to the stock through the grass, the "disease" disappeared.

Certain trace elements are now added to fertilizers, but we still lack knowledge of all such elements that play a vital role in our metabolic well-being. Progress is being made, however, in determining the roles played by trace elements in normal and abnormal metabolism. A deficiency of copper may cause defects in the connective tissue forming the walls of blood vessels, leading to hemorrhages; chromium seems to play a role in the sugar metabolism of the body; and vanadium may inhibit cholesterol formation (the fatty substance deposited on blood vessel walls).

We cannot assume that all the vitamins and trace elements for which we have physiological need are automatically provided by food from our supermarkets. Yet as the population continues to expand, demanding more food from less land, it is absurd to insist that only organically grown food is fit for human consumption, to imply that super phosphate from a bag is somehow unnatural if not subversive. In this age of additives (see Chapter 14), it is too easy to cut corners by adding something to take the place of something milled, processed, boiled, or frozen out. The pluses and minuses may balance, but no one seems to be certain that they do. To assure a reasonable intake of vitamins and nutrients one need not become a health food faddist and subsist on carrot juice and wheat germ. But a widely varied diet, including green and yellow vegetables, fruit, and grain, as well as protein and carbohydrates, *is* essential. The problem is nicely summed up by Dr. Walter Mertz of the Walter Reed Army Medical Center:

> Nobody can state for certain that all the elements with an essential function are known, and nobody can take for granted that our diet contains sufficient quantities of these. Careful animal experimentation must attempt to define deficiency symptoms and to clarify the biochemical and nutritional role of trace elements. Only then, can we hope to apply this knowledge to the detection and prevention of deficiency states in man and to their possible correlation with the many slow, long-term degenerative disease processes that concern us.[1]

Perhaps our general lack of concern with soil results from our increasing remoteness through urbanization. Man is rarely able to exact a yield directly from soil. Only through the medium of vegetation are we able to stabilize soil and control its yield of water and minerals. While we are more or less aware of what happens to vegetation as a result of insect attack, industrial or automotive fumes (see Chapter 10), fire, or drought, the soil is still too often regarded as a kind of incinerator capable of absorbing and decomposing all and every material developed by man. For most of man's history this has been true. The fungi, bacteria, algae, protozoa, and arthropods have had an enormous capacity to cope with organic materials. But in the past few decades, the production of materials unknown in nature has overwhelmed the resources of the soil. Plastics and biocides persist in the soil because so few organisms can break these materials down.

---

[1] Conference on trace substances in environmental health. 2nd. University of Missouri. 1968. D. D. Hemphill (ed.), 1969. *Trace substances in environmental health, II.*

If soil is to be maintained as a resource, we must give much greater priority to understanding its cycles, to untangling its structure and function, and to understanding its fauna and flora and their relationship to both natural environmental factors and those imposed by man's use. Until then, we run the grave risk of destroying a resource and permanently upsetting a balanced system before its full potential or basic importance to man is fully appreciated.

FURTHER READING

Anderson, E., 1952. *Plants, man, and life*. Little, Brown and Co., Boston, Massachusetts. Presents some alternatives to monoculture and its problems.

Bormann, F. H. and G. E. Likens, 1967. "Nutrient cycling." *Science*, **155**, pp. 424–429. Explores the movement of nutrients in the ecosystem.

Cole, L. C., 1966. "Protect the friendly microbes." *Saturday Review*, **49** (19), p. 46. Sound treatment of nutrient cycling, especially nitrogen, and of the contribution of soil microorganisms.

Eyre, S. R., 1968. *Vegetation and soils, a world picture*. 2nd ed. Aldine, Chicago, Illinois. A comprehensive coverage of soils—their nature, origin, and interaction with climate and vegetation.

Gourdou, P. and E. D. Laborde, 1966. *The tropical world*. 4th ed. Wiley & Sons, New York. A good discussion of tropical soils and their limitations.

Hyams, E., 1952. *Soil and civilization*. Thames and Hudson, London, England. Intriguing account of the effects of soil destruction upon history, with examples from the classical world, Far East, and United States.

McNeil, M., 1964. "Lateritic soils." *Scientific American*, **211** (5), p. 96. Explores the results of laterite development in tropical ecosystems.

Pendleton, R. L., 1941. "Laterite and its structural uses in Thailand and Cambodia." *Geographic Review*, **31**, pp. 177–202. Interesting account of the impact of agricultural practice, soil development, and building materials upon culture.

Richards, P. W., 1952. *The tropical rain forest*. Cambridge University Press, Cambridge, England. The most comprehensive treatment of the rain forests of the world.

Wagner, R. H., 1964. "The ecology of *Uniola paniculata* in the dune-strand habitat of North Carolina." *Ecol. Mon.*, **34**, pp. 79–96. Discusses the important reciprocal relationship between a dune grass and its sand dune habitat.

CHAPTER 4

NATURE IN CAPTIVITY

The forest and sand-dune ecosystems have been especially vulnerable to man, who has, first by necessity and then through habit, viewed nature as an adversary against which he must constantly struggle. Even today to many of us who would think nothing of shopping in a large city, oblivious to the very real perils of smog, dust, taxis, and muggers, the word "wilderness" still evokes images of hostility and danger. In fact, the mere suggestion of a wilderness hike may start us thinking about protection against the sun, rain, insects, snakes, and bears. Is it any wonder, then, that most wilderness has been made "safe" by intensive use or urbanization?

PARKS

While much wilderness has been destroyed, some has been transformed into parks. This transformation is most clearly seen in densely populated England. The thick oak forest that once covered England was methodically cut for ship timbers and lumber and replaced by cultivated land. By the seventeenth century, forests were limited to those selected as crown lands to be reserved for royal hunting parties and the forests which frequently surrounded the country manor houses of the aristocracy. By the eighteenth century, these woodlands, managed for both timber and game, lost much of their wildness. Often too, they were grazed, which tended to open the forest by curbing tree reproduction. Over the years these hunting parks became fixed in the public mind as representing natural vegetation. When the wealthy members of the rising middle class began to construct country estates, landscaping often reflected the influence of the hunting parks of the older estates. Landscape architects like Capability Brown manipulated the lands masterfully, damming streams to create ponds and arranging clumps of trees on large greens to soften the landscape and to provide vistas of distant fields and

Figure 4.1 This natural looking landscape at Blenheim Palace, England, including a bridge and pond, was planned and planted by the famous English landscape gardener, Capability Brown.

hills (Figure 4.1). In *Pride and Prejudice,* Jane Austen describes a walk in such a park:

> They entered the woods, and bidding adieu to the river for a while, ascended some of the higher grounds; whence, in spots where the opening of the trees gave the eye power to wander, were many charming views of the valley, the opposite hills, with the long range of woods overspreading many, and occasionally part of the stream. Mr. Gardiner expressed a wish of going round the whole Park, but feared it might be beyond a walk. With a triumphant smile, they were told, that it was ten miles round.

In the American colonies, New England towns and villages had their green or common, but these were obviously intended as places of public ownership where sheep and cattle could be conveniently pastured. The idea that parks were for public pleasure grew with surprising slowness. Besides, pleasure in an outdoor setting, or in any setting for that matter, was anathema to the established church of the late eighteenth and early nineteenth century in New England. This attitude began to change when orthodoxy gave way to the more liberal Congregational and Unitarian Churches in the nineteenth century. Indeed, moral fiber became so relaxed that one liberal man of some wealth donated a large tract of land to the City of Worcester, Massachusetts, as a public park for the *enjoyment* of the people of that city. Shortly thereafter, New York City became the second American city to establish a public park. Elaborately landscaped by Frederick Law Olmsted, Central Park set the tone for the creation of public parks from coast to coast, until few towns were without some patch of green with statue, fountain, or flagpole (Figure 4.2).

Figure 4.2 A typical town square in Pennsylvania includes a Civil War monument, flagpole, cannons, and flower beds.

The National Park

The first explorers of the West, hardy trappers and Indian traders, brought back tales of fantastic mountains and valleys, geysers of steam, gleaming caves, and huge trees. Subsequent expeditions corroborated even the wildest of these tales. Because of the Victorian glorification of the romantic, there developed a demand that some of these natural marvels be preserved for the wonderment and enjoyment of future generations (Figure 4.3). Surprisingly, many such monuments of nature *were* preserved; and this at the height of an era of westward expansion when nature was considered by railroad, mining, lumbering, and cattle interests to have been created for man's exploitation. Between 1860 and 1900, four of the nation's best-known national parks were established: Yellowstone (1872), Yosemite (1890), Sequoia (1890), and Mount Rainier (1899). The National Park Service, which inherited these natural wonders when it was founded in 1916, had as its prime duty their protection and preservation.

Problems in the Parks

Preservation of major geologic features like the Zion, Bryce, or Grand Canyons is relatively easy. They change so slowly that we have the illusion in our short lifespans of no change at all. But lesser environmental features such as plants and animals are constantly changing, at a rapid rate. One such feature common to our national parks was the stately

Figure 4.3 Thomas Cole and William Cullen Bryant view the sublimities of nature in a painting by Asher B. Durand. (Collections of The New York Public Library, Astor, Lenox, and Tilden Foundations)

Figure 4.4 An open grove of ponderosa pine in Washington. Maintenance of the parklike aspect of this grove depends upon periodic ground fires to remove competing species, thin ponderosa pine seedlings, and prepare the soil for the germination and growth of the next generation of trees. (Photo by H. Weaver, courtesy of H. H. Biswell)

groves of trees—douglas fir, ponderosa pine, redwood—their massive trunks suggesting the columns of great cathedrals (Figure 4.4). Beneath, like a primordial garden, the ground was carpeted with grasses and flowers. The coincidental resemblance of these forests to the European concept of a well-tended park surely gave early impetus to their preservation as parks, but ironically, it very nearly brought about their ultimate destruction.

Park officials seemed to ignore the fact that trees grow old, die, and are replaced by other trees, all in a dynamic balance with various environmental factors. Instead they excluded fire, removed diseased trees, and even tried to control natural predation to protect the "good" animals from the "bad." All efforts were made to stop time—to preserve the scene just as the early settlers saw it.

But in a few years things began to go awry. With the elimination of fire (see Chapter 5), twigs, needles, branches, and fallen trees began to pile up (Figure 4.5). Decomposition of these fallen materials proceeded slowly in the hot, dry western forests. Without fire to remove litter,

Figure 4.5 Accumulated litter representing years of protection from fire. Should this ignite, the entire forest would be lost. (Photo by H. Weaver, courtesy of H. H. Biswell)

Figure 4.6 Protection from fire allowed these weed trees to obscure the more scenic sequoias and increase the danger from wildfire. (Photo by H. Weaver, courtesy of H. H. Biswell)

seedlings of the desirable species were often unable either to germinate or, if they did, to survive the shade.

Under a regime of fire elimination, the desirable trees were unable to reproduce in national parks, and seedlings of shade-tolerant species, such as conifers, began to invade, forming a dense undergrowth that obscured the beautifully spaced trees that the parks were set up to preserve (Figure 4.6). This not only impaired the view, but because of the deep layer of highly combustible litter and the equally combustible understory of conifers, parts of some national parks became veritable tinder boxes. No longer could a fire be localized near the ground in needles and dry grass. Once the litter was ignited the fire would quickly jump into the understory and then into the crowns of the giant ponderosas, redwoods, or douglas firs. A crown fire is the end of any forest, and if many of them occurred, the recreational, aesthetic, and scientific qualities of parks would disappear.

That this is not altogether a recent problem can be seen in an 1887–1888 report of the commission charged with the responsibility of maintaining Yosemite Valley. In part the report stated:

Several times during the period of my labors on your behalf, it required suddenly almost the entire force of twenty or thirty men to divert the all consuming course of forest fires on the floor of the valley. Since the annual practice of the Indians in burning off the dried grasses and leaves has been discontinued, and even forbidden by law, the accumulation of vegetable matter beneath the trees has been practically undisturbed, until a growth of young pines has sprung up all over the valley, and destroyed much fine meadow land. A campfire carelessly left, or a match thrown among the leaves, has caused several fires within the past two or three years that could not be extinguished. They burned until the walls, the roadway, and streams defined and determined their course.[1]

This was true of Yosemite in the 1880s and it is now true of Lassen, Sequoia, and King's Canyon. The expense of hand-clearing trash species and removing forest floor litter from huge areas of rough country is enormous. To attempt this with bulldozers would severely injure tree roots, expose the forest floor to erosion, and be impossible altogether in inaccessible areas. Fire is too long overdue to be used now without grave danger. What can the Park Service do? The only practical solution, regardless of expense, seems to be hand-clearing, acre by acre, and removal of combustibles, followed by controlled burning every few years to keep litter to a minimum and to eliminate trash trees when they are still small seedlings. But because this hand labor is so terribly expensive, the task will take many years—years filled with the risk that large tracts of forests will suddenly explode in a grand conflagration, should a fire ever get out of control some hot, windy day in late summer.

The preservation of forests that began ten, fifty, or one hundred years ago is clearly impossible if we try to extend their existence by merely erecting a protective "fence." A forest *can* be maintained in the way that a hedge can be kept at a desired height—with active management requiring constant care. If we are to manage effectively, we must through ecological research devise management techniques and then unrelentingly apply them.

Wilderness Areas

The idea of preservation in a time-stopping sense dies hard. Many who have been disillusioned by the reality of changes brought about by increasing use of national parks have pressed for wilderness areas—remote, inviolate, unspoiled. But the idea of a brooding, mysterious, and awesome wilderness is finished. Man now views the wilderness in an entirely different light. In an age of crowded, dirty cities, the wilderness has come to symbolize a refuge, the last place where man can breathe clean

[1] Gibbens, R. P. and H. F. Heady, 1964. "The influence of modern man on the vegetation of Yosemite Valley." *Calif. Agric. Exp. Sta. Manual* 36.

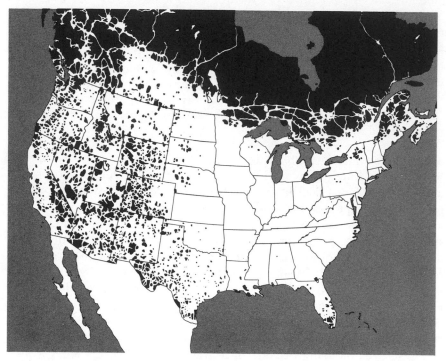

Figure 4.7   The areas in black are over five miles from the nearest road, railroad, or navigable waterway. This is America's remaining wilderness. (From C. Tunnard and B. Pushkarev, *Man-Made America,* Yale University Press, 1963.)

air, drink freely from streams, and get away from other people. Unfortunately, wilderness has become a symbol viewed in romantic anthropocentric terms, its cycles of destruction and renewal ignored because they extend beyond the lifespan of man, an attitude that will surely lead to the same difficulties encountered in the national parks. A true wilderness should be viewed biocentrically; its forests must be free to burn, free to be attacked by insects, free to be blown down by storms, and free to be carried away by floods, all because these are natural events to which the forest is adapted to respond. When left alone, forests regenerate themselves. The new forests may be different from the old, but that is the way things change in a natural ecosystem.

The idealized wilderness is a myth. We must have the wilderness, but not for selfish gratification. At stake is the survival of ecosystems, free to change in accordance with natural variations in the environment, without interference from man with his notions of good and evil, a last refuge of absolute freedom in a world of increasing technological control. To allow some measure of this ecological freedom, a wilderness must consist of at least 100 thousand acres in a fairly compact unit, without usable public roads or conspicuous signs of human activity. A block of land of such size is often free from the intrusive effects of the surrounding man-dominated countryside. In 1965 there were sixty-four tracts meeting these standards in the United States, and some have since been designated as reserves following the passage of the Wilderness Act in 1966 (Figure 4.7). Despite their size, however, wilderness areas share the plight of the national parks.

People, Parks, and Wildernesses

If people lived within national parks and wilderness areas by hunting and gathering, their populations limited by the availability of food, these areas could last indefinitely. But they will not last, because people, sustained from outside sources, are pouring into these areas in increasing numbers, destroying by their very presence the natural landscape that the parks were set up to preserve. On popular summer weekends Yosemite National Park experienced bumper-to-bumper traffic, which generated a pall of smog in that lovely valley. Finally, in 1970, its roads were closed to vehicular traffic. As the population has increased and more people have had leisure to visit parks and wilderness areas, the situation has worsened.

One solution might be to build just outside each national park a large, comfortable reception center which would display the wonders of the park, using lifelike dioramas, slide shows, films, and perhaps closed-circuit color television. A wag has suggested that if a three-color bumper sticker were available at such a center, proclaiming that a visit had been made, many visitors might be satisfied by that alone. For those who insist upon venturing beyond the safety and comfort of their automobiles, a system of monorails or cable cars could be installed to allow close observation of plants, animals, and earth formations, affording spectacular views unavailable from the ground, while protecting the ground from human traffic. For the minority (probably less than 5 percent) who want direct physical interaction with the park, a system of backpack campgrounds, connected by trails, rotated periodically to avoid overuse, might be arranged. Such a program would relegate roads, buildings, and parking lots to the park perimeter and at the same time would satisfy the three major types of park users.

Although it is unlikely that such a program will soon be implemented, certain features of the proposal are already in use or are being considered. During the summer, parts of Death Valley National Monument are closed to the public because of the excessive heat, but views of the barred area can be seen via slides and films from a visitors' center. Other parks are considering removal of facilities to outside locations, and in others, road construction has been halted and consideration is being given to the removal of some of those already in existence.

As the demand for park space increases on every level—national, state, and local—there is a real possibility of reconstructing many natural environments dispersed or destroyed by man's activities. A cypress swamp cannot be constructed in South Dakota or a rain forest in Vermont, but with proper space, research, and funding, a previously existing biological community can be reassembled. This suggestion is offered not as a license to destroy existing ecosystems, but as a means for restoring those long gone.

New York's Central Park is a striking example of just such an environmental restoration. Around 1850 the area now occupied by the park was covered with squatters' huts and shacks, and was on the outskirts of New York City. Period drawings and woodcuts show almost no trees or shrubs. In fact, there was little vegetation of any kind (Figure 4.8). The designer of the park, Olmsted, was obliged to start from scratch. Walking through the park over one hundred years later, it is difficult to realize that almost every tree and shrub was planted by hand (Figure 4.9). Naturally most of these species are ornamental, but despite the great odds against them, some wild species have returned, brought by bird droppings and random introductions. If some sort of a natural environment can be recreated in the center of a megalopolis, why not recreate prairies, forests, and marshes?

In an exceedingly interesting paper A. S. Leopold suggests that ecosystem reconstruction could be managed in four basic steps:

1. Historical research to determine what the area looked like before man began to alter things.
2. Ecological research to find out how the plants and animals are related to each other and to the ecosystem, and which environmental factors are most important in maintaining the area as we wish to see it.
3. Pilot experiments, based on the information gained from ecological research, to establish the feasibility of proposals.
4. Full-scale management of large ecosystems based on the more workable of the pilot studies.

This kind of ecosystem reconstruction is already being attempted in certain parts of the Great Plains where native plants and animals are being reintroduced in an attempt to reconstitute the tall grass prairie that was plowed into history.

THE HIERARCHY OF NATURAL AREAS

Many types of natural ecosystems—indeed the most scenic ones—are found in national parks and wilderness areas. But there are also immense

Figure 4.8 In 1859, squatters' huts occupied the land that later became Central Park in New York. (Museum of the City of New York)

Figure 4.9 Virtually every tree and shrub in this contemporary photo of Central Park was planted by hand. (Thomas Airviews)

portions of the West held by the government as national forests that are used for timber, watershed protection, and recreation, where a fine variety of natural ecosystems is found. While the scenery is not usually as spectacular as that in the parks, there is far more opportunity to enjoy the outdoors free from crowds of people. In the future, as the population of the United States increases, more and more national forests will probably have to be managed strictly for recreation. If this can be done, the pressure on existing parks may be reduced. National forests will also serve in future years as the last source of national parks, for most suitable private land will soon be either developed or priced beyond reach.

State parks fill an important role in making the natural world available to people who cannot travel to the great western national parks. Quite generously scattered across the country, state parks are within an hour's drive for most people. Finally, local parks provide contact with some semblance of nature, if only in the form of grass, trees, and squirrels.

As one descends the hierarchy of reserved natural environments from national park to city park, their size and scenic qualities decrease, but their *use* increases dramatically. Central Park, although huge by city park standards, is literally being worn to the bedrock by the enormous pressure exerted upon its facilities by New Yorkers (Figure 4.10). Unless alternate recreational facilities are provided in the form of smaller vest-pocket parks, or far more maintenance is provided for Central Park than is currently available, or both, the park will succumb like an isolated field of wheat to a swarm of locusts.

Figure 4.10 When tree roots are exposed the bark thickens. More than one foot of soil has eroded from the base of this tree in Central Park since its roots were first exposed. (Photo by Dan Jacobs)

Figure 4.11 This landscape along the Blue Ridge Parkway in the Southern Appalachians is purposely being managed to provide an interplay of forests and cultivated fields that is aesthetically pleasing. (National Park Service)

Scenic Easements

We conventionally think of natural reserves as separate, specified areas. There are many other possibilities. One of the most useful of these is the device of *scenic easement.* Easement allows a property holder to retain ownership of his land, yet he can sell or lease the right-of-way, air rights, or the right to develop. This means that land of great scenic value could remain in private hands, but the right to change the landscape by development might be sold or leased to a governmental body. The government, of course, would not exercise this right, but would leave the land in its natural state. The property owner would get some financial return for his sacrifice in not developing his land, and the public would gain by not having to buy the land outright to preserve its beauty.

A classic example of the use of scenic easement was the securing of easements on the development of land across the Potomac River from Mount Vernon, thereby preserving a view that Washington himself must have seen. Without such an arrangement the opposite bank would doubtless have sprouted high-rise apartments, marinas, and shopping centers, destroying the illusion of stepping back into history that is so carefully and beautifully fostered at Mount Vernon. While we have preserved a view, we have also saved a stand of irreplaceable riverbank forest.

Parkways

Easements are especially useful in the construction of parkways and the preservation of riverscapes. Their scenic value is a tremendously important aspect seldom considered in plans to acquire and reserve open space. Driving along the Blue Ridge Parkway in Virginia and North Carolina gives the feeling of intimate contact with nature. Although the Park Service bought most of the land bordering the Parkway, much has been leased to local farmers (Figure 4.11) to preserve the interplay of woods and fields. Thus the monotony that might otherwise have resulted from long tunnels through the trees has been relieved.

Rivers

A combination of federal, state, and local parks together with scenic easements has been proposed to retain the relatively unspoiled beauty of the Connecticut River in New England, the Hudson in New York (Figure 4.12), and the Potomac in the mid-Atlantic states. Ironically, the pollution of these rivers has prevented the commercial development that has already spoiled many cleaner streams. But time is running out: as the pollution in these three rivers is reduced, steps must be taken to forestall the usual haphazard development projects with ecologically sound master plans that will assure the continued beauty of the rivers commensurate with recreational and economic uses.

Trails

In 1937 the Appalachian Trail was opened (Figure 4.13), running 2000 miles from Mount Katahdin in Maine to Springer Mountain, Georgia. Like the proposed scenic rivers, the trail is a composite of federal, state, local, and private ownership, which has worked well over the years, allowing public access to some superb natural areas. Other shorter foot trails have been established around the United States and many are in the process of development. In the planning stage, for example, is a north-south trail through Florida. Since trails can be rapidly destroyed by overuse, there must be adequate alternatives to spread the traffic more equitably.

Nature Preserves

Among the smallest natural reserves are so-called nature preserves. Usually they are small-scale, under 1000 acres, often under 100 acres. They are composed of ponds, salt marshes, woods, prairies—features that are in danger of disappearing altogether. Mostly privately owned by schools, clubs, and conservation organizations, they serve a vital purpose in preserving natural areas too small to be considered for national or state parks. The Nature Conservancy, a private organization, has been instrumental in saving many of these small areas from bulldozers or sanitary landfill operations and holding them for future enjoyment and study.

Other areas capable of multiple use are golf courses, rights-of-way of various types, and even cemeteries. All preserve open space that can be used for alternative purposes.

Nature in an Urban Setting

All such categories represent occasional use by man, at best. The nature many of us live with is in our backyards, or if we are apartment dwellers, the trees in the street, the ivy growing on the wall, or the geraniums growing on the windowsill. The connection of man and nature, however tenuous, can be maintained in the dreariest apartment by a window full

Figure 4.12 A river valley does not have to be a virgin wilderness to be beautiful, as this view of the Hudson River Valley from Bear Mountain State Park in New York demonstrates. (Photo by Harry Thayer)

Figure 4.13 The Appalachian Trail runs for 2000 miles along the crest of the Appalachian Mountains from Mt. Katahdin in Maine to Springer Mountain, Georgia.

of house plants as carefully nurtured as the oldest sequoia tree in Sequoia National Park. To a confirmed urbanite, the sight of an ailanthus tree growing behind a tenement, or pushing its way through the soil at the edge of a parking lot, is a treasured thing indeed (Figure 4.14).

The greatest problem in the effort to save some fragments of a once-natural world has been the failure to recognize the hierarchy of natural systems from parking lot ailanthus to redwood; each community has its place in the world and each is equally important. We must have as broad a spectrum of natural environments as can be assembled, from entire forests, deserts, mountains, and lakes down to ponds, bogs, woodlots, to single trees. Our spectrum should also include a wide range of uses, from inviolate wilderness available only for scientific study, to open space so heavily used that its vegetation must be continually renewed. Man, despite the cries of impassioned romantics, is a part of the natural environment just as the natural world is part of man. While the opportunities still exist, land must be earmarked for *all* categories of natural areas. Once a salt marsh has been filled with waste, a sand dune leveled for houses, or a valley flooded by a reservoir, the options for preservation and continued renewal are forever lost, for there are limits to environmental reconstruction (Figure 4.15). Failure to act now can only impoverish future generations, which will have problems enough to bear. Let us give them at least a glimpse of how planet earth once looked.

Figure 4.14 Nature in the city is often limited to a window full of house plants, and scraggly ailanthus trees in unlikely places.

Figure 4.15 There will be no renewal for this salt marsh being "reclaimed" along the Connecticut coast.

FURTHER READING

Abbey, E., 1968. *Desert solitaire: a season in the wilderness.* McGraw-Hill, New York. A beautifully written account of Abbey's experience as a park naturalist in Utah.

Darling, F. and J. P. Milton (eds.), 1966. *Future environments of North America.* Natural History Press, Garden City, New York. Papers and superb discussions about the future of the natural world in North America.

Darling, F. and N. D. Eichhorn, 1967. *Man and nature in the national parks.* The Conservation Foundation, Washington, D.C. A brief survey of the problems arising from overuse of the National Park system.

Huth, H., 1957. *Nature and the American.* University of California Press, Berkeley, California. A leisurely look at our changing attitudes toward nature.

Leopold, A. S., 1963. "Study of wildlife problems in national parks." Transactions of the Twenty-eighth North American Wildlife Conference, pp. 28–45. A very perceptive study that points out the basic ecology behind the problems.

McHarg, I. L., 1969. *Design with nature.* Natural History Press, Garden City, New York. A very personal view of the landscape by a landscape architect with exceptional ecological awareness.

Nash, R., 1967. *Wilderness and the American mind.* Yale University Press, New Haven, Connecticut. Traces the wilderness concept through American history.

Shepard, P., 1967. *Man in the landscape.* Random House, New York. A literary review of man's past and present relationship with nature.

Webb, W. P., 1931. *The Great Plains.* Grosset & Dunlap, New York. Still a classic statement of how a landscape has influenced not only its own development, but the course of the American nation.

Whyte, W. H., 1968. *The last landscape.* Doubleday, New York. A fine sourcebook on what to do about the disappearance of open space.

U.S. Department of Commerce, 1966. *A proposed program for scenic roads and parkways.* Washington, D.C. Handsomely illustrated guide to the proper integration of the highway into a scenic landscape.

U.S. Department of Interior, 1966. *Potomac valley, a model of scenic and recreational values.* Washington, D.C. Outlines the opportunities of preserving the scenic and recreational values of the Potomac.

Part II

NATURAL TRAUMAS
BECOME POLLUTANTS

CHAPTER 5

FIRE AND MAN

The eighth of October, 1871, had been especially hot and dry in Peshtigo, a small lumber and wood products town in eastern Wisconsin. The continuous burning of slash from lumbering operations and railroad construction in the white pine forests surrounding the town had kept a pall of smoke hanging over the area all summer. Toward evening the smoke thickened and a rosy glow appeared in the sky. Soon sparks began falling and a roaring wall of flame swept through the town, consuming everything in its path.

It was a seething, searing hell, and the hurricane it was riding traveled almost as fast as light itself. It swept in so suddenly that no man could say for certain what happened in the next few moments. What is assuredly and horribly known is that some forty people, their senses blown away in that first blast of flames, rushed into the Company's boarding house, and there were burned to cinders.

John Cameron and many others were fleeing down the east bank of the river below the dam, and here many drowned and others lived it out. On the way they saw things they never forgot. Never. They saw horses and cattle, yes, and men and women stagger a brief moment over the smoking sawdust streets, then go down to burn brightly like so many flares of pitch-pine. Forty years afterward, an ancient man's voice choked as he told of watching pretty Helga Rockstad as she ran down a blazing sidewalk, her blond hair streaming, and of seeing the long blond hair leap into flame that stopped Helga in her tracks. Looking at the spot next morning, he found two nickel garter buckles and a little mound of white-gray ash.[1]

[1] Holbrook, S. H., 1943. *Burning an empire, the story of American forest fires.* Macmillan, New York.

Six hundred people died in the town of Peshtigo, over 1100 in the area swept by the fire storm, and one million acres of white pine forest, some of it virgin timber, were destroyed. Subsequent fires in Hinckley, Minnesota in 1894; in Yacolt, Washington in 1902; and in Bitter Root, Montana and Coeur d'Alene, Idaho in 1910, confirmed the public and professional feeling that fire and forests were incompatible. What is this natural phenomenon that has had such an enormous impact on the environment?

FIRE AS A NATURAL PHENOMENON

Fire has always been a part of the natural environment, and although man can greatly affect its incidence and spread, it is a normal oxidation reaction. Most metals oxidize in the presence of air and become covered with a thin coat of metal oxide, for example, copper oxide, the green patina on copper-sheathed roofs, and iron oxide, the rust on unprotected iron. These oxidation processes occur rather slowly, taking hours, days, or even years. But fire is an oxidation reaction which goes to completion in seconds, once initiated. All that is required is fuel, oxygen, and something to spark the reaction.

Most fuels that we commonly associate with burning are organic, that is, they are combinations of carbon, hydrogen, and oxygen. As the temperature of a fuel increases, bonds between carbon and hydrogen atoms begin to break. Oxygen, by acting as an acceptor of hydrogen and carbon, forms two products in this rapid oxidation process: water and carbon dioxide. But only a small portion of the original bond energy is transferred to the water and carbon dioxide; the rest is transferred into the environment as heat and light. It is the heat and light energy that makes fire useful to man.

Except for the driest of deserts, most terrestrial ecosystems contain plants and plant remains, which at some point in the year become combustible. When enough of these combustibles have accumulated, even a rather barren-appearing landscape may catch fire. In forests or grasslands, spectacular fires may be carried for great distances. Compared with an uncontrolled prairie or forest fire, the burnings of Rome, London, San Francisco, and Dresden were localized events.

Although *uncontrolled* fires are destructive, there is no reason to think of fire as an evil to be abolished, any more than an occasional deluge resulting in a flood is reason to ban precipitation. Certainly only controlled burning is truly useful to man. But people often view fire of any sort in the natural environment as an out-of-place phenomenon that must be curbed or eliminated at all costs. The roots of this belief are only partly explained by disasters like Peshtigo.

FORESTRY AND FIRE

Attempts have been made many times and by many cultures to manage and regulate forests for water supply, grazing, and the cutting of timber. However, formal education in forestry did not develop until 1825, when schools were established in Germany and France. The concept of forest management came late to North America's seemingly inexhaustible forests. It was not until the superb white pine forests of New England and then the Great Lakes area were gone that people began to consider the possibility of managing the remnants and second growth for sustained yield. Since there were no schools in the United States, professional training had to be obtained abroad in the highly respected schools of Europe, particularly in Germany.

The major emphasis of European forestry in the nineteenth century was protection of the forest. Considering the population density of Europe even then, the demands made on forests for kindling were great, and although gleaning rights to this forest by-product were often passed from father to son for generations, forests had to be protected from "poaching." But European forests also had to be protected from fire. Fire was considered an enemy in the carefully tended conifer forest plantations, and any role that it might have played in the ecology of natural vegetation was thereby blurred. However, what was useful in the forest plantations of Europe was not necessarily useful in the incredibly large and ecologically diversified forests of North America. Because fire was considered intolerable in European forests, it was not surprising that this view was extended by European trained foresters to *all* forest types in North America. Fire exclusion came to be regarded as a basic article of forest management, first by private foresters, and later by the government's Forest Service, set up to manage the huge tracts of public land still covered by timber.

Longleaf Pine and Fire

After much of the best white pine had been cut in the Northeast and the Great Lakes area, attention passed to the hitherto neglected longleaf pine forests distributed in a band along the coastal plain from Virginia to Texas. The trees were well spaced, with a light cover of grasses on the forest floor. After the initial logging of longleaf stands, it was noticed that reproduction was poor, leading to replacement by much less valuable hardwoods or other species of pines. It was supposed that ground fires, quite prevalent in the region, and the foraging of hogs destroyed the seedlings, and that if both of these factors were controlled and enough seed trees were left to provide a seed source, the forest would soon regenerate itself.

A young forester by the name of Chapman took a close look at the ecology of longleaf pine and realized that the species, far from being eliminated by ground fires, actually seemed to have many features suggesting long association with fire as a part of its environment. He noticed that the young seedling at the ground surface develops a fat bud, surrounded by a tuft of needles (Figure 5.1). Then, for the next three to seven years, the seedling puts most of its energy into developing a deep taproot. Because of its close resemblance to a clump of grass, this phase of longleaf pine development is called the grass stage. By the end of seven years the seedling suddenly begins active vertical growth, and in a few years' time grows ten to fifteen feet. Thereafter the sapling grows at a more conventional rate but develops thick fire-resistant bark. Chapman concluded that periodic ground fires may singe the tuft of needles but leave the terminal bud and taproot intact. By using energy stored in the taproot over the three to seven years spent in the grass stage, the seedling is able to make exceptional growth in height over a period of

Figure 5.1 In the young seedling or "grass" stage of the longleaf pine, the total resources of the seedling go into production of a deep taproot and a fat terminal bud. Food is stored in both of these areas for the active growth phase that follows in a few years.

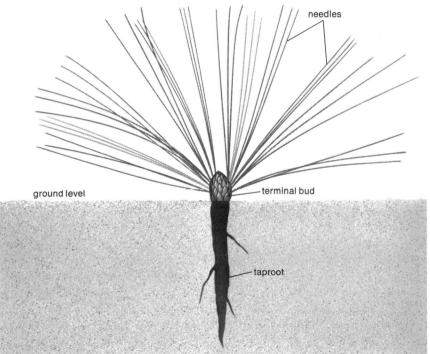

two or three years, time enough to get the more sensitive new branches up and away from the danger of fire. Then, development of thick fire-resistant bark protects the young tree from further ground fires.

Drawing on his observations as early as 1909, Chapman noted that "fire always has and always will be an element in longleaf forests, and the problem is not how fire can be eliminated, but how it can be controlled so as, first, to secure reproduction; second, to prevent the accumulation of litter and reduce the danger of a really disastrous blaze."[2] This sound appraisal of the situation was not taken seriously at first.

Data versus Dogma

Because he was not affiliated with the Forest Service, Chapman was free to speak out. The Forest Service seemed to have already made up its mind about the role of fire in all forests, contending that southern forests were no different than any other in their need to be completely protected from fire. An experiment was set up in Louisiana in 1915 to demonstrate the effect, both combined and singly, of foraging pigs and ground fires in eliminating longleaf reproduction.

An acre of young seedlings was divided into quarter-acre plots and treated as follows: one plot was burned and grazed, a second was unburned and grazed, a third was burned and not grazed, and a fourth was neither burned nor grazed. By 1918 the two grazed plots were completely cleared of longleaf pine seedlings. By 1924 the Forest Service reported that in the third plot some seedlings were killed and were not growing as well as the larger number of seedlings in the fourth plot, so fire was condemned for damaging longleaf reproduction and growth. Its fire policy thus affirmed experimentally, the Forest Service terminated the experiment and landowners were strongly urged to eliminate fire from their pine stands.

Chapman, however, continued to follow the now officially abandoned experiment and found that by the late 1920s the trees in the third or fire-thinned plot began making much faster growth than the unthinned trees in the fourth plot where the crowded trees began to stunt each other's growth. The Forest Service, however, though informed of these results, never got around to publishing them and continued in its well-established dogma that fire must be eliminated from longleaf stands.

Too frequent burning of the Southeast reinforced the now rigid attitude of the Forest Service toward fire. Seeing no possible value in woods burning, the Forest Service in the late 1930s hired a young psychologist named Shea to explore the psychological reasons behind the

[2] Schiff, A. L., 1962. *Fire and water: scientific heresy in the Forest Service.* Harvard University Press, Massachusetts.

Figure 5.2 The familiar Smokey Bear and some of his slogans. (McGraw-Hill Films)

stubborn tendency of the local people to burn the woods. As Shea put it, "it was hoped that here might be found a point of vaccination that with an improved educational serum would reach the germs of the woods-burning desires."[3] Reinforcing the dogma that no good could be found in a destructive practice, Shea concluded that:

> the sight and sound and odor of burning woods provide excitement for a people who dwell in an environment of low stimulation and who quite naturally crave excitement. Fire gives them distinct emotional satisfactions which they strive to explain away by pseudo-economic reasons that spring from defensive beliefs. Their explanations that woods fires kill off snakes, boll weevil and serve other economic ends are something more than ignorance. They are the defensive beliefs of a disadvantaged culture group.[4]

In a final effort to modify these inherited beliefs, the Forest Service invented a talking bear to symbolize the grave danger that fire held for

[3] Shea, J. P., 1940. "Our pappies burned the woods." *Amer. For.* **46**, p. 159.
[4] Ibid.

Figure 5.3 (a) This Texan pine stand is crowded with hardwood invaders that shade out pine seedlings. (b) Periodic controlled burning removes the fire-sensitive hardwoods. (c) This pine forest in Mississippi is kept "clean" by periodic prescribed burning. (Photographs from U.S. Forest Service)

all forests and their inhabitants. To this day a solemn, reproving, or imploring Smokey the Bear asks the customers to "please help prevent forest fires" in national parks and forests from coast to coast (Figure 5.2).

Then in the late 1930s a children's novel, *Bambi*, was made into a feature-length cartoon by Walt Disney. Who of that generation can forget the climatic scene where Bambi, Thumper, and all their friends flee from that worst of all evils, the forest fire. By now the man in the street had it firmly in mind that there was no place for fire in the forest.

But by the early 1940s, clearly something was wrong in the great longleaf pine forests in the Southwest. Both private and public forests which had been, with great labor, protected from fire were afflicted. There was no reproduction. There was a very damaging brown spot fungus on the needles of the grass stage seedling. There was massive invasion of worthless hardwoods which shaded out the longleaf seedlings. And worst of all, there were catastrophic fires fueled by twenty- to forty-year accumulations of combustible trash. In some forests more wood was lost from a single fire than was supposed to have accrued from all those years of fire protection. Beginning to doubt the advice of the Forest Service, private owners set up controlled burning to reduce the danger of wholesale conflagration in their forests. By the late 1940s the Forest Service was finally forced to admit that controlled burning, at the right time and under the right conditions, not only was effective in reducing dangerous litter accumulation, but was absolutely necessary to the survival of longleaf pine stands. Controlled burning would assure that the longleaf pine seeds had a mineral soil seedbed free of litter, that the brown spot disease would be reduced by occasional singeing of the grass stage leaves, and that hardwood invasion would be reduced (Figure 5.3a, b, c).

Today the Forest Service, as well as most foresters, realizes that the belief that fire in the forest environment is *always* bad is far from correct. No one suggests that indiscriminate use of fire by anyone, anytime, is

appropriate; rather, that in the right time and place, with the proper conditions, for certain species, fire can be an exceedingly useful management tool. Where evidence exists that fire has always been a regular part of the environment, as in the ponderosa pine forests of the Southwest, in the longleaf pine forests of the Southeast, or douglas fir in the Northwest, it makes little sense to exclude fire because it is damaging to spruce forests in Bavaria or white pine forests in Wisconsin. But the Forest Service is still left with a dilemma: on the one hand it must ask people to follow the admonition of Smokey the Bear not to set forest fires; yet on the other hand, it must deliberately burn many of those same forests as a necessary management practice upon which the very existence of many western forests and national parks depends (see Chapter 4).

FIRE IN THE GRASSLANDS

There is increasing evidence that the grasslands of the Great Plains of North America, and perhaps in other parts of the world as well, were established as man evolved and have been maintained indirectly by fire ever since. There is little evidence that the earlier theory—that grasslands resulted from inadequate rainfall to support the growth of trees—is correct.

A striking feature of many parts of the Great Plains is the sudden breaking of the land surface into a steep slope or scarp leading to another flat surface. These scarps trail in irregular lines all over the Great Plains. They often support trees, which is unusual; therefore, this was long regarded as reflecting better or unusual soil, augmented supply of moisture, or shelter from direct sunlight (Figure 5.4). But since trees are found on both the sunny and the shady sides of scarps and in soils derived from sandstone, shale, limestone, and even basalt, this theory cannot be correct. The trees seem able to reproduce in the thin and often stony soil of the scarps even though the soil on the nearby level plains may be richer.

Among the species found on scarps today are the drought-resistant juniper and ponderosa pine. Presently, the plains of Texas are the most arid region of the Great Plains, but fossil pollen grains from dried lakes indicate that during glacial times pine trees were common in this part of Texas, and probably in much of the rest of the Great Plains. Why are the trees restricted to the river valleys and rocky scarps? The reason is hinted at by the lack of trees on gentle scarps that are similar in every other way to the heavily wooded, steep scarps. Apparently the lack of grass under the trees and the steep slope act as firebreaks. On a level, grass-covered prairie, there is nothing to stop a wind-driven fire from running for hundreds of miles until it comes to a natural firebreak. Whereas grass recovers quickly from fire because its buds are protected by the ground, the exposed buds of woody species are killed by fire. Therefore recurring fires will destroy most trees on the prairie. Viewing man as an agent of fire, it seems likely that the incidence of fire on the Great Plains was greatly increased when man appeared, and that the grasslands have been with us ever since.

The importance of fire in eliminating trees can be seen in the Nebraska National Forest. A forest of ponderosa pine was planted around 1880 as an experiment to see whether trees could grow in former grassland if protected from fire. In thirty years the trees grew to be 22 feet tall and 7 inches in diameter before being destroyed by fire in 1910. Replanted and again protected, the forest grew until partially destroyed once again by fire in 1967. This is not to say that without fire North Dakota or West Texas would be a continuous pine forest, but without fire, the Great Plains would be likely to support occasional patches of shrubs or trees on the open plains away from the scarp faces and bottomlands to which they are limited today. This is seen in the large areas of former grassland in Arizona, New Mexico, and Texas, which when protected from fire, grew up to shrubby junipers and mesquite of no timber value, severely limiting the grazing value of these lands. Today some areas are being systematically cleared by bulldozing and replanted in grass. But unless these restored grasslands are periodically burned, shrubs will re-invade, negating the effort of their initial removal.

Figure 5.4 These ponderosa pines survive in Nebraska only because the steep slope on which they grow does not carry fire as well or as frequently as the level prairie above. (U.S. Forest Service)

Although the Indians increased the frequency of burning and so helped to maintain and even increase the area of prairie in North America, the pioneer practice of converting grassland into farmland or range began to reverse the process by creating broad firebreaks that stopped ground fires just as effectively as scarps. Today much of the eastern edge of the Great Plains is laced with shelter belts of drought-resistant trees and shrubs, which manage quite nicely when protected from fire.

FIRE-MAINTAINED NICHES

Animals, because of their mobility, are less obviously adapted to fire; however, the survival of two species of birds and the abundance of moose in parts of Alaska are both related to the incidence of fire in their habitats.

The Condor and Chaparral

The California condor, one of the largest vultures in the Western Hemisphere, has been restricted by man-induced disturbances to a horseshoe-shaped range in the coastal mountains of California from Coalinga south to near Ventura and north again to east of Bakersfield. The center of this range has been protected by its inclusion in the Sespe Wild Area of the Los Padres National Forest. Despite its name, Los

Figure 5.5   Chaparral-covered mountains in the Los Padres National Forest in California are the last refuge of the California condor.

Figure 5.6 One of the sixty California condors left, soaring in search of food in its home range. (U.S. Dept. of the Interior, Sport Fisheries and Wildlife)

Padres National Forest, set up to protect part of Los Angeles watershed, contains more chaparral than trees (Figure 5.5). Large areas of easily burned chaparral, before fire protection, presented a mosaic of burned and open patches surrounded by a dense growth of shrubs. When fire protection was begun with the formation of the National Forest, large tracts became solidly covered with dense chaparral.

A recent estimate by Koford suggests that the condor population is rather precariously stabilized at sixty individuals, for the condor, though protected by law, is a conspicuous target, supporting its twenty pounds in flight with a ten-foot wingspread (Figure 5.6). Although condors may live up to forty years, they do not begin to breed until they are five years old, and then produce only one young every other year thereafter. The birds are restricted to their present small range mainly by urbanization, since the condors feed upon carrion, which, as a result of small, well-cared-for cattle ranches and the trend toward grain farming, has become scarce.

Because of its enormous wingspread, the condor needs room to approach its food supply, and, more importantly, to take off after heavy feeding. Thus, open space is vital to the condor's survival. Since fire elimination tends to close chaparral stands, controlled burning helps the condor in some respects. However care must be taken to avoid over-burning, since the heavy brush helps to keep people out of nesting areas. Perhaps the best way to provide feeding space *and* protective isolation would be to spot burn unconnected small areas of the chaparral on a rotation basis. In this way it may be possible to increase the survival of the young condors and so increase the number of condors to a point where no single disaster could exterminate the species.

Figure 5.7 A female Kirtland's warbler feeding her young. Notice the protection provided by the jack pine whose branches can be seen at the top and bottom of the
picture. (Photo by G. Ronald Ausking from National Audubon Society)

Jack Pine and Warblers

A much less conspicuous bird, Kirtland's warbler breeds only in an 85-by
100-mile area in the north-central region of the lower peninsula of
Michigan (Figure 5.7). The nesting area is limited to dry, porous, sandy
barrens covered with patches of jack pine, a scrubby pine of minor
commercial value. After the white pine forests originally covering this
area were cut or burned, jack pine was one of the few plants able not
only to survive but to thrive in the frequently burned barrens. This is in
part because of the pine cones of the jack pine, which remain tightly
closed and attached to the tree, protecting the seeds, often for years.
When seared by fire the cones pop open, scattering several years' supply
of seeds over the ground. These quickly germinate in the fire-cleared
mineral soil and a stand of young trees results. Although Kirtland's
warbler nests on the ground, it requires the protection of living pine
boughs in thickets near the ground, which allow the birds to enter and
leave the nest unobserved. This arrangement of branches occurs only
when the trees are about six feet tall and disappears when the trees get
much over fifteen feet tall. Trees shorter than six feet are too far apart
and lack the necessary thick lower branches. In older trees the lower
branches are shaded out and die, reducing cover.

With the extensive fires following logging in the 1870s, the number
of trees with the proper characteristics increased enormously and the

population of Kirtland's warbler was probably at its greatest. When fire was eliminated or restricted in this part of Michigan after the turn of the century, and large areas of jack pine began to exceed the size required for warbler habitat, the warbler population started to fall. In 1951 it was estimated that there were less than 1000 birds remaining. A program of rotational controlled burning of patches of jack pine forest at least forty acres in extent (the minimum breeding territory for a pair of Kirtland's warblers) has recently begun. Hopefully this will ensure a stable warbler population and reduce the chance of extinction.

Moose and Spruce

Before the advent of white men, the Kenai Peninsula in Alaska was primarily a forest of spruce and fir with lichen-covered open patches of ground (a lichen is a primitive plant that in temperate forests usually grows on rocks or tree trunks). Large herds of caribou depended on lichens for their winter food supply. When man appeared in the late nineteenth century much of the forested area was burned and the lichen destroyed. In a few years, deprived of their winter food, the caribou were gone.

The spruce and fir killed by fire were succeeded by willow, alder, and aspen. Since moose require the buds and bark of these successional species as their winter browse, in the same way the caribou needed lichens, the large fire-generated stands of these hardwoods allowed a significant increase in moose survival. In fact, so spectacular was the increase in the moose population that the government reserved a very large portion of the Kenai Peninsula as a National Moose Range. But as the short-lived alder, willow, and aspen were replaced by spruce and fir in the natural sequence of events, the moose population declined again, raising the prospect of a mooseless Moose Range.

But when unauthorized fires, for this was a time when fires were forbidden, swept through the Moose Range, restoring successional browse, the moose population quickly recovered. Now that the lesson had been demonstrated, the size of the moose population in the Kenai could be controlled by planned fires to maintain a certain proportion of the forest in a fire recovery stage, thereby providing moose with the winter food supply necessary for their survival.

Because of its emotional connotations and potential economic impact, fire has been regarded for too long solely as a destructive force. Fire is also a natural part of most ecosystems, indeed of critical importance in some. Properly applied and controlled, ground fires can maintain the necessary balance of open mineral soil and forest cover to allow tree reproduction, reduce populations of destructive insects and diseases, and maintain certain types of vegetation in a stage suitable for use by rare species of animals adapted to fire-swept ecosystems. Had the relationship of fire as an environmental factor been objectively examined in the early days of forest management, many of the expensive mistakes of the last fifty years could have been avoided. If such mistakes are to be avoided in the future, comprehensive policies governing the management of ecosystems must be based on research by scientists, free from political restraints or preconceived conclusions.

FURTHER READING

Ahlgren, I. F. and C. E. Ahlgren, 1960. "Ecological effects of forest fires." *Botanical Review,* **26,** pp. 483–533. A comprehensive overview of the role of fire in ecosystems.

Cooper, C. F., 1961. "The ecology of fire." *Scientific American,* **204** (4), p. 150. Well-illustrated account of the role of fire in maintaining ponderosa pine stands in the West.

Gibbens, R. P. and H. F. Heady, 1964. "The influence of modern man on the vegetation of Yosemite Valley." *Calif. Agric. Exp. Sta. Manual* 36. Many before-and-after photographs showing graphically the impact of man on Yosemite National Park.

Holbrook, S. H., 1943. *Burning an empire: the story of American forest fires.* Macmillan, New York. An absorbing account of the spectacular "big burns" of the past.

Koford, C. B., 1966. *The California condor.* Dover Publications, New York. Details of the life history of this slowly vanishing species, with some suggestions of how fire might be used to increase survival rates.

Mayfield, H., 1960. *The Kirtland's warbler.* Cranbrook Institute Sci. Bull. 40, Bloomfield Hills, Michigan. The relationship between this rare warbler and fire is carefully discussed.

Schiff, A. L., 1962. *Fire and water: scientific heresy in the Forest Service.* Harvard University Press, Cambridge, Massachusetts. Fascinating detailing of a classic "biopolitical" situation.

Stone, E. C. and R. B. Vasey, 1968. "Preservation of coast redwood on alluvial flats." *Science,* **159,** pp. 157–161. The role of fire in maintaining redwood in California.

Wells, P. V., 1965. "Scarp woodlands, transported grassland soils, and concept of grassland climate in the Great Plains region." *Science,* **148,** pp. 246–249. Presents interesting evidence connecting fire with the treelessness of the Great Plains.

CHAPTER 6

"WATER, WATER, EVERYWHERE, . . ."

The summer of 1965 found New York City as closely described by Coleridge's verse, "Water, water, everywhere,/Nor any drop to drink" as its inhabitants would ever want it to be. Picture the irony of a city surrounded by water, yet plagued by low water pressure. In New York car washing or lawn watering was prohibited, whereas Los Angeles in a desert, comparatively speaking, had plenty of good water for every possible use. The 1960–1967 drought in the Northeast underscored the seeming paradox of a well-watered East suffering recurrent shortages of water while a dry West luxuriated in its abundance.

Once people become accustomed to drawing water from a tap they quickly forget that their faucets are connected by long miles of pipe to a supply somewhere. A continuous flow of water is assumed to be an inalienable right like the "right" to toss an empty beer can into a lake or a woods. This seemingly inexhaustible supply of water must be replenished ultimately by precipitation falling somewhere.

THE HYDROLOGIC CYCLE

All water is locked into a recycling process called the hydrologic cycle. Solar energy, especially in the tropics, evaporates water from the ocean surface, filling the air mass above with large quantities of water vapor. When these warm, moist maritime masses conflict with cool, dry air over large land areas, some of the water vapor precipitates out. The reason temperate latitudes (contrary to what the name implies) produce frequent storms, is that contrasting warm and cool air masses mix between the high and low latitudes. Although it may seem, when caught in a convectional thunderstorm triggered by this meeting of air masses, that *all* the water vapor is quite abruptly being dumped from above, no more than 20 percent of the water held by maritime air masses is lost over land.

Once it was felt that most of the rainfall over land was the result of evaporation from the land surface together with transpiration from vegetation, evapotranspiration. If this were true, land use might have a profound effect upon precipitation both in quantity and distribution. So it was hoped by some and feared by others that by manipulating evapotranspiration, precipitation could be brought under control. It was later determined that almost 90 percent of the precipitation falling over land areas is oceanic in origin, with perhaps 10 percent derived from land sources. This is not to say that moisture released by a magnolia tree in Louisiana into a moist air mass from the Gulf of Mexico will not fall on a corn plant in Ohio, but the amount would be insignificant. The principal effect of ocean-generated precipitation is to moderate the effect that man through his land use practices can have on continental precipitation.

Water Management

Precipitation that has fallen can be controlled to some extent by various management practices. J. E. Church, one of the early pioneers in hydrology, once made the observation, "forests catch the falling snow directly in proportion to their openness, but conserve it, after it has fallen, directly in proportion to their density." Entrapment of snow by fences or sheltering belts of trees can also significantly increase the water content of adjacent soil.

In dry areas, treatment of the watershed itself can affect the yield of water. By replacing chaparral or forest with grassland, evapotranspiration can be sharply reduced and more runoff water made available for man's use. This has been successful in some watersheds. Going a step further, small areas of desert have been experimentally blacktopped to obtain close to 100 percent runoff from summer storms. This works well for a while, but the paving is difficult to maintain under the extreme conditions. Another less drastic approach is to remove trees and shrubs that line the banks of many southwestern streams. These plants, salt cedar, cottonwoods, willows, and mesquite, absorb close to twenty-five million acre-feet of water per year (an acre-foot is one acre flooded to a depth of one foot) and transpire it into the dry air. The expense of their removal, however, has slowed implementation of such management.

Water Demand

The chief point to be learned by examination of the hydrologic cycle is that the water supply of the earth is fixed. Approximately 97.3 percent of the earth's water is in the oceans, of which a very tiny amount, 0.007 percent, is distributed annually as precipitation over the land areas.

Much of the remainder is locked up for the present in ice and snow; glaciers and ice caps cover 11 percent of the land area and pack ice and ice floes cover 25 percent of the ocean. The amount of fresh water immediately available to people is a small fraction of the total water found on earth. Of the earth's total water supply, 3.59×10^{20} gallons,[1] the United States receives a rather generous 4.3×10^{12} gallons a year in rainfall and its rivers and lakes contain another 6.5×10^{11} gallons. Since this is about fifty times the average yearly demand of 1.12×10^{11} gallons, there should be an ample supply of water for years and years of unrestricted growth.

As population increases, however, the supply slowly becomes critical—not because it is used up, but because of the tendency for abundant water sources (rivers or lakes) to be used for waste disposal and the unrealistic "need" for absolutely pure water for *every* human use. The problem, put quite simply, is not so much short supply, as it is thoughtless and unnecessary contamination of the supply coupled with irrational demands for purity.

Metabolic Need

Most terrestrial animals have a limited water need; many satisfy this need from their food with a very small supplement of water. Some desert animals, such as the gerbil, fulfill all their water needs from their food supply and are able to exist without an independent supply of water. Man's needs are quite modest. Although composed of about 65 percent water, man requires only a few quarts a day for maintenance, and, since much of this is supplied by food, often a pint or two is adequate. This innate ability to conserve water evolved over a very long time and probably had certain survival value where water was scarce, as in desert areas inhabited by primitive man.

Cultural Use

Culturation changed this irrevocably, however, and although today's civilized man *needs* little more water than his ancestors of 20,000 years ago, he *uses* incredible quantities in his daily activities. Flushing a toilet requires three gallons, while a single shower uses thirty gallons. Food preparation, laundering, and house maintenance add further to a total of about 180 gallons per typical residence a day.

Public Water Supply

Supplying this quantity of water to huge urban areas has required long-range planning by all communities. Founded in 1781, Los Angeles man-

---

[1] Exponentials are a handy way of expressing huge numbers. For example, 10^2 is shorthand for 100; 10^3 for 1000. So 10^{20} would be 100,000,000,000,000,000,000.

aged to supply a small population with water derived from the annual
precipitation of about fifteen inches. But a drought in 1862–1864 put an
end to the huge cattle ranches that surrounded the town and began the
inevitable subdivision process. Further population growth accompanied
the arrival of the Southern Pacific Railroad in 1877 and the Santa Fe in
1885. By 1900 it was clear that future growth of the Los Angeles basin
would require a large dependable water source, available only in the
Sierra Nevada some 200 miles away. The most convenient and economi-
ically feasible Sierran source was the Owens Valley on the Nevada
border, but Owens Valley was already occupied by ranches and small
settlements that were naturally unwilling to become ghost towns for the
benefit of distant Los Angeles. The ensuing struggle ended with the
construction in the 1920s of a 200-mile-long aqueduct which assured Los
Angeles' continued growth and the abortion of similar hopes for Owens

Figure 6.1 Los Angeles' water supply has had a long history of careful planning: in
the past, the Colorado River Aqueduct; for the present, the Los Angeles Aqueduct
from the Sierra; and in the future, the California Aqueduct from northern California.

Valley. The Colorado River was tapped in the 1930s after a long, contentious struggle by the watershed states over water rights to the Colorado's flow. By the 1960s even the combined water resources of the Colorado River and Owens Valley seemed limited, so the Feather River in underpopulated northern California is now being tapped (Figure 6.1). This tremendous project, costing well over a billion dollars, will assure southern California of adequate water for continued growth until limited by other factors—perhaps air pollution—but that is another story.

Water and the Engineer

Encouraged by the enormous demand of arid regions for water, a plan of continental dimensions, called the North American Water and Power Alliance (NAWAPA), was prepared by an engineering firm to "utilize the excess water of the northwestern part of the North American continent and distribute it to the water deficient areas of Canada, the United States, and Mexico." Making use of the unusual geographical, geological, climatological, and hydrological features of North America, this plan would take water now being poured into the Arctic Ocean and store it in an interconnected system of reservoirs at high elevation (Figure 6.2). This water could then be redistributed by means of a reservoir-canal-

Figure 6.2 The projected North American Water and Power Alliance (NAWAPA): the implications of technology on the environment are enormous in this scheme; the ecological implications are unknown. (Courtesy Ralph M. Parsons Co.)

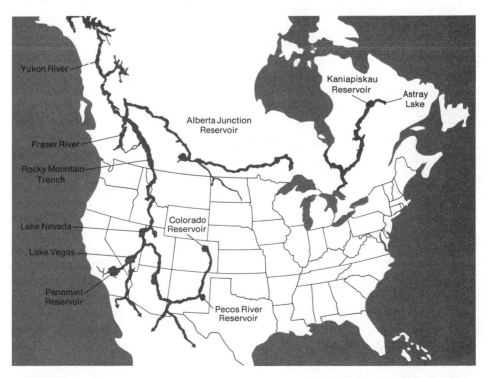

river system throughout the continent, producing not only a network of navigable waterways but generating power as the water flows to the sea. The cost has been estimated at 200 billion dollars and it would require twenty years to construct.

Considering the problems Los Angeles had with one small valley, one doubts that British Columbia would agree to put some of its most level and cultivatable land under water for the sake of the dry areas of the southwestern United States and Mexico. But neither technology nor money are real obstacles when a water supply is at stake, as has been demonstrated time and time again. We are learning just enough about long term weather cycles to wonder what would happen if rivers which warm the Arctic Ocean by flowing into it, were diverted. Would the currents carrying cold water back to the tropics dwindle, making the poles cooler and the tropics warmer? No one knows, but for the first time people are wondering.

USE AND REUSE

The paradox of plentiful water in dry California and periodic shortage in the humid East can now be explained. California cities, fearing a lack of water, long ago made plans for future water needs, based on the current rate of usage and the expected population increase. Eastern cities with plenty of rainfall grew beyond their long-range supplies, buoyed by the illusion of abundant water. When drought occurred and the supply failed, officials wrung their hands. When the drought was broken, illusion once again replaced planning.

It is more than a little surprising, then, to realize that in a very real sense *there is no shortage of water in the United States nor is there likely to be in the future,* no matter how many people live in the country and make demands on its water supply (see Table 1). Except for irrigation, no major water use consumes more than a fraction of the total water used.

Table 1   United States intake and net consumption of water (10^9 gal/day 1965)[a]

	Intake	Net consumption
Irrigation	142	116
Steam generated power	111	1
Industry	74	12
Public water	23	1
Rural domestic	6	4

[a] Dykes, D. R., T. S. Bry, and C. H. Kliner, 1967. "Water management: a fashionable topic." *Env. Sci. & Tech.* **1** (10), pp. 780–784.

All of the water now "in captivity" is to be found in reservoirs, water supply systems, and industrial and municipal waste systems. If all of this water were efficiently recycled, the annual increase in demand for new water would be exceedingly small. Loss to the atmosphere through evaporation from soil, rivers, reservoirs, plants, and animals, as well as metabolic demands by population increase, would easily be replaced by annual precipitation.

A Domestic Recycling Scheme

To be sure, a society which completely recycles its water will be a long time in coming. But such a plan is not as visionary as it might seem. At the present time, all water piped into our homes is grade A drinking water regardless of the use we make of it. It makes little sense to throw away three gallons of grade A water just to dispose of a small volume of wastes in a toilet, or thirty gallons to get rid of a little detergent with its dirt and lint from clothes. A priority system of water use could easily be established, reserving grade A water for drinking, cooking, washing, and bathing. Grade B water reprocessed from grade A uses would be quite adequate for outdoor uses: car washing, lawn watering, filling swimming pools (these would still be cleaner than most public bathing beaches), and perhaps a grade C water for flushing toilets.

Grade A water would be supplied from a well or reservoir direct to the faucet. After grade A uses, the waste water would be filtered and stored in a basement tank for grade B and grade C uses. In a recent pilot experiment, water from filtered laundry and bath water was simply stored in a basement tank and subsequently pumped into toilets for flushing, reducing water consumption by 39 percent in a typical suburban home. Even under prototype conditions, which are far more expensive than mass produced versions, the capital cost was only $500 and the savings in water and sewer rental costs $20 a year, which over the lifetime of the house (twenty-five years) would pay for the cost of installation. Such a simple recycling device could cut domestic water consumption by almost 50 percent if widely adopted.

A more comprehensive system would be to recycle carefully treated and purified sewage effluent into the water supply, drastically curtailing the need for new water. Doubtless this strikes a repugnant note to most of us who feel that the water that comes out of our tap is or should be as pure as distilled water. On the other hand, communities that draw their water supplies from rivers are already simply recycling the effluent from the town above, particularly when under drought conditions many rivers continue to flow only because of the effluent contributed by towns along the banks.

Drinking our Effluent

A classic case of intentional, probably desperate, reuse of effluent for water supply took place in Chanute, Kansas in 1956–1957. The drought of 1952–1957 was the most severe in Kansas history. By the fall of 1956 the Neosho River, which was Chanute's water supply, simply stopped flowing. Shutting off the water supply to industry was considered but rejected, because so many people would lose their jobs. Wells were then suggested, but well water was too saline for use. Hauling was too expensive and logistically complex. Finally recycling was adopted as the best solution. As luck would have it, a new sewage disposal plant constructed in 1953 produced an effluent lower in its bacterial count than the Neosho itself. The effluent was pumped into a holding pond for a number of days, then prechlorinated and recycled. The resulting water, carefully monitored by the state board of health, met minimum standards, although it had a musky taste and odor, a pale yellow color, and produced foam when agitated. Some people were concerned about the effectiveness of chlorination in eliminating viruses, particularly the polio virus (this was before the various vaccines) and infectious hepatitis. And their concern seemed justified, since 20–40 thousand cases of hepatitis were suspected to have resulted from similarly treated water distributed by the city of New Delhi, India a few years before.

After a few months the drought was broken by rains, the river resumed flowing, and Chanute returned to its previous water source. A doctors' survey conducted after resumption of normal water supply showed there were no illnesses that could be related directly to the use of recycled water. Apparently the major problem was the increasing levels of various salts. If recycled water is to be used in the future, it will be necessary to have rigorous filtration systems to remove odor, taste, and color and precipitation processes to remove salts. The virus problem will be more difficult to solve.

Recent evidence suggests that Down's Syndrome (mongolism) may be related to cases of infectious hepatitis contracted by mothers in early pregnancy, in the same way that German measles infections during early pregnancy have been associated with birth defects. Unless a fail-safe means of removing or killing viruses in sewage effluent can be perfected, full recycling of effluent cannot take place. It is only a matter of time, however, before population pressure requires the extensive recycling of *all* available water resources, including those which have been traditionally regarded as unfit for human use.

Industrial Recycling

Water reuse is not limited to the individual user. Industry must also make a greater effort to recycle and reuse some of the enormous quantities of water it uses in processing various products and services. As an example of what can be done, Kaiser Steel in Fontana, California has employed various recycling methods to cut the water requirements for manufacturing steel from the industry average of 65,000 gallons per ton to 1400 gallons per ton.

With the added incentive of an even tighter water supply and more stringent state and federal pollution standards, even such gains as these can be exceeded. Finally a point will be reached when, beyond initial charging of a system, very little additional water will be needed. This means not only more but cleaner water will be available in the environment for other uses.

DESALINIZATION

Many articles have been published about the coming wonders of desalinization, complete with glowing promises about severing our dependence upon natural precipitation, making deserts bloom, changing climate, and other age-old fantasies of mankind. Any desalinization process, and there are many, requires energy, but only one process, electrodialysis, removes the salt directly. While desalinization is usually thought of as a sea-edge process, any brackish water is potentially desalinizable; the lower the concentration of salt, the more efficient and less expensive the separation. However, the need is often most critical in dry climates near an ocean, which provides the nearest source of water.

There are many methods of desalinization, some rather complex, and all ingenious (Figure 6.3). The older processes, stills, use heat energy;

Figure 6.3 Desalinization of water is now a technically feasible possibility. Different methods of desalinization are shown schematically: (a) a multi-stage flash distillation process; (b) an electrodialysis cell, and (c) an electromagnetic separation. (Modified from *Chemical and Engineering News,* June 8, 1970)

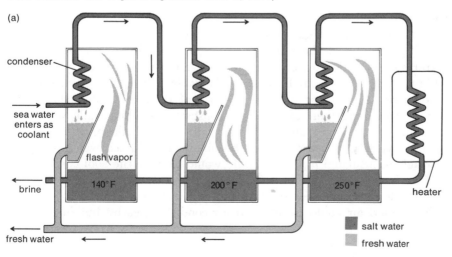

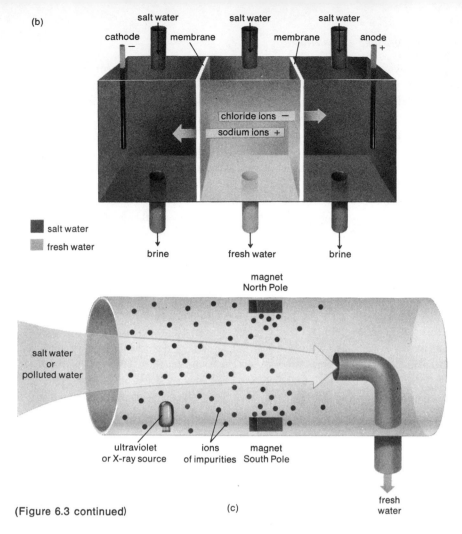

(b)

salt water   salt water   salt water

cathode   membrane   membrane   anode
 —   +

chloride ions —
sodium ions +

■ salt water
▨ fresh water

brine   fresh water   brine

magnet
North Pole

salt water
or
polluted water

ultraviolet   ions   magnet
or X-ray source   of impurities   South Pole

fresh
water

(Figure 6.3 continued)   (c)

the more recent processes, electrical energy. With the development of nuclear power plants it has been suggested that the two be combined, producing both electric power and fresh water. Such a plant projected for an artificial island off the coast of southern California was planned to produce 150 million gallons of water a day.

Perhaps one of the most ingenious schemes for augmenting fresh water supplies was suggested by Gerard and Worzel a few years ago (Figure 6.4). Many small coastal islands in the tropics have grossly inadequate fresh water supplies because of their tiny catchment area, but by exploiting their exposure to constant prevailing trade winds of high moisture content and their proximity to cold, deep, nutrient-rich water, a variety of benefits could be obtained. Cold sea water pumped through an insulated pipe to a baffle would provide the cool condensing surface for condensation of moisture from the warm, moist trade wind. The condensed water could be stored in a cistern below, and the cool, dehumidified air could be used to air condition nearby buildings. The sea water coolant, rich in nutrients, could be released into a lagoon whose productivity would be increased, thereby supplementing the protein resources of the island. At least a part of the power needed to pump the water from offshore depths could be provided by a windmill powered by the same trade winds.

The biggest problem with desalinization is the cost. Although costs are being reduced constantly by building larger and more efficient plants, desalinization still compares unfavorably with competing sources of fresh water. Even if desalinization costs were reduced to those comparable with other water sources, there remains the problem of transport from the coast to the inland valleys and deserts. Moving anything against gravity is extremely expensive (all of the existing far-flung water supply systems work for the most part *with* gravity flow). It has been said that if the quantity of water needed is large enough, it is cheaper to bring water from Baffin Bay to Baja California than to desalinate and transport sea water to the place where it is needed. The very existence of the Feather River scheme in California or the projected NAWAPA lends credence to that statement.

Desalinization will surely come to play a much larger role in the water supplies of major coastal communities in the future, but unless we develop some low-cost energy source to push water uphill, desalinized water will not quickly transform the arid regions of the earth into lush, tropical gardens. Indeed, to do so might have major climatological implications. Perhaps by the time this is feasible we will have developed the ecological insight to predict the consequences of such drastic changes.

Figure 6.4 Proposed water recovery plant. (1) Large-diameter pipe to deep water. (2) Pump. (3) Connecting pipe. (4) Condenser. (5) Freshwater reservoir. (6) Windmill electric generator. (7) Baffles to direct wind. (8) Small turbine to recover water power. (9) Lagoon receiving nutrient-rich water for agriculture. (10) Community enjoying cooled dehumidified air. (Modified from R. D. Gerard and J. L. Worzel, *Science*, **157**, 1967. Copyright 1967 by the American Association for the Advancement of Science.)

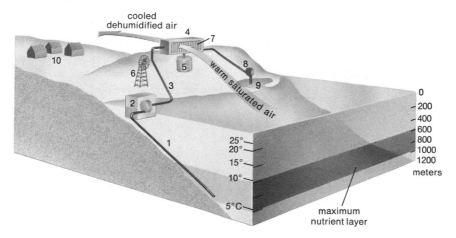

WEATHER CONTROL

Besides the management of rainwater after it has fallen, there is the possibility of weather control. Ever since artificial snow was produced in a deep freeze some years ago, a great deal of interest has been generated in seeding clouds to generate rain and break up hurricanes. The simple fact of the matter is that there is no weather data available for over two-thirds of the world's land area and nearly all of the water area. This lack of information makes weather forecasting inexact and widescale manipulation impossible. While reconnaissance satellites are extremely useful in helping to cover weather information gaps—for example, in various polar, tropical, and oceanic areas—the ability to use satellite information to project and predict weather is making slow headway. Weather modification of large areas is therefore not yet possible. Indeed, to undertake any broad scale experiment in the atmosphere aimed at making permanent changes with our present level of knowledge would be irresponsible in the extreme. However, a great deal of experimentation has been carried out on a local basis, and the results, which range from inconclusive to promising, bear further examination.

Seeding certain types of clouds with silver iodide crystals provides water-attracting nuclei for condensation of the tiny supercooled droplets of water that make up the cloud. As these droplets condense on the silver iodide nuclei, drops of water are formed that ultimately may grow to precipitable size. Evaluation of seeding attempts is extremely difficult because of the great variability of cloud types. While some experiments have increased rainfall around 10 percent, the increase is rarely more than 20 percent, and, in some instances, a decrease in rainfall has probably occurred. Other problems include the accurate calculation of lag time (the time from seeding to the time rain falls), for clouds cannot be seeded directly over the desired rainfall area. Seeding must take place as much as 50–100 miles upwind to allow for raindrops to form. Their trajectory from cloud to target area is especially important if the precipitation is to be snow. Also, most successful seeding attempts have used cumulus clouds or orographic storms, that is, unstable clouds formed by air being forced up over mountains. Efforts to seed large cyclonic storms, that is, those associated with low pressure areas, have not been very successful.

Hail suppression has been reported by the Russians who have fired silver iodide directly into potential hail-producing clouds using antiaircraft shells or rockets. But because of the extreme variability in the formation and fall of hail compared with rain, proof of success is elusive.

Lightning suppression is being attempted by introducing silver iodide into clouds, in the hope that the large number of ice crystals formed would relieve the electrical potential before enough charge had accumulated to result in a lightning stroke. Metallic chaff dipoles—small pieces of foil—dropped into thunderheads might have the same effect.

Hurricane and tornado modification is making slow progress. Recent experiments with a hurricane in the Gulf of Mexico have succeeded in lowering the wind velocity around the eye by seeding with silver iodide (Figure 6.5), but reducing the force of tornadoes by direct action has not yet been achieved. Progress has been slow because there is no precise way to predict conditions likely to spawn tornadoes. Until tornado modification becomes feasible, a more accurate warning system is our only defense.

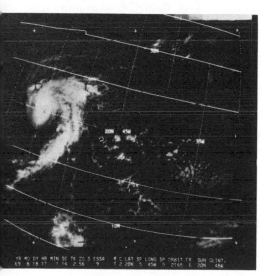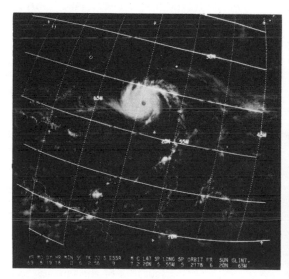

Figure 6.5 Satellite photos of Hurricane Debbie (1969). The photo on the left was taken three hours after the hurricane was seeded. The photo on the right was taken the next day. Seeding apparently accomplished some temporary reduction in the storm's wind velocity. (Environmental Science Services Administration)

The action necessary to change the weather or climate over large areas may never be within man's grasp, but there are probably many instances where the various controlling factors are delicately balanced and man may already have the ability to effect radical change. The question is, as always: does man have the insight to predict the consequences of his manipulations, and then can he exercise restraint where he must?

FURTHER READING

Anonymous, 1968. "Reusing storm runoff." *Env. Sci. & Tech.*, **2** (11), 1001–1005. Ingenious scheme for using a presently wasted resource.

Dykes, D. R., T. S. Bry, and C. H. Kliner, 1967. "Water management: a fashionable topic." *Env. Sci. & Tech.*, 1(10), pp. 780–784. Overview of the water supply-demand problem.

Gerard, R. D. and J. L. Worzel, 1967. "Condensation of atmospheric moisture from tropical maritime air masses as a fresh water resource." *Science*, **157**, pp. 1300–1302. A very clever scheme to utilize the environmental features of a tropical island for water and energy supply.

Hickman, K., 1966. "Oases for the future." *Science*, **154**, pp. 612–617. A review of many of the problems in increasing the supply of fresh water.

Hodge, C. (ed.), 1963. *Aridity and man.* AAAS Publ. 74, Washington, D.C. A fine collection of papers dealing with various aspects of water supply.

Maass, A., 1951. *Muddy waters: the Army Engineers and the nation's rivers.* Harvard University Press, Cambridge, Massachusetts. A critical examination of the role of the Corps of Engineers in manhandling the environment.

Nace, R. L., 1967. "Water resources: a global problem with local roots." *Env. Sci. & Tech.*, 1(7), pp. 550–560. An excellent summing up of the water supply problem.

Okun, D. A., 1968. "The hierarchy of water quality." *Env. Sci. & Tech.*, 2(9), pp. 672–675. Suggests an approach to a meaningful reuse of water.

U.S. Department of Interior, 1968. *The a-b-seas of desalting.* Office of Saline Water, Sup. Doc., Washington, D.C. A concise, well-illustrated discussion of the various techniques of desalinization.

CHAPTER 7

". . . NOR ANY DROP TO DRINK."

Traditionally speaking, pollution, like prostitution, is any departure from purity. But in the environment it has come to mean departure from a normal rather than a pure state; otherwise we would have to say that water is polluted with algae, fish, or fowl.

Pollution is, of course, a subjective and often emotionally charged term, one that man has used to describe the environmental traumas of fire, overproductive or deoxygenated water, smoke- or dust-filled air, and background radiation that have always existed. Natural traumas become pollution when man's activities overwhelm the capacity of an ecosystem to handle them, thereby causing an imbalance in the system. This imbalance requires an enormous input of energy to correct. Further, because of man, environmental traumas are now regional instead of local, continuous rather than episodic. Water and, as we shall see in Chapter 10, air, are the two most polluted areas in our environment.

Because of its unique bipolar properties, water readily dissolves a great variety of substances; those substances that remain insoluble are at least dispersed. Where human populations are small and widely scattered, dilution *is* a reasonable solution to waste disposal. Rivers and streams are, in a few miles, able to carry out the work of breakdown, absorption, oxygenation, or consumption, and small amounts of pollutants are effectively dispersed.

In primitive times, by the time waste water from a village of a few hundred people had been carried a few miles downstream, the water had been purified naturally—at least, enough to be drinkable in the next village. When villages became towns, and towns cities, rivers could no longer break down the quantity of pollutants poured into them, and the effluents from each population center added to the water supply of the next. Running water almost everywhere became sewage water. This is an old, old story, as stated in the Bible: "and the fish that was in the river died; and the river stank, and the Egyptians could not drink of the water of the river, and there was blood throughout all the land of Egypt."[1] The blood color may have been due to blooms of red algae or perhaps purple bacteria; the stench is all too familiar.

CATEGORIES OF WATER POLLUTANTS

Toxicity.   Heavy metals such as lead, mercury, copper, arsenic, and chromium which are used in tin plating, galvanizing, and chromium plating solutions can act as metabolic or respiratory blocks in many organisms, inhibiting or destroying enzymes that are essential to life processes. Since metals may affect the decomposer organisms as well as the higher forms of life, their excessive addition to an aquatic system can be quite destructive. These pollutants are actual poisons which may be absorbed by the bottom mud and be released whenever bottom deposits are disturbed.

Salinity.   Occasionally, salt brines from mines or oil wells are released into normally fresh water. Whereas some organisms can tolerate a certain range of salt concentration, many disappear as fresh water becomes brackish. If the salt level remains constant, salt-tolerant species often appear and reproduce successfully. A marine species of the green seaweed, *Enteromorpha,* became quite abundant on the bottom of a New York stream that was badly polluted with salt brine. However, because of constantly fluctuating levels of most pollutants, organisms that might otherwise adjust to the new conditions rarely have the opportunity to become naturalized, so neither old nor new organisms can survive.

Many large rivers normally keep brackish sea water out by their continuous outflow of fresh water. During times of low runoff, the 1962–1967 northeast drought for example, river currents cannot resist the pressure of tidal flow from the sea, and tide lines may advance many miles upstream, with severe effects on freshwater organisms.

Philadelphia, which obtains much of its water supply from the Delaware River above the tide line, was threatened in 1965 when the slow current of the Delaware allowed sea water to advance almost to the city's water supply intake. New York City also had a pressing need for the water from reservoirs on tributaries of the upper Delaware, but interstate watershed compacts compelled New York to release badly

[1] Exodus vii:21.

needed water from its reservoirs to maintain a minimum flow in the Delaware. This saved Philadelphia's water supply by keeping salt water safely downstream.

Acidity. In 1803, T. M. Morris made one of the first references to acidic drainage in America when he noted, "the spring water issuing through fissures in the hills, which are only masses of coal, is so impregnated with bituminous and sulphurous particles as to be frequently nauseous to the taste and prejudicial to the health."[2] Morris was referring to natural seepage, which has apparently always been a minor problem. But since the opening of the nation's first coal mine in Pennsylvania in 1761 the problem has become far more widespread. Of course, acidic water from mines increases with production of coal, but it also continues long after mines are abandoned.

Joshua Gilpin visited a coal mine in 1809 and stated that "above the coal is several feet of a mixed kind of bad coal and iron abounding in sulphur and in vitriol effervescing in white and yellow crystals."[3] The materials Gilpin was describing were iron pyrites which enter streams following strip mining. Certain bacteria are able to obtain energy by changing iron from one oxidation state to another (ferrous to ferric). They use the energy to respire and reproduce. With an unlimited source of food they multiply greatly; the by-products are the orange-red precipitate of ferric hydroxide that stains river banks and channels, and the acidity caused by the release of sulfuric acid.

This set of reactions not only drastically lowers the pH of the stream and contributes huge quantities of iron and sulfate to the water, but it reduces the availability of oxygen in the stream as well. In the low-water season of summer and early fall, the pH may rise to 4.5, but with the high water of winter and spring, flushing of wastes from strip mines and mine shafts by tributary streams may lower the pH to 2.5. At pH levels below 4.0 all vertebrates, most invertebrates, and many microorganisms are eliminated. Most higher plants are absent, leaving only a few algae and bacteria to save the stream from total sterility.

Acid mine drainage is a severe problem wherever coal is mined. In the United States, the most severely affected areas are parts of Pennsylvania, Ohio, West Virginia, Kentucky, Illinois, Missouri, and Tennessee; in Europe, parts of the British midlands, France, Germany, and Poland. Most coal mining areas now have strict regulations which have greatly reduced pollution from contemporary sources. Regulations, however, do nothing for those shafts or stripped areas already abandoned which continue to discharge acid wastes into the nearest drainage.

[2] Braley, S. A., 1954. "Acid mine drainage: I The problem." *Mechanization,* **18,** pp. 87–89.
[3] Ibid.

Control of these abandoned workings will have to come bit by bit, adding enormously to the ultimate cost of acid pollution control.

Although coal mining operations add the greatest amount of acid to streams, other industries produce wastes which affect pH too. Continuous casting and hot rolling mills use sulfuric acid (pickling liquor) to clean oxides and grease from the processed metal. When the pickling solution becomes diluted with iron salts and grease, it must be replaced with fresh solution. Since over 50 percent of all steel products are treated with pickling liquor at the rate of fifteen gallons of liquor per ton of steel, a considerable volume of spent liquor is disposed of every year. After pickling, the metal is rinsed to stop the pickling action; every ton of steel requires another eighty gallons of water which picks up about 15 percent of the acid. Because of their acidity these spent pickling and rinse solutions are a most difficult waste water problem.

Turbidity.   Relatively inert and finely divided materials are easily suspended in water, cutting down light transmittance enough to inhibit photosynthesis of both macro- and microscopic water plants. The china clay industry, which washes its raw material to remove grit, is one gross offender (Figure 7.1a, b), as is the steel industry whose new hot-strip mills produce very fine particles that give a red or black color to the effluent stream. Ultimately, the suspended particles settle out, smothering all life at the bottom of the body of water with a fine ooze.

Figure 7.1 (a) A clay pit in Cornwall, England. Waste clay from the tailing pile above the pit enters the surface drainage, turning it opaque milk white. (b) When the suspended clay reaches the sea it discolors the inshore waters for miles.

One pollutant which does not settle out is an oil made soluble by stabilizing its emulsification with detergents. When these used emulsified oils are released into water they impart a milky turbidity which is relieved only by dilution.

Many streams go through changes of bottom flora and fauna as the soil and vegetation of the watershed are disturbed. Severe fires or clearance for development release soil and clay particles into streams, producing silt which destroys the habitats of the larvae of mayfly, caddis fly, and stone fly—all important sources of food for game fish. After stabilization of the watershed by the regrowth or replanting of vegetation the amount of silt in streams is reduced. Seasonal flushing by storms can then scour the bottom, restoring in a few years the pebbly substrate required by the original fauna. With the return of animal life the food chain again becomes balanced.

Deoxygenation. The most common pollutants are organic. These materials are not poisonous to stream life, nor do they affect pH, necessarily. Their effect is more subtle. Most organic materials are attacked by bacteria and broken down into simpler compounds. To do this, bacteria require oxygen. The greater the supply of organic food, the larger the population of bacteria that can be supported, and the greater the demand on the oxygen supply in the water. This demand for oxygen by the bacteria is called *biological oxygen demand*, BOD. The BOD is a

useful index of pollution, especially that related to the organic load of the water. Because all stream animals are dependent upon the oxygen supply in water, the BOD is of particular importance in determining which forms of life a polluted stream is capable of supporting. Fish have the highest oxygen need; usually the cold-water fish require more oxygen than warm-water fish. Invertebrates can tolerate lower concentrations of oxygen, and bacteria still lower.

Two invertebrates are able to tolerate such a low oxygen level that they have become recognized as a kind of biological index of oxygen depletion. One is a small worm called *Tubifex*, and the other is a bright red larva of a tiny midge called *Chironomus*. Both live in the bottom mud in numbers that are inversely proportional to the oxygen content of the water. In badly polluted water *Tubifex* populations have been estimated at twenty thousand individuals per square foot. In unpolluted water *Tubifex* may be absent altogether.

Sometimes when the organic load is especially great and the water warm, oxygen is so depleted that the usual oxygen-requiring or aerobic decomposers are replaced by species that don't require oxygen (anaerobic), with a quite different group of end products (Table 1).

Table 1 **Comparison of decomposer end products under differing conditions**[a]

Aerobic conditions	Anaerobic conditions
$C \rightarrow CO_2$	$C \rightarrow CH_4$
$N \rightarrow NH_3 + HNO_3$	$N \rightarrow NH_3$ + amines
$S \rightarrow H_2SO_4$	$S \rightarrow H_2S$
$P \rightarrow H_3PO_4$	$P \rightarrow PH_3$ and phosphorus compounds

[a] Klein, L., 1962. *River pollution Vol. II. Causes and effects.* Butterworth, London.

While methane, CH_4, is odorless, amines have a fishy smell, hydrogen sulfide, H_2S, smells like rotten eggs, and some phosphate compounds have a wormy smell. When added to the smell of decaying fish or algae, the shift to anaerobic conditions is not a pleasant one.

ORGANIC WASTES

Now and then severe pollution from excess organic matter occurs in natural situations, as when heavy leaf fall temporarily overwhelms a small stream creating a large BOD, and possibly killing fish. But most organic wastes are related to man's misuse of his environment by the release of untreated human and animal wastes as well as those from industrial processing—blood, milk solids, grease, pulp liquor, washings from fruit and vegetables.

The Human Factor

Human waste disposal is not a recent problem, though it certainly has been made more difficult by the sharp population rise of the past century. The Old Testament met and solved, for its time, the question of human waste disposal: "and thou shalt have a paddle upon thy weapon; and it shall be when thou wilt ease thyself abroad, thou shalt dig therewith and shalt turn back and cover that which cometh from thee."[4] Millions of farmers in Asia still utilize a similar approach, fertilizing their fields and rice paddies with "night soil." The people of medieval towns and cities, being unable to return waste to the soil, dumped it from their windows into the street, requiring well-bred gentlemen of the day to defend their sensitivities by carrying an orange stuck with cloves to mask the foulness of the streets. Urban people today prefer sewage treatment plants in which organic wastes can be broken down by bacteria, so that the effluent does not deplete the oxygen in rivers and lakes.

But with our increasing level of affluence, this ecologically sound pathway is being short-circuited. For example, people are spending more and more of their leisure hours aboard a great variety of pleasure craft (Figure 7.2), many with toilets but few with means to handle the waste

[4] Deuteronomy xxiii:12–13.

Figure 7.2   Most of the boats in this marina are large enough to have onboard toilets, which dispose of their wastes directly into the water. (Photo by Mario Marino)

properly. In most instances waste is dumped overboard with little or no treatment. In addition to the sanitary wastes there is litter (one pound of paper, cans, and bottles, one-half pound of garbage per capita per day of boating); bilge water containing lubricating oil leakage; and in large ships, ballast. The average freighter carries 1000 tons of ballast (usually water taken from one polluted harbor and released in another) making ships pollution carriers par excellence. In 1964, of the fifty thousand ships clearing United States customs, sixteen thousand released sixteen million tons of polluted ballast water. Other sources of vessel waste include wash water from cleaning operations, usually containing high concentrations of detergents, and accidental cargo discharges through negligence, collision, grounding, or sinking. The Torrey Canyon disaster, which will be discussed in Chapter 9, is just one example of this.

The United States Navy until quite recently disposed of its shipboard wastes over the side, causing an enormous sewage problem in large ports when the fleet was in. Consider how much waste is generated in a single day by a nuclear aircraft carrier with five thousand men aboard!

Some possibilities for aboard-ship sewage disposal include holding tanks, incinerators, biological treatment systems (BTS) and maceration-disinfection. Holding tanks are adaptable to any size vessel and minimize mechanical failure, but they are large and heavy. They are also suitable only for human wastes, and require shore support facilities that are nonexistent even in some large ports. Incinerators are lightweight, but have a high power requirement. Like holding tanks, they are limited to human waste, and, in addition, can be a fire hazard. BTS can handle all kinds of sewage, since it is a miniature secondary sewage treatment system, but it is quite large and heavy and requires trained personnel for proper operation. Maceration-disinfection is an easily installed, lightweight system, occupying little space, but it assures no reduction in BOD, or mineral and organic content: its disinfectant efficiency is not always certain and it is subject to failure. Perhaps the most promising approach is a variant of the recirculating flush toilet. The airline model can handle 80–100 uses before servicing; larger models, capable of servicing fifty people for 7–10 days, have been developed for ships.

Animal Wastes

Humans are not the only waste producers, of course. Kansas, which is not overpopulated, has such a water pollution problem that no stream or river in the state is safe for drinking or swimming. The cause is the more than 200 feedlots scattered across the state (Figure 7.3). The 5.5 million

Figure 7.3 The animal wastes from this midwestern feedlot enter the nearest surface drainage system without treatment. The waste from one cow equals the wastes of ten people. (U.S. Dept. of Agriculture)

cattle and 1.3 million hogs daily excrete the sewage equivalent of seventy million people. Most of this goes directly into streams and rivers with little attempt at control. Because of this pollution load some towns have had to use ten times the usual level of chlorination to purify their drinking water. New regulations will require feedlots, which process millions of livestock a year, to treat their wastes before releasing the effluent. Even so, it will take five to ten years before pollution from this source can be reasonably controlled.

Treatment of Human and Animal Wastes

Once the organic materials are broken down into their components by sewage treatment, the minerals must be removed to avoid overfertilizing the environment. Precipitation removes only some of the minerals; those like nitrates, which are highly water soluble, are more difficult to deal with. A useful approach, first used by a large food processor in New Jersey and later developed and refined by a research team at The Pennsylvania State University, uses the soil and its plants and animals as a living filter. Effluent is piped from a sewage treatment plant and sprayed on a field, pasture, or forest at a rate that is commensurate with soil ab-

Figure 7.4 Year-round spraying of treated sewage effluent onto forest and crop lands at Pennsylvania State University fertilizes the soil and filters the effluent, returning clean water to the water table. (Photo by L. T. Kardos)

sorption (Figure 7.4). Applied at a rate of two inches a week, 129 acres could filter a million gallons of effluent a day. The plant roots and soil microorganisms are especially efficient in removing phosphates and nitrates, the two nutrients which cause the most problems in aquatic systems.

Gratifyingly, experimentally grown crops benefited from the fertilizing effect of effluents; hay yield increased 300 percent and the nitrogen content of wheat increased by 30 percent. Some problems with this approach are the requirements of no surface drainage, which would short-circuit the process, and deep soil to allow maximum filtration.

A recent extension of this technique has been its application to revegetation of strip mine spoil banks. Because of their dark surface and exposed position, spoil banks become excessively hot and dry in the summer, exceeding even the broad tolerance of weeds. In addition the soil is usually both infertile and acid, hence, establishment of seedlings is impossible.

Application of sewage effluent to these banks ameliorates all these problems. Moisture, which also cools the ground surface by evaporation, allows a broad range of plants to become established, and the mineral nutrients which abound in the effluent encourage rapid and luxuriant plant growth. Finally, the addition of two inches of water a week helps leach some of the acid into lower soil levels, which allows the establishment of species requiring less acid soil. Since transport of the effluent is expensive, the spoil banks and sewage disposal plants should be as close together as possible. The potential benefits of the scheme might even justify the location of sewage plants in the center of spoil areas. Once vegetation is established and the vicious cycle of excessive heat and inadequate moisture is broken, the pipes distributing the effluent could be shifted to another barren area. When the area surrounding the sewage plant is revegetated, the effluent could be used indefinitely to increase the productivity of the new forest.

Industrial Waste

Ironically, industry is its own pollution victim: the greater any one industry's pollution of water and air, the greater the costs to all industry of pretreating or cleaning these two resources before they can be used. Of the many industries which contribute organic wastes to aquatic systems, adding to the load already present from human and animal wastes, we will examine three: pulp, oil, and food processing, which typify both the variety of the wastes, and the magnitude of the problem.

Pulp and paper. Pulping is the process of breaking down wood into its component fibers, which are then formed into paper. The cellulose wood fibers are held together with lignin and a complex sugar or hemicellulose which can be dissolved under high pressure and temperature with bisulfite and either sulfurous acid or sulfur dioxide. This treatment yields 50 percent of the dry weight of the wood as cellulose fibers and a waste sulfite liquor containing the remaining 50 percent in the form of lignosulfonate and various sugars hydrolyzed from the hemicellulose. Until recently much of the 2.5 million tons of spent sulfite liquor produced each year was simply dumped into the environment. Today, more and more is being recycled.

The production of paper involves reuse of waste paper. Some paper *is* reused despite the enormous quantities strewn about the landscape. But reuse requires de-inking treatment which also extracts the chemicals used in paper-making and the clay filler used in sizing glossy paper. Naturally, the ink and filler must be disposed of. The desirable whiteness of paper is obtained by bleaching the fibers with hypochlorite, requiring 3–5 washing cycles. The wash water contains dissolved solids and unreacted chlorine. Finally, preparation of paper stock requires sizing, filler, resins, and coloring; all are diluted in water. These processes require huge quantities of water, from the floating of the log to the mill to the final sheet of paper. Without reuse of water, few plants could continue operation, for virtually none have unlimited supplies of water. But the scale of operation is such that leaks, spills, rejects, and equipment cleaning produce an effluent with much organic and inorganic material that is injurious to organisms and to ecosystems (Figure 7.5).

Petroleum refining. Petroleum is a rich mixture of organic compounds; the refining process simply sorts these out and refines and purifies them for many general and specialized uses. Crude oil is emulsified with water to separate the salts and other impurities which, when settled out, are released along with the water. Then, by a complex series of fractionations, cracking, and various refinings, products ranging from heavy grease to high-octane fuel are obtained. Water used in these various processes amounted to 16.8 million gallons per day in 1964; however, much of this was recycled. Perhaps a more meaningful index of pollutant discharge would be gallons of waste water given off per barrel of crude oil—which varies from 50 gallons in new plants to 200 in old ones. Ninety percent of this water is used for cooling, but the principal source of pollutants in the oil industry is inefficient or sloppy operating procedures, leading to small leaks.

Fruit and vegetable processing. In 1964 the packing of 944 million cases of canned and frozen produce used 76 million gallons of water. The use of water in processing starts with the cleaning, grading, and sorting process. Machine-picked crops, which are gradually replacing hand-picked crops, usually include more soil and other foreign material

Figure 7.5 Paper pulp waste released into a small stream without treatment in North Carolina. (U.S. Dept. of the Interior, Sport Fisheries and Wildlife)

and water is used to rinse them. Chemical peeling—dipping of fruit and vegetables into a hot lye solution to dissolve the skin—produces a wash water that is hot and very alkaline, containing not only minerals but large amounts of dissolved organic matter from the skins.

In the freezing process most foods have to be blanched, that is, dipped into hot water or steam briefly to denature enzymes that might cause them to deteriorate after freezing. This blanching waste is also hot and contains much dissolved organic matter.

When the steam used in the water-removal phase of making juice concentrate is condensed to avoid air pollution, a liquid waste with various dissolved materials remains. Because of the seasonal nature of this and other food-processing industries, waste water is produced at a time when streams are often at their lowest and water demands highest. This factor also greatly increases the pollution potential of these wastes.

The industrial processes described paint a bleak picture of the misuse of a resource. But most industries are finally realizing that water resources are inexhaustible only when they are recycled, and some progress is being made in other areas of waste treatment as well.

Industrial Waste Treatment

The waste water from some industrial plants, food processing, textiles, rubber and plastics, machinery, and transportation equipment, can usually be accommodated by the plant's own sewage treatment equipment or by nearby municipal facilities for a fee. Wastes from paper, chemicals, petroleum, coal, and primary metals are not so easily handled because of the acidity, toxicity, or resistance of the wastes to bacterial breakdown.

Reuse and by-products. Recycling and reuse of water in the pulp and paper industry has increased tremendously since 1950 and will continue as water supplies and pollution standards tighten. Lignosulfonates, because of their dispersal and sequestering characteristics, are being used in other industries as additives to drilling mud in oil wells and as stabilizers of unpaved road surfaces in the summer. Even though 425 million pounds of lignosulfonates are used yearly, these markets are small compared with the supply.

There are other possible uses of pulp wastes. Over 40 percent of the domestic vanillin supply (artificial vanilla flavoring) is synthesized from lignosulfonate, as are glacial acetic and formic acids. The sugar xylose, which is quite abundant in pulping liquor, is very sweet but has few calories, making it a potential substitute for cyclamates. The complex sugars from hemicellulose can be converted to ethanol (grain alcohol), for which there is a steady industrial demand. Hemicellulose can also be eliminated from the effluent by culturing a yeast, *Torula,* which can be harvested as a valuable protein source. However, competition with established chemical industries already producing these materials is great, and these by-product activities are not as profitable as they may appear.

Consolidation. The economic elimination of the small processor and his replacement with much larger operations is already reducing pollution loads in industry. Food processing plants that are large and diverse enough to afford their own disposal plants can effectively reduce the organic content of their wastes and ultimately the BOD of the aquatic environment. This has been demonstrated with canning plants, meat packers, and dairy product processors.

Most industrial problems, then, can be handled by recycling, reuse, by-product utilization, more efficient and effective equipment, and biological breakdown of organic wastes. For those problems remaining, more radical methods have been proposed.

Deep well injection. When oil wells are drilled, large quantities of salt brine are brought to the surface with the oil. One approach to disposing of the brine has been to pump it, under pressure, down dry wells or new wells drilled for this purpose. In the past forty years over 40,000 brine injection wells have been drilled or utilized. The chemical industry and others have followed this example, using deep injection wells for the disposal of highly toxic or difficult-to-handle wastes generated in various processes. As surface disposal becomes more and more restricted, other industries will doubtless use this technique as the cheapest and most convenient way to get rid of wastes. But deep injection wells cannot be put down anywhere; there must be permeable sedimentary rock layers such as limestone or sandstone, capable of transmitting fluids, and porous enough to hold large quantities of liquid in the spaces between the grains of rock. To contain this liquid properly the porous layer must be bounded above and below by impermeable layers. The desired stratum must also be below the water table to prevent contamination (Figure 7.6). Although these requirements seem rather specialized, over half the land area of the United States appears to be suitable for injection, especially the southeastern coastal plain and the interior lowland of the Midwest. Wastes put into deep injection wells must have a low enough solid content to avoid filling the pores of the rock stratum, thereby reducing its permeability. Potentially dangerous is the generation of heat by radioactive wastes which may produce explosive steam, and the precipitation of salts or heavy metals which may contaminate water supplies.

Despite these restrictions, wells are used by petrochemical, chemical, steel, plastic, pulp, and photoprocessing industries. The effluents they dispose of include phenols, cyanides, phosphates, chlorides, nitrates, chromates, sulfates, alcohols, chlorinated hydrocarbons, acetates, and ketones. Although most of the wells are between 1000 and 6000 feet deep, some are shallower than 1000 feet, and others as deep as 12,000 feet.

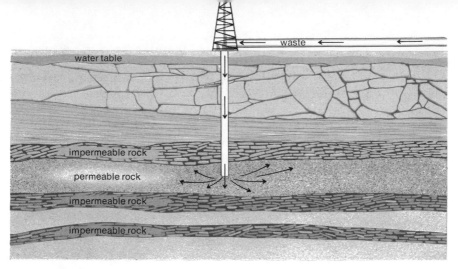

Figure 7.6 A deep well injection must deposit the waste fluid in a permeable layer of rock bounded by impermeable layers. The water table lies much closer to the surface.

One such deep injection well near Denver, Colorado, has stirred up much controversy. This well was drilled through the flat strata into the crystalline bedrock 12,000 feet below. Fluids were both injected and deposited by gravity flow at rates varying from 2 to 6 million gallons a month, beginning in March 1962 and terminating in February 1966. Since a number of minor earthquakes had been recorded in the Denver area during this period, it was thought that the deep injection well might have caused them. A comparison of earthquake frequency and waste injection (Figure 7.7) does indeed show an interesting correlation. The chances

Figure 7.7 When earthquake frequency is compared with the volume of fluid injected into a disposal well at the Rocky Mountain arsenal near Denver, Colorado, a cause and effect relationship is strongly suggested. (Modified from J. H. Healy et al., *Science*, **161**, 1968. Copyright 1968 by the American Association for the Advancement of Science.)

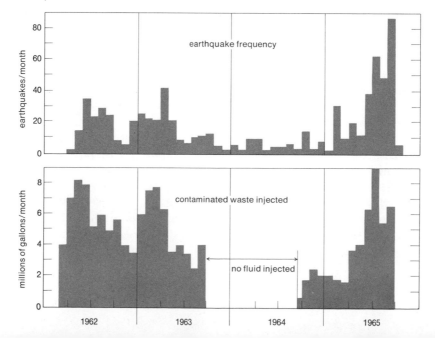

were calculated to be only one in 2.5 million that the earthquakes associated with the injection well in both time and space were solely due to
coincidence. Apparently the fluid injected into the basement rock lubri-
cated fracture lines, allowing tension built up over the years to be released through a series of slippages along these fault lines, which resulted
in earthquakes. As the volume of fluid injected increased, either there was
more slippage or a greater area was affected. The only solution beyond
cessation of injection would be to remove some of the fluid already
present, or barring that, to wait for a natural decrease in the resistance
to diffusion as the fluid percolates further and further from the well.

This example raises several questions about deep injection wells and
liquid waste disposal in general. One wonders, in light of the Denver
example, whether such wells should be used at all in areas of seismic activity. In addition, the qualifications outlined previously suggest that
storage capacity is limited and perhaps should be reserved for wastes
difficult or awkward to handle by conventional means. Injection wells
certainly should not be used for mere convenience or economy. The
presence of faults even in the most stable rock casts doubt on the ability
of supposedly impermeable rock layers to contain wastes permanently
and to prevent their introduction to the water table. Finally, we know
little about confining the lateral spread of liquids in a stratum; this information would be of great significance should we wish to retrieve these
wastes in the future. There are many instances of yesterday's garbage becoming today's food, and there may come a time when it would be most
useful to recover as resources the wastes that we discard today.

The near future will doubtless see a sharp increase in by-product
recovery and in the reuse and recycling of water. As this takes place,
industry's role as a major polluter of water will steadily decrease.

TOO MUCH PHOSPHORUS

Deoxygenation is just one problem from organic wastes. Another of perhaps even greater magnitude is the fertilizing effect of nutrients released
as organic wastes are decomposed.

Of all the minerals cycling in the environment (see Chapter 3), the
one which most often limits the growth of plants and animals is phosphorus. Too little phosphorus results in very low productivity of aquatic
systems; too much leads to population explosions of plants and microorganisms.

The concentration of phosphorus in aqueous environments is normally low, about 0.02 parts per million (ppm), because most phosphates
are insoluble in water. If the input of phosphorus into aquatic systems is
increased much beyond this figure, it is either organically fixed by organ-

Figure 7.8   A mass of filamentous algae in the nutrient-rich water of a trout hatchery in Pennsylvania.

isms and ultimately deposited in sediments, or precipitated directly. Either way, excess phosphorus is incorporated into sediments eventually, but the pathway is of critical importance.

When a nutrient normally in short supply is added to a system it may trigger a rapid growth of organisms in that system (Figure 7.8). Phosphorus is just such a trigger factor in most aquatic systems. Microorganisms are often held in check by a lack of phosphorus, but when it is made available they use it immediately, and with their rapid reproductive rates they can quite suddenly become extremely abundant. This population explosion, usually of one or more species of algae, is called a *bloom*. Most freshwater ecologists or limnologists call an aquatic population a bloom when there are 500 individuals of a species per milliliter of water. If a given amount of phosphorus caused a bloom of a certain intensity, we could easily set up maximum allowable concentrations of phosphorus and control the bloom. Unfortunately, the information available on phosphorus and algal blooms indicates that there is no one-to-one relationship between phosphorus concentration and algal blooms. In some Wisconsin lakes, algal blooms have been recorded with a phosphorus concentration of 0.01 ppm. In other lakes, blooms have been observed where the phosphorus concentration was 0.001 ppm and yet other lakes have not had blooms with a phosphorus level of 0.05 ppm, an increase of fifty times. Clearly other factors are involved too—nitrates, trace elements such as manganese, boron, cobalt, vanadium, as well as vitamins (B_{12} especially) or hormones—but phosphorus seems to be critical.

What are the sources of phosphorus in the environment, both natural and man-made? The data in Table 2 show that domestic waste and agricultural runoff are the largest contributors. Although there is some phosphorus in domestic sewage, a much greater source is household detergents.

Table 2   **Sources of phosphorus in water supplies**[a]

Source	Production, million lbs/yr	Phosphorus discharge, ppm
Domestic waste	200–500	3.5–9
Industrial waste	†	†
Agricultural runoff	120–1200	0.05–1.1
Nonagricultural runoff	150–750	0.04–0.2
Urban runoff	11–170	0.01–1.5
Rainfall	3–9	0.01–0.03

[a] Task Group Report, 1967. "Sources of N and P in Water Supplies." *JAWWA* 59, pp. 344–366.
† No estimate available.

Detergents

The unusual cleaning powers of detergents are not the result of a single magic molecule engineered by far-seeing chemists. There are many components, each tailored for a specific cleansing function, yet working together so that the result is greater than the potential of any single part. Typically, a detergent contains a surfactant, a builder, a silicate, and carboxymethylcellulose.

The surfactant or sudsing agent (30–50 percent of the total) is an organic agent that is able to penetrate between the soil particle and cloth or fabric. This is because the surfactant molecule is bipolar; one end is attracted to the dirt particle and the other is attracted to water. This bipolar action combined with machine agitation quite effectively removes dirt and, by lowering surface tension, grease as well.

Once in the wash water, dirt must be kept in suspension; this is one function of the builder. Another function is to soften water, eliminating the scummy precipitate soap produces in hard water, that is, water containing a large amount of calcium and magnesium salts. While varying greatly from brand to brand, most detergents contain up to 50 percent builder. The remaining detergent ingredients, though present in small amounts, complement and complete the cleaning action of the surfactant and builder. Sodium silicate is added as a corrosion inhibitor and carboxymethylcellulose is an anti-redeposition agent that helps keep removed dirt in suspension until the wash water is flushed away.

At one time the surfactant caused problems because of its tendency to keep sudsing in the environment beyond the washing machine. A change in the structure of this component in 1965 alleviated this problem. But an even greater problem has since been recognized. The builder, which comprises up to half of the packaged detergent, contains an impressive amount of phosphorus in the form of tripolyphosphate. Since this as well as the other detergent ingredients must be water soluble to perform its cleansing function, it is also available to microorganisms when it enters the environment in sewage effluent. Initially the potential fertilizing effect of tripolyphosphate was probably not considered important enough to worry about, but with the enthusiastic acceptance of detergents by the public, millions of pounds of phosphorus were soon being added to the environment per year. The result was algal blooms in increasing frequency across the country. Up to a certain point bloom organisms increase the useful productivity of streams and lakes, making available a more abundant food supply for higher links of the food chain. But in very large numbers these blooms cause problems.

Since many bloom organisms are photosynthetic, the oxygen level in the water during the daylight hours is sharply augmented as oxygen is released in the photosynthetic process. But during the night when there is no photosynthesis to balance respiration which consumes oxygen, the oxygen content of the water may fall below the level necessary for respiration of higher forms of animal life; fish kills often result. This is a particular problem in the still water of lakes. In addition, windrows of algal mats washed up on lake shores decompose, producing hydrogen sulfide and other by-products, often making bodies of water unsuitable for use of any kind.

The connection between detergent builders and algal blooms was eventually recognized, and the industry is presently dealing with its second unanticipated problem. Unfortunately we cannot assume that once phosphates were somehow removed from detergents in manufacture or processed out of sewage effluent, algal blooms would quickly disappear, since detergents are only one source of man-generated phosphorus. Nor can we assume that replacements such as NTA (nitrilotriacetic acid) have no environmental impact until they are thoroughly tested.

Agricultural Runoff

Despite the enormous quantity of phosphorus being added to the environment via detergents, we are adding several times this quantity in the fertilization of farmlands. As the world's population continues to increase, the need for greater productivity in agriculture becomes critical, and increased fertilization is one of the chief means of accomplishing this. But when highly soluble fertilizer is used, 10–25 percent is leached away into the surface runoff before the plants are able to use it. This represents a loss to the farmer and a burden to the aquatic ecosystems, which are being grossly overfertilized. Another agricultural source of phosphorus is animal waste. Livestock, increasingly grouped in feedlots, contribute seventy million pounds of phosphorus per year, almost as much as detergents.

While the phosphorus problem will never be completely solved, there are several ways by which the various sources of man-generated phosphorus can be brought under control.

SOME SOLUTIONS TO THE PROBLEM

Where it exists at all, sewage treatment consists of one or two stages. Primary treatment simply removes suspended solids; after the filtering process, the remaining liquid is released into the environment. Secondary treatment goes a step further and utilizes bacteria to break down into inorganic compounds, sludge or organic wastes that were settled out in the primary process. The water is then filtered, chlorinated to remove the bacteria, and released. The BOD is greatly reduced because of the in-plant decomposition of organic material, but the effluent still contains large amounts of phosphates, nitrates, and most other nutrients known to promote plant and animal growth. Primary and secondary treatment of sewage is only partly effective. Some kind of tertiary treatment to remove nutrients *must be included* if artificial enrichment of surface water is to be controlled.

There are several possible tertiary treatments using existing technology. The effluent can be prebloomed, that is, held in shallow tanks or pools for a time to allow algal growth to remove nutrients. Algae can then be filtered out, leaving the water considerably reduced in its nutrient content. Problems with this approach are the space necessary for the holding ponds and the need for supplying heat to keep algae growing at a high rate in the winter in most northern states. Perhaps heat generated from the activated sludge phase of secondary treatment could be used to heat the water to allow a longer season of algal growth. Also, the algae filtered out of the final effluent could be processed and sold as a protein feed for livestock, generating some economic return. A more promising solution uses lime or alum to take phosphorus out in the form of insoluble precipitates that can then be treated to regenerate the precipitator and perhaps provide some form of phosphorus that has agricultural or industrial use. Unfortunately nitrates in sewage are much more difficult to remove because unlike phosphates, most nitrates and ammoniates are water soluble.

As for the other man-generated sources of phosphorus, there is a growing tendency for livestock and poultry to be concentrated in feedlots or various controlled environments, raising the possibility of treating animal wastes in the same manner as domestic sewage. In contrast, fertilization of agricultural land, which is by far the greatest source of man-generated phosphorus in the environment, presents far more serious management problems. If highly soluble fertilizers, which are immediately available regardless of the plant's ability to make use of them, could be replaced by less soluble fertilizers, which release nutrients more slowly and keep pace with crop needs, there would be a substantial decline in nutrient leaching into the water table or surface water. The problem of industrial phosphorus sources and the phosphorus contained in bottom muds is still unsolved.

One thing is certain, however: if phosphorus-stimulated algal blooms are to be significantly reduced, ways must be found to reduce the release of man-generated phosphorus into the environment. But even when this is accomplished, natural sources which are impossible to control will continue to supply phosphorus, often in critical amounts.

WHERE SOLUTIONS FAIL

Most problems of excessive aquatic productivity, however unpleasant, have been limited in scope to streams, segments of rivers, and small lakes. But with the appropriate conditions, there is no upper limit on the size of overproductive waters: large lakes, seas, and ultimately even oceans can become overfertilized with catastrophic results.

The Aging of Lakes

It was long thought that lakes aged gradually, progressing through a sequence of events from their birth as very pure unproductive or *oligotrophic* bodies through slowly increasing numbers of plants and animals and greater nutrient levels (*eutrophic*) that left the former lake bed a marsh, and finally, a forest. But things are not quite so simple. Studies in the lake district of England seem to indicate that this simplistic picture of lake development should be re-examined. Apparently the thickness of bottom deposits in some lakes bears little relationship to the internal productivity; it is the drainage basin that is the major source of sediments. When erosion in the watershed of a lake is high, minerals are removed and sealed away in lake bottoms before nutrients locked in these minerals have a chance to be leached away in the process of soil formation. During periods of less intense erosion when more leaching of soil takes place, the nutrients pass into the lake and stimulate biological productivity. When the watershed lacks vegetation, erosion is rapid; when it is covered with trees, erosion is much reduced. The history of a lake and its watershed can often be read by analyzing the bottom mud and tracing the abundance of various nutrients layer by layer. When this was done at Esthwaite Water in England, sodium and potassium levels were found to be quite low soon after the lake was formed. As the watershed became covered with vegetation and erosion was checked, the levels of sodium and potassium rose. At the time of forest decline associated with man's activities about 5 thousand years ago, these nutrients decreased again and have continued declining up to the present. A similar picture was obtained from work done on Linsley Pond in Connecticut—one of the most studied bodies of water in the world.

The implication of this work is that lakes may be eutrophic or oligotrophic from the very beginning and may retain their initial state for thousands of years. Or they may change rather suddenly when man intervenes. It now seems unlikely that all lakes are in the inexorable process of gradual change from oligotrophy to eutrophy. If the old pattern no longer represents the true situation, we can no longer predict what will happen. Detailed studies of just what is happening in grossly overfertilized lakes right now are necessary. From this, we may be able to make some predictions about the future of the lakes. No remedial action can be taken until the base lines are drawn.

Death of a Great Lake?

Of all the overfertilized lakes in the world, Lake Erie is one of the largest and best known. In fact, the western basin of this lake is so polluted that discussion has arisen as to whether the lake is merely dying or already dead.

Lake Erie, like the four other Great Lakes, is a glacially scoured lowland that was filled by the melting continental glacier around twelve thousand years ago. Although Lake Erie is not the smallest Great Lake, because of its shallowness (mean depth of fifty-eight feet) it has the smallest volume. This shallowness is at the root of its subsequent problems.

Once clear and filled with valuable fish and game, Lake Erie today is the embodiment of all that can go wrong in an aquatic system. The more desirable species of fish have disappeared; the once clear water is filled with excessive numbers of microorganisms; mats of filamentous algae at times cover whole square miles of lake surface; swimming is

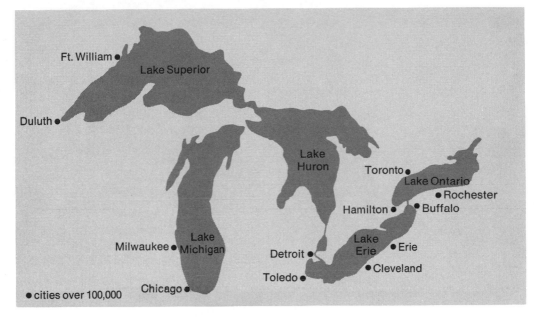

Figure 7.9 When Great Lakes cities with a population of 100 thousand or more are spotted on a map, the clustering around Lake Erie helps to explain the critical nature of that lake's pollution problem.

impossible in many places because of the quantity of untreated sewage in the water and the decaying vegetation covering the beaches; oil scums often cover harbors and coves, making boating and water sports unpleasant.

Of all the Great Lakes, the shore of Lake Erie has the largest number of population centers, each having varied industries with effluents to be disposed of. Erie produces paper; Buffalo, flour and chemicals; Cleveland, petrochemicals and steel; Toledo, glass and steel; and Detroit, automobiles, steel, and paper (Figure 7.9). Naturally, large population centers contribute quantities of sewage both treated and raw. And the agricultural land between the cities adds pesticides, herbicides, and fertilizers, giving no respite. Bear in mind this is (or was) a "great lake," 240 miles long and 50 miles wide, with a volume of 109 cubic miles!

Fish as a pollution indicator. The first symptom of lake-wide problems in Lake Erie was seen in the 1920s, when the fish crop—fifty million lbs annually of cisco, whitefish, pike, and sturgeon—began to fall. By 1965 the catch of these commercially valuable fish had fallen to just 1000 lbs. This doesn't mean that Lake Erie is devoid of fish, for there are still about fifty million lbs of other types of fish caught every year; but the more valuable species have been replaced by catfish, carp, smelt, and other less valuable types (called "rough" fish by commercial fishermen).

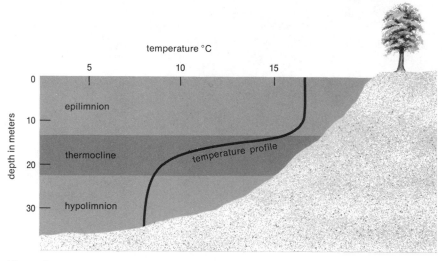

Figure 7.10 In the summer deep lakes stratify into a cold hypolimnion, a zone of rapid temperature change or thermocline, and an upper layer of relatively warm water. The dark line describes this temperature profile.

Why did this happen? During the warm season of the year all but the most shallow lakes stratify into three layers (Figure 7.10), a warm surface layer, *epilimnion,* a transition layer of rapidly changing temperature, *thermocline,* and a cold deep layer, *hypolimnion.* Under normal conditions whitefish, cisco, and other cold-water fish move to the deeper, colder hypolimnion in summer, and return to the surface in fall when storms and colder weather mix the layers and the lake again becomes thermally homogeneous.

With increasing eutrophication, problems develop. Normally, a certain number of algae are found in the warm surface water during the summer. At the end of the growing season, they die and fall to the bottom where they are broken down by bottom bacteria into inorganic nutrients which are held in the bottom mud. By the next summer, enough nutrients have leached from the watershed into the lake to support another population of algae. No more algae are able to grow than accumulated nutrients are able to support, and so a balance is maintained. When nutrients pour in from man-generated sources, the balance is upset. Great blooms of algae released from their dependence on natural nutrients fill the epilimnion during the summer, sinking to the bottom during fall and winter. When the lake stratifies the following summer, extra thick organic layers at the bottom so stimulate the growth of bacteria that the oxygen level of the hypolimnion begins to fall, often falling below the minimum requirements of fish. Because cold water can hold more oxygen than warm water, cold-water fish such as trout or cisco have developed, through evolution, higher oxygen requirements than warm-water fish such as carp or catfish. Hence, when the oxygen level of the cold deep water begins to fall, the valuable cisco, whitefish, and pike are eliminated.

The Future of Lake Erie

The deoxygenation of the hypolimnion previously discussed is far more serious than the loss of valuable fish. Under normal oligotrophic conditions, when the organic sediment load is light enough to allow reasonable oxygenation of the hypolimnion, the bottom mud is covered by a thin layer of iron in the insoluble or ferric oxidation state, usually ferric hydroxide. This compound not only absorbs much phosphorus from the water, but because of its insolubility seals the bottom mud, preventing exchange of nutrients in the mud with the water above.

When oxygen is exhausted, the iron changes to the soluble ferrous state. This not only puts phosphorus into solution as ferrous phosphate but exposes the bottom mud, allowing free entrance of nutrients into the lake water. The problem in Lake Erie, then, is simple and overwhelming. Wherever pollution stimulates algal blooms, which later decompose and deoxygenate the bottom, some 30–120 feet of nutrient-rich mud is exposed which is capable of contributing its nutrients into the lake water, stimulating still more algal growth, deoxygenating more lake bottom, thus closing a vicious cycle. Even if all man-generated pollutants or nutrients were prevented from entering the lake, there might still be quite enough nutrients already in the lake to continue algal blooms with concomitant problems far into the future.

This misuse of Lake Erie is of critical interest because we can now see that pollution of aquatic systems not only extends far beyond a local pond or stream, but can have widely ranging effects which were totally unexpected. If water resources are to be available in the future for man or any other organism, we must rethink our traditional view of water as being a free waste disposal system. While the supplies of water are unlimited, the ability of water to dilute or carry off wastes is not. Since water pollution is basically a reflection of our sophisticated technology, it is time for that sophistication to be applied to the environmental problems it has caused.

Whether Lake Erie can ever be restored to full usefulness is uncertain, for we don't know enough about the system to say. But we have very little time to find out.

FURTHER READING

Bigger, J. W. and R. B. Corey, 1969. *Eutrophication: causes, consequences, correctives.* National Academy of Sciences, Washington, D.C.

Boyle, R. H., 1969. *The Hudson River: a natural and unnatural history.* W. W. Norton, New York. A personal account through the eyes of a sport fisherman of the natural history of one of America's most famous rivers.

Eliassen, R. and G. Tchobanoglous, 1969. "Removal of nitrogen and phosphorus from waste water." *Env. Sci. & Tech.,* 3(6), pp. 536–541. A state of the art report.

Healy, J. H., W. W. Rubey, D. I. Griggs, and C. B. Raleigh, 1968. "The Denver earthquakes." *Science,* **161**, pp. 1301–1310. This paper reports on the correlation of deep injection of wastes and the frequency of earthquakes.

Hynes, H. B. N., 1960. *The biology of polluted water.* Liverpool University Press, Liverpool, England. A fine account of the organisms involved in water pollution, with good illustrations.

Mackethun, K. M., 1965. *Nitrogen and phosphorus in water: an annotated selected bibliography of their biological effects.* U.S. Public Health Ser. Publ. No. 1305. An excellent review of what has been published (to 1965) in this area.

Mackethun, K. M. and W. M. Ingram, 1967. *Biological associated problems in fresh water environments: their identification, investigation, and control.* FWPCA, U.S. Department of Interior, Washington, D.C. A good source book for anyone concerned with water quality problems.

Parizek, P. R. *et al.,* 1967. *Waste water renovation and conservation.* Pennsylvania State University Studies No. 23. Admin. Committee on Res. P.S.U. Description of the technique of spraying liquid effluent on crop or forest land to remove nutrients.

Pearl, I. A., 1968. "Waste product use helps paper industry control pollution." *Env. Sci. & Tech.,* 2(9), pp. 676–681. A look at how the paper industry is attempting to deal with its pollution problems.

Task Group Report 261OP, 1967. "Sources of nitrogen and phosphorus in water supplies." *Jour. Amer. Water Works Assoc.,* **59**, pp. 344–366. Fine analysis of where waste nitrogen and phosphorus come from in the environment.

U.S. Department of Interior, 1968. *Lake Erie report: a plan for water pollution control.* FWPCA, Sup. Doc., Washington, D.C. Well-documented breakdown of the sources of Lake Erie's pollutants.

CHAPTER 8

THERMAL LOADING

Although water is the most abundant compound on earth, it has several strikingly different characteristics: one of the most unusual and certainly most useful is its high specific heat. By absorbing and releasing heat slowly, water becomes an excellent medium to effect heat exchange, hence, water has been used for heating and cooling purposes for thousands of years. When limited, this loading of waste heat into a natural aquatic system caused few problems, but when industrial plants began to use almost the entire flow of a river for cooling purposes, environmental problems resulted.

WATER AS AN INDUSTRIAL COOLANT

With the rise of industry in the eighteenth century, water came to play an increasingly important role as a coolant in many manufacturing processes. With continued growth and diversification of industry, the demand for cooling water has soared. Today 13.1 trillion gallons, over 80 percent of the water used by industry in the United States, is used for cooling. Of all industrial water users, the electric power industry is the largest, requiring 111 billion gallons per day.

Water was first associated with power generation in a mechanical rather than electrical sense, in the form of water mills. Dams were constructed across small streams or rivers and the head of water they captured was used to turn a wheel which, through an ingenious if inefficient series of gears and levers, performed mechanical work, for example, grinding corn or wheat to meal and flour (Figure 8.1). Later, by means of belt drives, crude machinery for more diverse uses was operated. With the development of electricity, water was used to drive a turbine which generated power. Despite the publicity that has surrounded dam con-

Figure 8.1 This old water mill still grinds corn and wheat alongside the Blue Ridge Parkway in Virginia. (National Park Service)

struction, expense and a growing shortage of natural sites have limited the growth of hydroelectric power. Today only a small portion of the power needs of most countries is met from hydroelectric sources. Although quite the cleanest means of generating power, there is a growing feeling among large numbers of people that rivers have uses other than for forming a series of pools for hydroelectric power generation (Figure 8.2a, b).

Power Plant Coolant

The major alternative method of power generation is based on the same premise as hydroelectric power, the turning of turbine blades. But instead of using the energy of water falling from the dam to the power plant below, steam is the energy source. Until recently, steam was produced by heating water with coal, fuel oil, or natural gas (all called fossil fuels). The steam turns the blades of the turbine, which generates mechanical energy. This is converted by a generator into electrical energy, which is conducted away by power lines. Unfortunately, because of all the energy transformations involved—heat to mechanical to electrical energy—only a small portion of the energy contained in the carbon-carbon bonds of the coal is converted to usable electricity. The energy leaks in the system are enormous. After the steam passes through the turbine it must lose enough heat to condense into water for return to the boiler. By passing

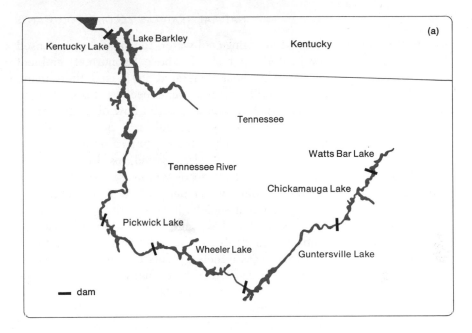

Figure 8.2 (a) The Tennessee River has been almost totally converted from a river into a series of pools. (b) These huge reservoirs have caused the once-free Missouri River to become a series of lakes in the Dakotas.

the spent steam over a coil containing cool water, the steam is condensed and the cooling water heated up to 20°F above its normal, ambient temperature. For each kilowatt hour generated in a coal fired plant, 6000 Btu[1] must be dissipated. The medium that receives this no-longer-needed heat is the coolant, which may be the entire flow of a small river.

In the early years of this century, hydroelectric power produced much of the power needed in the United States, but as dam sites grew fewer and more remote and power needs skyrocketed, fossil-fueled plants were built. Until recently these plants were widely scattered and not especially large so that their cooling water demands were not excessive. However, with a growing, affluent society buying more and more air conditioners, television sets, dishwashers, and many other electric appliances, power demands began doubling every six to ten years. In response to this surging demand, fossil-fueled plants four or five times the size of older plants were constructed. These plants generated more power, but they required still more water for cooling. There was closer spacing of larger plants demanding more water, and finally the cooling ability of naturally running water became inadequate. This was evident when the unrestricted dumping of heated effluent by power plants and various heavy industries raised the temperature of the Mahoning River in Ohio to 140°F, forcing industries to slow or terminate operations during periods of low water.

The Rise of Nuclear Power

Through efforts of the Atomic Energy Commission (AEC) exercising its mandate to explore the peaceful uses of atomic energy, and of industry eager to use new technology to reduce costs and increase profits, a new energy source for power production has been developed—nuclear fuel. There is a popular misconception that nuclear-fueled plants are radically different in their mode of power generation. They are not. Certainly, heat is generated from a radioactive source rather than a fossil fuel, but the rest of the power train is the same. The heat produces steam, which turns turbine blades, which run a generator, which produces electricity. The major improvement is the elimination of the combustion products of fossil fuels, which have contributed greatly to air pollution (see Chapter 10). The price paid for this boon was an increase of almost 50 percent in the already heavy demand for cooling water in the condensing system, and in the various radioactive wastes, liquid, solid, and gas that were released (see Chapter 11). Use of nuclear fuel requires lower steam pressure and the less efficient use of steam results in more waste heat. Hence, nuclear plants produce 10,000 Btu per kilowatt-hour rather than

[1] One British thermal unit or Btu raises the heat of one pound of water by 1°F.

the 6000 produced by fossil-fueled plants. Breeder nuclear units, which reuse some of their energy to create new fuel, will be coming into use in the next decade and are expected to reduce this disparity.

In addition, because of this lower efficiency, nuclear-fueled plants must have a quite large generating capacity to be economically competitive with fossil fuel plants. This demands an even greater supply of cooling water. Although nuclear power has been on the scene for a mere twenty years, some indication of its future role and of its demands can be appreciated by reviewing a large power utility's cost analysis for its various sources of fuel: uranium, 22 cents per million Btu, coal, 29 cents; fuel oil, 33 cents; gas, 40 cents. In the next 30 years the power demand is projected to be 2×10^6 megawatts, producing as waste 20×10^{15} Btu per day. To dissipate this incredible quantity of heat energy would require *one-third* of the daily average fresh-water runoff of the United States. Unless steps are taken to find alternate means of dispersing or utilizing this heat, there is a distinct possibility that all major rivers in the United States will reach the boiling point by 1980 and then evaporate entirely by 2010!

HEAT AS POLLUTION

Putting aside for the moment this dire though unlikely denouement, why is addition of heat energy to an aquatic system even considered pollution? Although heat loading of a stream causes it to depart from its normal thermal character, which was our definition of pollution in Chapter 7, there is a more specific reason to view thermal loading as a form of pollution. As the temperature of water increases, its ability to hold oxygen decreases (Figure 8.3). Since dissolved oxygen is the key to

Figure 8.3   The quantity of oxygen dissolved in water is related to temperature, as shown in this graph. The higher the temperature, the lower the oxygen content of the water. Thermal pollution can lower the oxygen content below the point necessary to sustain many animals.

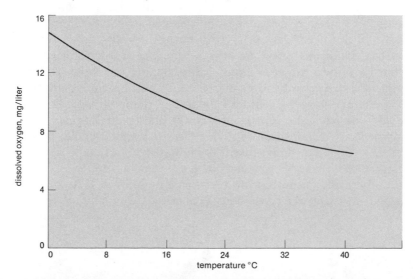

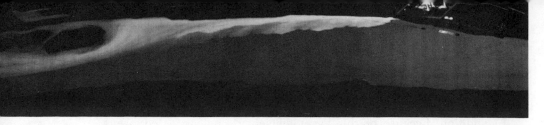

Figure 8.4 An infrared aerial photo of hot coolant water from a coal-burning power plant being dumped into a river. Since infrared film is sensitive to heat, the lighter areas are those having the warmest surface. Note the slow rate of mixing of hot and cold water. (Environmental Sciences Branch, HRB-Singer Inc.)

assimilation of organic wastes by microorganisms, any activity that might impair this assimilation can be labeled as pollution.

The presence of dissolved oxygen is probably the single most important factor in the biology of aquatic systems, and a great variety of physical and biological interactions stem from it. Aeration or oxygenation is derived from two sources: exchange with the atmosphere, and photosynthetic release by green plants. Depending upon the temperature difference between a heated effluent and the body of water into which it is being flushed, the heated discharge can either mix vertically and horizontally, or stratify in a layer above the cooler, denser receiving water (Figure 8.4). If the temperature difference is small and the receiving water shallow, complete mixing usually takes place. If the temperature difference is great and the receiving water deep, stratification usually results: the colder water beneath is covered with a blanket of hot water, which not only carries less oxygen because of its higher temperature but is less likely to exchange with the cooler water beneath because of its different density. Biological reduction of the oxygen content of the atmospherically unreplenished water below might then produce anaerobic or oxygenless conditions.

With a given thermal load and assuming complete mixing, a stream or lake might have its temperature elevated by around 20°F, which could lead to a complete shift from cold- to warm-water forms of life. Whereas such a replacement might cause little disruption to the productivity of a body of water once the adjustment had taken place, in almost no instance would the production of heated effluent be constant for long enough periods to allow either adaptation or replacement to be effective. Shutdown of a power plant or factory for just one day in midwinter would have a disastrous effect on warm-water species introduced and maintained by the artificially high water temperature. Such an occurrence, perhaps extreme in its regular periodicity but illustrative of the problem, is seen in Figure 8.5. Heated effluent is dumped periodically into the River Lea from the Rye House Power Station in England, making adjustment by organisms adapted either to warm *or* cold conditions extremely difficult. Only those organisms without specific temperature preferences can survive such a change of temperature.

Much more unusual but occasionally observed is a cold effluent released into a warm river. The Fontana Dam, a unit of the Tennessee

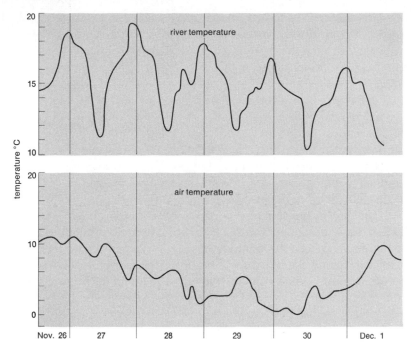

Figure 8.5 Periodic pulses of hot water flowing down the River Lea in England from a power plant make any adjustment of flora and fauna to either warm or cold water impossible.

Valley Authority on the Little Tennessee River in North Carolina, releases cold bottom water from its reservoir downstream into the river, eliminating for many miles the warm-water fish usually characteristic of southern rivers.

Effect of Temperature Change on Organisms

As temperature increases, dissolved oxygen content decreases, as we have seen, but respiration and oxidation rates double for every 10°C temperature increase (Figure 8.6). In order to meet the respiratory needs

Figure 8.6 Occasionally hot summer weather naturally heats shallow water to the point where it cannot hold enough oxygen to meet the needs of fish, causing them to suffocate. This is what killed these shad in the Anacostia River near Washington, D.C. (U.S. Dept. of the Interior, Sport Fisheries and Wildlife)

of warm-water fish, the dissolved oxygen content of water should be at least five ppm during sixteen of every twenty-four hours and should not fall below three ppm at any time. Even carp, which have an exceptionally low oxygen requirement for fish, 0.5 ppm at 33°F, must have 1.5 ppm at 95°F.

But temperature interacts with other factors as well. At 34°F, carp are able to tolerate a carbon dioxide concentration of 120 ppm, but at 86°F a concentration of 55–60 ppm is lethal. Similar type interactions are seen with pH, salinity, and toxicity. All organisms have temperature limits to their survival ability, but for most if temperature change is gradual, acclimatization is possible to some extent. It is the sudden change either up or down which is most often lethal.

The life cycle of many aquatic organisms is closely and delicately geared to water temperature. Fish are often distributed, migrate, and spawn in response to temperature cues. When water temperature is artificially changed, the disruption of normal activities and patterns can be catastrophic. Some animals spawn as the temperature drops in the fall; conversely, some spawn as temperature rises in the spring. Shellfish, such as oysters, are so delicately attuned to temperature change that they spawn within a few hours after their environment reaches a critical temperature. Trout eggs take 165 days to hatch at 37°F, only 32 days at 54°F, and will not hatch at all above 59°F. Lifespan is also affected by water temperature; water fleas (*Daphnia*) live for 108 days at 46°F, but only 29 days at 82°F.

Even more subtle effects involve competition between various species, predation rates, incidence of parasitism, and spread of disease. Any one or a combination of these more covert factors can destroy a species as effectively as the more obvious thermal shock.

MAN SIMPLIFIES THE ECOSYSTEM

Plainly, adding hot water to the environment is not the simple act of dilution it might appear to be. To put this subtle disturbance into perspective, we must view an organism together with its biotic and abiotic environment as an integrated system, an *ecosystem*.

Natural ecosystems, whatever their size, are complex; the larger they are, the more complex. Man's chief effect on ecosystems is to try, consciously or unconsciously, to simplify them. A potato field serves as a good example of this. To channel the productivity of a piece of land as directly as possible into a useful product—potatoes—man simplifies the ecosystem by removing all plants or animals that might compete with the potato plants and thereby decrease the yield.

Cultivated plants were selected to survive the stress that is amelio-
rated by the diversity of the natural ecosystem; drought, direct sunlight,
flooding, and wind are some of these stresses. Most native plants have
requirements too specialized to tolerate such stress conditions. However,
one group of plants, weeds, is not only able to tolerate stress but is able
to outcompete man's plants and become superabundant. The appearance
of weeds in a field is the first step in a natural process (succession) that
reduces stress and ultimately leads to the return of a complex ecosystem
capable of sustaining itself.

A typical nonpolluted stream may support as many as two dozen
species of fish. After pollution, although there may be only three "weed"
species, carp, killifish, and shiners, the *numbers* of fish may remain the
same. Through the simplifying stress of pollution, the productivity, which
in the normal stream was shared by small numbers of many species, has
now been channeled into large numbers of a few species.

Even more basic are the effects of stress on food chains. The basis
of all higher life is the green plant or, in aquatic systems, various forms
of algae. Because of their ability to fix energy from sunlight, carbon
dioxide, and available mineral nutrients (nitrogen and phosphorus especially) without dependence upon other organisms, we call them primary producers. The primary producers are in turn eaten by a series of
consumers leading in a chain from tiny crustaceans and larvae to fish
and, perhaps ultimately, to man. The productivity of a body of water
depends upon the kinds and numbers of algae present.

There are three major groups of algae in fresh water from a food
chain point of view: diatoms, greens, and blue-greens. While all types
are eaten by some organisms, most consumers prefer diatoms to greens
or blue-greens. Blue-green algae are eaten by few consumers, and they
can be toxic to some aquatic and terrestrial consumers. These three algal
groups differ in their temperature tolerances (Figure 8.7). With an in-

Figure 8.7 The three most important groups of common freshwater algae, (a)
diatoms, (b) greens, and (c) blue-greens, have radically different temperature tolerances. (Modified from J. Cairns, Jr., *Scientist and Citizen*, **10,** 1956.)

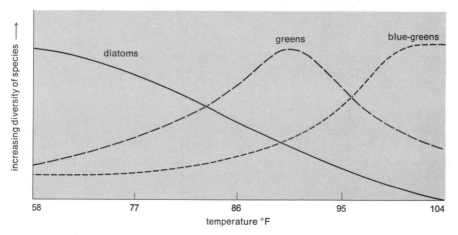

crease in temperature the number of the more desirable diatom species decreases as the number of the greens and blue-greens increases. At a certain level the greens also decline, leaving the blue-greens most abundant at the higher temperatures. As these primary producers shift in their relative abundance, the consumers depending upon them for food are also affected. A water flea, for example, which might be able to tolerate the thermal extreme of 95°F, would probably starve to death if the diatoms on which it fed were unable to survive at that temperature. In turn, fish feeding on water fleas would be similarly hard pressed to survive, regardless of their tolerance or adaptability to the high water temperature.

Another factor affecting food chains involves the large volume of water used in cooling systems. A plant circulating 500 thousand gallons a minute may intercept and enclose a considerable portion of the flow of a small river. Because of the tendency of microorganisms to grow in the pipes and reduce the flow, periodic or continuous metering of chlorine is used to control such growths. This not only affects organisms where the coolant is returned to the river (the outfall) until the chlorine is diluted, but it may sterilize the water flowing through the cooling system. In some installations 95 percent of the organisms are killed during the cooling period. The thermal shock of a rapid 20°F rise is destructive to many species. If the cooling system's intake uses more than a small part of a river's flow, a considerable portion of its microfauna and microflora could be eliminated, with grave effects on the food chain downstream, even to the point of eliminating certain populations of fish.

A heated effluent as an environmental stress, then, not only affects organisms directly but also the entire interaction of factors that make up the ecosystem. This makes it extremely difficult to predict the effects of thermal loading. With hydrologic models an engineer may be able to predict the degree of mixing, thickness, and the extent of stratification, but assessing the total and often subtle biological consequences is far more difficult than determining the thermal death point of one or two fish.

COOLING THE COOLANT

There are several current approaches to removing heat from cooling water before it is reused or discarded.

Wet Tower

The most common practice is to run water in a thin sheet over baffles in huge hyperbola-shaped towers relying on drafts of air entering at the base

Figure 8.8 A set of four wet cooling towers used by a huge power plant in Pennsylvania. (U.S. Dept. of Agriculture, Soil Conservation Service)

to remove heat by evaporation (Figure 8.8). A variant is to spray the hot water inside the tower in a fine mist. Cool air rising through the tower condenses the mist, releasing heat, which is carried by the air column out of the tower (Figure 8.9). In either case the cooled water is either discharged into the environment or recycled. Two problems associated with this wet tower technique are water loss (20–25 thousand

Figure 8.9 In a wet cooling tower, the hot water is exposed to air circulating through the tower. As water evaporates, heat is lost. The cooled water is either recycled or released into the environment.

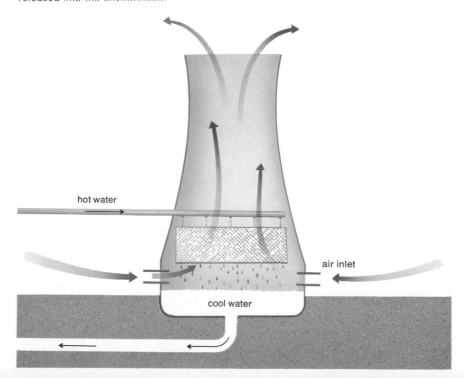

hot water

air inlet

cool water

gallons per minute for a 1,000-megawatt plant), and fog formation on cold days. When the surface temperature is less than 32°F, the fog freezes on contact, forming hoar frost or rime ice, potentially damaging to vegetation and extremely dangerous on highways. The water loss not only vitiates somewhat the benefit of recycling but serves to concentrate whatever pollutants the effluent contains.

Dry Tower

In a dry tower the heated effluent is contained in a system of pipes much like the radiator of an automobile. Air is passed over the pipes by a large fan facilitating heat exchange by radiation and convection (Figure 8.10). Water loss and, to a lesser extent, fog are controlled by this method, but installation costs are much higher, perhaps two or three times the cost of a wet tower. Also, maintenance costs for one plant of 500-megawatt capacity were calculated at 3 percent of the power output or $500 thousand a year to keep the blower fans operating.

Figure 8.10 A dry tower resembles a huge car radiator. Contained in pipes, the water gives up its heat by radiation and convection to the air, avoiding evaporative water loss.

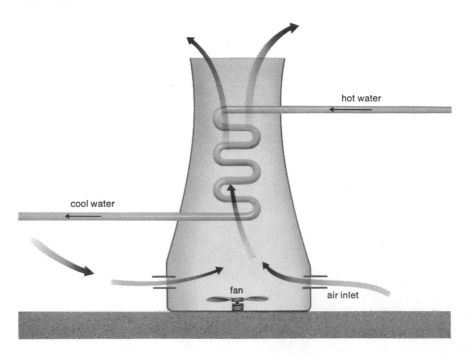

Cooling Ponds

This approach takes advantage of the few positive aspects of thermal loading. Warm-water outfalls in winter attract fish (and fishermen) because of the extended feeding period and ice formation, which often prevents adequate oxygenation of the water beneath, is prevented. By combining waste water from a power plant and effluent from a sewage treatment plant, the heat and nutrients, considered a nuisance in most environments, can be put to good use. By supplying heat-tolerant algae as primary producers and introducing warm-water fish to harvest the algae, water would have time to cool, have its nutrient content sharply reduced, and its algae harvested in the form of useful protein, before being returned to the natural environment. One tropical species that might be used, milkfish (*Chanos*), is one of the few palatable fish that feeds on blue-green algae. This type of operation would probably have to be limited to northern latitudes, since heat tolerance of even tropical species would be exceeded in areas where heated effluents are released during long, hot summers.

In terrestrial ecosystems, heated water could also be put to good use by applying it to crops. Preliminary evidence indicates that growth of some crops is accelerated by irrigation with warm water, suggesting the possibility of double cropping in areas of mild climate. Frost protection can also be achieved by spraying fruit trees with warm water. There are difficulties, however: frost is a problem in most orchards on only a few days each year, and irrigation is necessary for only a few months in the summer. A large power plant must cool 500 thousand gallons per minute every day of the year. Also, the cooling effluent from nuclear powered plants contains slight amounts of radioactive minerals, which might be concentrated in crop plants irrigated with such an effluent.

These are problems that can be worked out; it is becoming increasingly evident that we cannot continue to inject our waste heat either into the water or into the air without causing serious environmental problems. There is a tremendous potential for ingenious uses of this heat energy, as these examples indicate—uses that must be exploited if the natural ecosystems are to survive until the day when more efficient power sources are found.

CAYUGA LAKE AND THERMAL LOADING

To understand what thermal loading means when applied to a specific body of water, let's examine a recent and quite typical proposal for a power plant on a northeastern lake.

Cayuga Lake, one of the Finger Lakes of New York, is thirty-eight miles long, averages 1.7 miles wide, and is up to 435 feet deep. The average residence time of water in Cayuga Lake is nine years. A drop of water entering the lake from rainfall or a tributary stream would take nine years to find its way out, via the lake outlet. During the summer the lake stratifies into an epilimnion or upper layer, 35–50 feet deep at 50–73°F, and a hypolimnion or lower layer at 40–43°F. This stratification lasts from May to November. As we have seen in Chapter 7, biological production is confined to the epilimnion and decomposition to the hypolimnion, where as a result of this activity and respiration of higher organisms, oxygen decreases until the autumnal mixing.

In 1968 it was announced that an 830-megawatt nuclear-fueled electric power plant would be constructed north of Ithaca on the shores of Cayuga Lake. Cooling water would be withdrawn from the hypolimnion at around 100 feet and cycled at the rate of 9 thousand gallons a second through the cooling system. Withdrawn from the lake at around 45°F, it would be returned to the epilimnion at 65–70°F.

During the May-October period, 10 percent of the average volume of the lake would be so heated. This supplement of heat would delay both cooling and mixing of the lake in the fall and promote earlier summer stratification, thereby extending the growing season. This, plus the addition of warm water, could significantly increase the productivity of the lake. The prolonged period of stratification, together with the greater productivity of the lake, would increase the amount of organic matter settling out at the bottom of the lake. As this organic matter decomposes, oxygen will be used at a greater rate, lowering the amount available for cold-water fish when they move into deeper water during the summer. One commentator suggests,

> Maintenance of steam-electric power plants such as the one proposed for Cayuga requires periodic flushing treatments with such compounds as chlorine (for controlling growths in the pipes), hydrochloric acid, detergents, and corrosion inhibitors. All of these would have undesirable effects on the lake, but again, the magnitude cannot be predicted now.
>
> Small aquatic crustaceans living in the hypolimnion are the basic food source for the young of fish species such as lake trout in Cayuga. It is not likely that these forage organisms could survive the extremely rapid temperature rise in the cooling system, once they are pulled into the plant's 1,200 cubic-feet-per-second water intake. The volume of water involved over a period of time suggests that this effect on the fish food supply could be considerable.
>
> In cooler months of the year, the current of warmer water entering the lake from the plant would tend to attract and concentrate fish in this area— a common behavior pattern . . . where daily (or more frequent) flushings of

chlorine might kill fish or other organisms. In addition, more subtle influences on behavior, reproduction, and survival could be involved.[2]

Altogether, this is a prime example of the possible ecological conse-quences of what seems to be a routine construction project for the public good. It is unusual only in that no consideration was initially made by the sponsoring utility company to provide an alternative to dumping heated effluent into Cayuga Lake. After much public protest the utility agreed to provide cooling towers. More characteristic of the increasing sensitivity of power companies to growing public concern about thermal loading was the announcement by a New England firm of its intent to construct dry towers to control its thermal output to the Connecticut River. Without these, two-thirds of the total flow of the Connecticut River at this site would have been heated by more than 20°F. The public will pay for this consideration in higher rates, but at least the attempt is being made to stop a pollution problem at its source.

ALTERNATIVE SOURCES OF POWER

Nuclear-fueled power plants and their successors, the breeder reactors, give the illusion of innovation and radical departure from conventional power sources, but they are only intermediate steps that will help bridge the gap until truly new pollution-free sources can be developed. Three possibilities, which are being actively studied, anticipating their wide use for power in the next thirty years, are magneto-hydrodynamics (MHD), fuel cells, and fusion.

MHD

A simple MHD generator resembles a piece of pipe in which are im-bedded electrodes. The pipe is placed between the poles of a powerful electromagnet. When hot gas is forced through the pipe at great velocity, the magnetic lines of force generate current that is tapped by the elec-trodes and fed into a power distribution system. Its advantages include an ability to withstand very high temperatures, its lack of moving parts, and at least 50 percent efficiency (50 percent of the energy applied to the device can be harvested as electricity).

[2] Eipper, A. W. *et al.*, 1968. "Thermal pollution of Cayuga Lake by a proposed power plant." *Citizens Committee to Save Cayuga Lake.* Ithaca, New York.

Fuel Cell

This device, given great impetus by the space program, consists of a positive and a negative electrode immersed in a conducting solution or electrolyte, which is able to convert chemical energy into electrical energy. The greatest potential use of such devices would be to supply the power needs of individual buildings, eliminating the need for wires, transmission lines, and all the other paraphernalia associated with centralized power production.

Fusion

While this is the most distant possibility of the three, it holds the most promise. Deuterium, a form of hydrogen, is found in great abundance in the ocean. When heated in its gaseous phase to about 5000°C, the electrons and nuclei that normally are related by electromagnetic forces to each other are so energized that they form a substance, called plasma, that lacks electrical charge. Plasma is simply a mixture of nuclei and electrons without relationship to each other. If this plasma is heated to 100 million°C, the nuclei of deuterium have such great velocity that, when they collide, they fuse, releasing huge amounts of energy. The only vessel that could possibly contain plasma at such a temperature would be a magnetic field. When the plasma at this state of excitement presses against the magnetic field containing it, power can be generated and trapped by electric circuits surrounding the magnetic field. In this way the energy sources can directly and quite efficiently be tapped.

The value of these projected power sources is that each has the potential of producing electricity directly, making unnecessary the crude and inefficient energy conversions that hobble present sources. Even more important will be the elimination of thermal loading, which has been the inescapable by-product of steam-generated power, and of air pollutants, either those products of combustion or radioactive pollutants. Despite the willingness of industry to invest in new power generating processes, the need to amortize already existing plants and meet pressing power demands builds in a lag period of twenty to thirty years before new technology can be economically absorbed by industry and put to full use. Hence it will be several decades before pollution abatement can be achieved by basic process changes. For our sake and for the sake of the environment, the continuation of power plant proliferation will require redoubled efforts to explore alternative types of power generation. Neglect will lead inevitably to environmental degradation of such magnitude as to make present problems an idyllic interlude by comparison.

FURTHER READING

Clark, J. R., 1969. "Thermal pollution and aquatic life." *Scientific American*, 220(3), p. 18. Well-illustrated popular review of the problem.

Mihursky, J. A., 1967. "On possible constructive uses of thermal additions to estuaries." *BioScience*, 17(10), pp. 698–702. A novel, somewhat optimistic approach.

Thermal Pollution. 1968. Hearings before the subcommittee on air and water pollution of the committee on public works, U.S. Senate, Ninetieth Congress. 2nd Sess., February 1968. U.S. Govt. Printing Office, Washington, D.C. Grand bonanza of materials related to thermal pollution.

CHAPTER 9

THE WORLD OCEAN: ULTIMATE SUMP

The line "how deep is the ocean" has become a cliché, but clichés are also truisms, and to most people the ocean is the personification of depth and immensity. For this reason its pollution seems impossible. Since people interact with the oceans at the edges, much the way a mold attacks cheese, most of what we have to say about ocean pollution describes the edges. Keep in mind, however, that the edges get wider every year. Thor Heyerdahl, who recently sailed from Morocco to the West Indies on a papyrus raft, reported seeing oil slicks much of the way.

THE ORIGIN OF COASTAL FEATURES

One hundred thousand years ago the level of the ocean was almost 500 feet lower than it is today. Extensive continental glaciers tied up tremendous quantities of water, exposing the continental shelves as broad coastal plains. As the ice slowly melted, the ocean level began to rise, first flooding the incised lower channels of large rivers forming *estuaries*. These estuaries were covered as the sea advanced over the flat coastal plain, making a series of barrier islands separated from the mainland by broad, shallow lagoons. Ultimately, when all the coastal plain is flooded and its uplands are encroached upon by the sea, another set of estuaries will be formed. Because there has been some crustal movement as well as sea rise, the effect in eastern North America is uneven (Figure 9.1). To the north all of the coastal plain is flooded; Maine and the Maritime Provinces of Canada have coastlines indented with fjords, the result of the sea flooding the coastal headlands. Further south, parts of whole river systems have been flooded, forming Delaware and Chesapeake Bays.

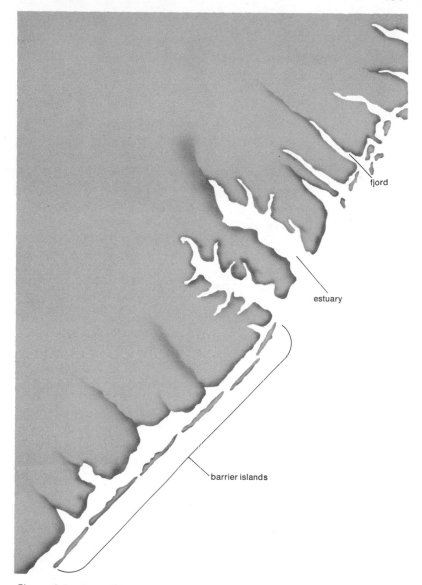

Figure 9.1 A combination of land subsidence and sea level rise has created a variety of land forms along the eastern coast of North America:
barrier islands and lagoons to the south, estuaries in the Middle Atlantic
States, and a fjord-like coastline to the north.

From Norfolk, Virginia, south to Yucatan, Mexico, the sea has advanced
only part way across the coastal plain, hence the characteristic barrier-
island-lagoon dominates. Should the sea rise another hundred feet or so,
the character of the southern coastline would change, with the flooding
of its flat coastal plain from barrier-island-lagoon to estuarine type. The

Figure 9.2   If all the ice now present at the poles should melt, the sea would rise by about 500 feet, flooding much of the eastern United States and most of the population centers of the Pacific coast.

sea is continuing to rise, as ice caps melt, at the rate of less than three feet per thousand years. We have about fifty thousand years to make plans.

If the sea level were rising very much faster than it is, all coastal cities would be in great danger (Figure 9.2). Why have so many great cities been built on or near estuaries? Since estuaries are usually drowned river mouths, they form natural harbors, protected from the oceans, yet connecting the ocean and a river-drained hinterland. Because of this vital connection, estuaries serve as a focus of transportation. In addition, estuaries, combining characteristics of both fresh and salt water, are extremely productive. So it is doubly ironic that such potentially productive areas are often considered wasteland and in many regions have been converted into convenient sewers.

Salt Marsh Productivity

Just how productive tidal marshes and estuaries are can be seen by comparing the amount of dry organic matter produced in a terrestrial system with an estuarine system. A typical wheat field produces 1.5 tons per acre per year (including straw and roots); a coastal marsh near Sapelo Island, Georgia produces ten tons per acre per year. The amount

Figure 9.3 A healthy stand of cord grass in one of the few remaining salt marshes in Connecticut.

of this production that is actually used, however, is low and diffused through a large food web, with man harvesting only a small fraction in the form of shellfish, crabs, and fish. Why the great fertility? Broadly speaking, the great productivity of salt marshes and the estuaries, bays, mud flats, and tidal creeks with which they form an inseparable unit, is based on three communities of plants: cord grass (*Spartina*) in the marshes (Figure 9.3), mud algae on the creek banks, and microscopic plants in the water called *phytoplankton*.

Nutrient Cycling

Less than 5 percent of the cord grass is eaten in place, since few insects are able to survive both salinity and tidal flooding. Even the few hardy grasshoppers sometimes seen in the salt marshes excrete, undigested, two-thirds of the material they eat. As the grass is crushed and fragmented in the fall and winter by storms, it is broken down by microorganisms into fine organic particles containing much protein, carbohydrates, and vitamins. These materials are distributed by tides, but are retained and cycled in the ecosystem by various organisms. For example, phosphate is more abundant in estuarine ecosystems than in rivers or the ocean itself. The Altamaha River in Georgia contains 0.1 micrograms per liter, while coastal tidal creeks contain nearly 2–4 micrograms per liter. The reason for the abundance is found in the ecology of one of the most common animals in the southern estuaries, the inedible horse mussel. By filtering the water for its food, the fine organic particles, which the mussel does not eat, are stuck together with mucus into little particles called pseudofeces (pseudo, because they do not pass through the digestive system of the mussel but are eliminated before entering the mouth). These particles, rich in phosphate, remain in the area and are fed upon first by fiddler crabs, then by other animals ultimately eaten by man, such as blue crabs and striped bass. Here is a classic demonstration of an economically worthless organism[1] performing an extremely valuable role in the cycling of phosphate, a most important mineral nutrient.

Food Chains

The mud algae include dinoflagellates—microscopic algae enclosed in sculptured jackets with a whiplike tail or flagellum—and diatoms, mentioned in Chapter 8. These organisms are usually so abundant that they give the muddy creek banks a characteristic brown or yellowish appearance. They are adaptable to the changing season and so are able to photosynthesize either at high tide in the summer when cooled by the tidal water, or at low tide in the winter when the creek banks are warmed by the sun. By reproducing at a rapid rate and washing off into the water, the diatoms and dinoflagellates provide a year-round food source not only for filter feeders such as oysters, clams, and mussels, but for tiny larvae and small crustaceans that are the basic food for most fish. The importance of the reproductive rate of these mudbank algae cannot be overestimated, for we see only a very small fraction of the total algal production per year sticking to the mudbanks. Although the population on the mudbank remains constant, huge numbers are "sloughed off." Much of the productivity of the ecosystem depends upon this process.

Because of the mixing of fresh and salt water in estuarine areas, mineral nutrients tend to be trapped. Also, daily tides continuously provide food and oxygen while removing waste products. This results in good production of phytoplankton and supplements the mudbank algae as a primary food source for the various consumer levels.

Utilization of Estuarine Productivity

The great productivity of estuarine areas has long been overlooked and underused because, unlike land areas, production and harvest in estuaries

[1] Organisms are normally considered useful if they are edible or directly provide food for edible organisms.

rarely take place in the same spot. These processes are constantly being moved by the tide here and there, and they are separated not only by space but by time as well. There are two approaches toward better use of estuarine production: we could overcome our narrow prejudices and eat more of the elements of the food web than we consider edible today, or we could simplify the ecosystem somewhat by channeling more of the production into acceptable harvestable units or by introducing species more efficient at harvesting producers than those now naturally available. Ecosystem simplification, however, as we saw with the potato patch in Chapter 8, has its dangers. Although formidable, they are not insurmountable.

The point is that tidelands, regardless of the way in which their productivity is used, are extremely valuable as producers of food and as spawning grounds for many economically valuable species. This is seen in the productivity values of the continental shelf (1.3 tons per acre per year) and the open ocean (0.5 tons per acre per year), compared to the 10 tons per acre per year of the estuaries. Too often a fisherman catching a striped bass sees his fish as existing in a biological vacuum, not remembering that fish must spawn to reproduce, that eggs must have the proper environment to hatch, and that the fry must have the right food sources to develop into catchable fish. Very often, too, he ignores the fact that these early stages in the life cycle of many ocean organisms may require radically different temperatures, salinity, mineral nutrient levels, and food supplies than are found in the open ocean—conditions that are unique to estuarine areas.

Unfortunately, the productivity and usefulness of estuaries is not as immediately obvious as good farmland or a stand of prime timber. So the estuarine complex remains wasteland to most people, to be tolerated until it can be made suitable for development.

Misuse of the Estuaries

Because of tidal flushing, estuarine areas are usually able to handle a level of mineral cycling that would turn a lake into a pea soup of overproduction we call eutrophication. But even estuaries are overwhelmed at times. Because of the regular removal of wastes by the tides, efforts to control pollutants are either not made or are initiated so late that recovery is extremely expensive. As a result, the future of estuaries and their adjoining wetlands as a continuing resource is mixed. In Florida, wetlands continue to be regarded as wastelands, which are to be dredged and used for housing developments. Large tracts of marshes on the west coast of Florida have been urbanized in this way, and the process is only beginning (Figure 9.4). If this type of development continues around

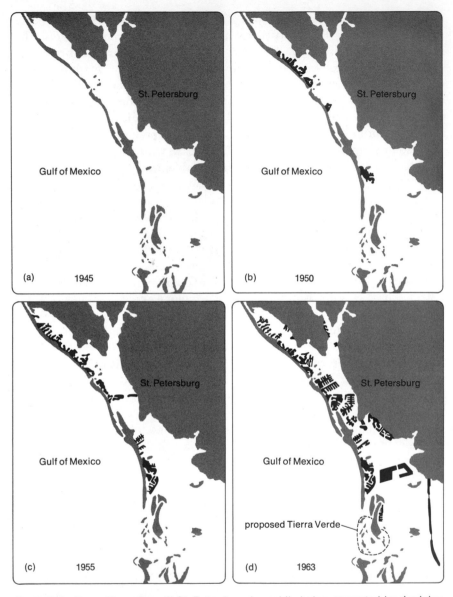

Figure 9.4   Boca Ciega Bay off St. Petersburg is rapidly being converted by dredging and filling from a productive estuary into a typical suburban sprawl. (Modified from G. H. Lauf (ed.), *Estuaries,* 1967. Copyright 1967 by the American Association for the Advancement of Science.)

the perimeter of the Gulf of Mexico without effort to preserve some marshes for wildlife "nurseries," the harvest of various types of seafood in the Gulf, particularly shrimp, is bound to decline sharply.

Other states, perhaps having undergone the first round of their population growth, have developed more progressive attitudes toward

wetland development. Several New England states with extensive areas
of coastal marshes and estuaries have passed legislation requiring con-
sultation with the state before development can take place. This is designed to assure that the value of the wetlands to the owner and to the
public is preserved. While a polluted river or bay can be restored by
channeling public concern into the proper substitution of money for talk,
there is no hope of restoring or renewing a marsh once it has been
"reclaimed."

ABUSE OF A BAY

Great South Bay, like Lake Erie, stands preeminent as a well-polluted
body of water. In this instance, however, a sea edge is being affected
with the clear implication that in time the open ocean itself could be
similarly polluted. Great South Bay, on the south shore of Long Island,
New York, is a lagoon some twenty-four miles long and up to three
miles wide, with an average depth of four feet, protected from the open
Atlantic by the barrier of Fire Island (Figure 9.5). Just a few years ago
most of Fire Island was incorporated into Fire Island National Seashore
to assure the survival of a wild and relatively untrammeled stretch of
sand dunes less than fifty miles from New York City. The bay, however,
has been less fortunate.

Because of its small area and porous soils, Long Island has few
streams of any size. The streams that there are flow from the Ronkon-
koma moraine, a line of rubble marking the southernmost extent of
glaciation, across an almost flat plain, and then into various bays, including Great South Bay, as a series of small estuaries. The combination of fresh water and tidal flushing makes these streams ideal for raising
ducks.

Figure 9.5 Great South Bay on the south shore of Long Island, New York, over twenty
miles long and three miles wide, is protected from the open Atlantic by Fire Island. Its
only connections with the ocean are Fire Island Inlet to the left and Moriches Inlet,
several miles off the right edge of the drawing.

The Long Island Duck

Although ducks have been raised on Long Island since its early settlement, wide-scale duck farming did not develop until the 1920s and 1930s. In 1965, thirty-two farms in the Great South Bay area produced over three million ducks, 60 percent of the national production in the United States, making the Long Island Duck a household word (Figure 9.6).

At first, ducks were penned on the rivers entering the bay and all duck wastes entered the water either directly or were flushed by rain from the stream banks. Later, small ponds or lagoons were constructed to settle out some of the solid wastes, but the soluble materials still drained into the streams and then into the bay.

The raw wastes of 1000 ducks per day produce a total of 5.7 pounds of nitrogen (in the form of uric acid which is converted into ammonia by bacteria), 7.6 pounds of phosphate compounds, and 3.6 pounds of soluble phosphate. To get rid of these wastes, 14–120 gallons of water a day per duck are required. Larger farms use 2–3 million gallons of water a day. Altogether, each day the duck industry on the bay produces 3300 pounds of nitrogen, 5600 pounds of phosphorus, and 55,600 pounds of suspended solids in a total effluent of 133 million gallons.

The solid materials give a gray turbidity to the stream water and the heavier of these settle out in ten-foot-deep deposits of sludge in some places, suffocating almost all the bottom life and greatly reducing the dissolved oxygen in the water. With oxygen in short supply, anaerobic conditions encourage sulfide bacteria to produce hydrogen sulfide. This gas then bubbles to the surface, buoying solids with it, forming lumps or rafts that drift about. Finer material is precipitated by contact with saline water in the river mouths. Soluble nitrates and phosphates, of course, are distributed widely in the bay.

Figure 9.6 Duck farms lining the shores of Great South Bay have seriously polluted the water. Because Great South Bay is closely bounded by Fire Island and the mainland, there is little flushing of the bay by clean ocean water and wastes accumulate.

A Eutrophic Bay

If Great South Bay behaved in a typical fashion, even this large quantity of mineral nutrients could be flushed by tidal action into the sea and diluted, sparing the bay the effects of overfertilization. But Great South Bay has only two inlets allowing entrance of sea water, Fire Island Inlet at one end and Moriches Inlet at the other. Flushing time is four to ten days and circulation is weak. Prevailing winds and currents from the southwest tend to confine the waste-enriched water to the east end of the bay farthest from the source of clean sea water. For these reasons, Great South Bay is unsuitable for the assimilation of large quantities of waste from any source.

Once in the bay, the wastes contribute nitrogen and phosphorus which stimulate the growth of a very small green alga, *Nannochloris* (*2–4 millionths* of a centimeter in diameter), previously so rare that plankton surveys in the early years of this century failed to notice it at all. Normally oysters feed on diatoms, particularly *Nitzschia*, which has quite different ecological requirements than *Nannochloris*. *Nitzschia*, for example, requires a salinity of at least fifteen parts per thousand, grows best at 50–68°F and is unaffected by changes in the nitrogen-phosphorus ratio. *Nannochloris*, on the other hand, grows well in a variety of salinities, prefers temperatures in the range of 50–78°F, and doubles its growth rate when the nitrogen-phosphorus ratio is decreased from 15:1 to 5:1, suggesting that phosphorus rather than nitrogen is a limiting factor. This simply means that because of poor circulation, bay waters have lower salinity, higher temperature, and higher nitrogen and phosphorus content than ocean water. Each of these factors selects for *Nannochloris* against *Nitzschia*. At times *Nannochloris* becomes so abundant that there may be 3–10 million cells per milliliter of bay water, reducing visibility through the water to less than one foot.

Oysters versus Clams

Oysters feed by straining water through their gills and directing the food to their mouth where it is ingested. But oysters feed efficiently only if the water is relatively clear. As the quantity of suspended particles increases, the feeding rate decreases. In plankton-rich water, the gills become covered with the tiny algal cells of *Nannochloris* that interfere not only with normal feeding but respiration as well, leading to starvation or suffocation. As a direct result of eutrophication, oyster production declined sharply between 1940 and 1960, from 600 thousand bushels a year to, in some years, zero bushels.

Although the duck farm wastes have had a very deleterious effect on the oyster industry, production of hard clams (*Mercenaria*) has soared, for this species is able to feed upon the pollution-induced *Nannochloris*, whereas the oyster is not. If the problem were simply the shifting of productivity from oyster to clam, we might simply dig for clams instead of diving for oysters. But, the problem is not that simple. Besides their nutrient content, duck wastes contain a variety of bacteria similar to those found in man. They are collectively called coliform bacteria, and are used as an index of pollution by the Bureau of Public Health. One of these bacteria, *Salmonella*, often found in clams growing in polluted water, causes an intestinal infection so violent that the victim may be cured of eating raw clams for life. As a result of such bacterial contamination, 40 percent of the clam beds in Great South Bay have been closed. When you consider that this closed area can produce 300 thousand bushels of clams a year, worth over two million dollars, the economic result of bay pollution comes into focus.

Pollution from People

Although we have emphasized duck farms, population growth in Suffolk County has also been a factor in bay pollution. Most of the housing disposes of its wastes through cesspools or septic tanks which percolate the mineral nutrient and bacteria-rich wastes through the very porous Long Island soil into the water table draining into the bay. In all probability, if the duck farms were to disappear tomorrow, the unsewered wastes of central Long Island could easily keep Great South Bay oversupplied with both mineral nutrients and bacteria. Further organic pollution is contributed by the increasing number of pleasure craft being used in the bay (see Chapter 7).

   Nannochloris is not the only alga to be stimulated by overfertilization. A filamentous green alga (*Cladophora*) forms great floating mats, at times almost thick enough to stand on. When these finally sink during the fall and winter they accumulate in the holes left in the bay bottom from sand dredged for beach stabilization. Under anaerobic conditions, hydrogen sulfide is produced, which then bubbles to the surface the following summer, often in such quantity as to blacken housepaint on shorefront homes. A third plant, eelgrass (*Zostera*) is also affected. Normally eelgrass is found in small clumps scattered over the bay floor. This not only stabilizes the sandy bottom but provides shelter for the highly prized bay scallop. But with overfertilization, eelgrass spreads rapidly, covering the bay bottom for miles. Storms and the changes of season loosen the blade-like leaves of this plant, which then wash up on shore to decay anaerobically because of their sheer volume; again hydrogen sulfide is produced, and the beaches are covered with a slippery organic slime (Figure 9.7).

Figure 9.7 Windrows of eelgrass washed up from Great South Bay. When these masses of dead plants decompose, hydrogen sulfide is produced in such quantities that the beach is unusable. (Photo by Dan Jacobs)

The Solution of the Bay Problem

The answer for Great South Bay is as simple as it is expensive. Duck wastes should be settled in basins isolated from surface and ground runoff, dried, and used for fertilizer. The supernatant liquid should be treated to remove mineral nutrients, chlorinated to kill bacteria, and then either recycled or returned to the watershed. Human wastes must be collected by sewers and centrally treated through the tertiary stage before being released.

Increased flushing of the bay is certainly desirable, but present inlets must be carefully studied from all ecological angles before being supplemented by man-made inlets. New inlets should not be constructed at the expense of organisms that cannot tolerate the salinity that full-strength ocean water pouring through them would supply. The solid wastes deposited as shoals of sludge in rivers and at their mouths must be dredged so that their mineral nutrient content is permanently taken out of circulation. Finally, onshore facilities must be developed at marinas, yacht clubs, and town docks to handle wastes generated and stored in holding tanks aboard the ever-growing fleet of leisure-time craft. Great South Bay and other estuaries with similar problems can be cleaned up, not easily or cheaply, but there is no inherent ecological reason why oysters, ducks, and people cannot coexist in an ecosystem satisfactory to all.

OIL ON THE WATER

Pollution of the ocean surface by oil slicks is a relatively recent problem; it was not until after World War I that oil began to replace coal as a ship fuel. The discovery of huge reserves of oil in the Near East and in South America stimulated a tremendous increase in the quantities of crude oil shipped to North America and Northern Europe for refining. A typical tanker carrying crude oil during World War II held 16 thousand tons of oil. By 1965, average capacity had almost doubled. New tankers today carry 150 thousand tons and some tankers are on order that will hold 312 thousand tons. Whereas it makes good business sense to carry as large a load as possible, the larger the tanker the greater the hazard from any single accident.

Oil Spills

Just such an accident occurred on March 18, 1967. The Torrey Canyon, a 970-foot tanker carrying 117 thousand tons of Kuwait crude oil from the Persian Gulf to Milford Haven, England, ran aground fifteen miles west of Land's End in Cornwall, seven miles northeast of the Isles of Scilly. The oil was contained in eighteen storage tanks each holding 6500 tons. As a result of the grounding, six of these tanks were ruptured, releasing by March 26 some 30–40 thousand tons of crude oil into the western end of the English Channel. Initially there was hope of refloating the Torrey Canyon, and of salvaging both ship and cargo, but heavy seas soon broke the back of the vessel, dooming any hopes of recovering either (Figure 9.8). For the next six weeks, until drained of her oil and able to

Figure 9.8 The Torrey Canyon, split in two, hangs on a reef off Cornwall, England, while thousands of tons of crude oil pour into the sea from her ruptured tanks. (Popperfoto)

sink, the Torrey Canyon continued to spill the remaining 75 thousand tons of oil she contained. For several days the oil slick was kept well off-shore by storm winds but with a change of the weather the oil began to pile up on the coasts of England and France, causing great consternation to governments and holiday-seekers alike.

But the Torrey Canyon episode was only a reminder that oily wastes are constantly being discharged into the sea from a great variety of sources, which together pose a sizable problem. Some pollution is produced by all ships large or small, whatever their cargo. Slop water from routine ship maintenance, which picks up oil from machinery lubrication, bilge water, oily wastes from engine rooms, all find their way into the ocean. But the tanker is the big problem. Let us look at a typical modern tanker carrying 300 thousand barrels of oil in thirty-three tanks in three rows of eleven tanks each. After emptying her cargo the ship still contains 1700 barrels of oil, 500 in the pipelines and 1200 coating the insides of the tanks. Oil in the pipes is drained into one tank and pumped ashore. In turning around for another load of crude oil, ballast water is pumped into several central tanks from the sea to stabilize the otherwise empty ship for the return run to the oil field. The mixture of oil and water in the ballast tanks soon separates; the water below is pumped into the sea and the oil above pumped into a collection tank. While this is going on, the other tanks are cleaned with steam. This mixture is also pumped into the collection tank. By the time the ship has traveled a few thousand miles her tanks are cleaned and filled with sea water, and the 1200 barrels of waste oil from the previous trip are in the collection tank to be pumped ashore at the next stop.

This is what modern tankers are supposed to do. Outmoded tankers, of which there are many, or modern tankers operating on a very slim profit margin, may not adhere scrupulously to this procedure and may simply dump the 1200 barrels of crude waste overboard at the first opportunity. The ocean is large, the nights are dark, and detection is difficult. Prosecution under existing antipollution laws is hampered by the need to prove willful negligence, a provision that is almost impossible to enforce.

Offshore Wells

A third source of oil waste is rapidly becoming a major problem. At one time most oil wells were located on land or in very shallow bays or estuaries. But with improvement of deep drilling techniques over water, off-shore wells are being drilled on continental shelves all over the world. The North Sea, the Gulf of Mexico, the Arctic Ocean, and the Pacific Coast of the United States have suddenly opened up to oil exploration

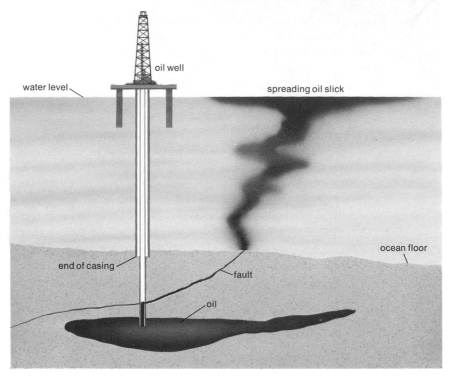

Figure 9.9 A diagram showing the probable explanation for the Santa Barbara oil spill in January, 1969. Incomplete casing allowed oil to escape to the surface along a fault line in the channel bottom.

and exploitation. One of the most recent areas to be developed is in the Santa Barbara Channel off the coast of Santa Barbara, California.

On January 28, 1969, workmen removing a drilling bit from a 3500-foot well were greeted with a gush of oil indicating a "blown" well. Normally, well holes are cased with pipe which allows the well to be capped should any accident threaten uncontrolled flow of oil from the well. In this instance the well casing extended only thirty-nine feet below the 200-foot channel bottom. Apparently under great pressure, the oil penetrated a nearby fault in the rock below the casing level, thus bypassing the pipe which was supposed to contain it, and flowing to the surface without restriction (Figure 9.9). While no one was sure how much oil was actually flowing from the well, flow rates of nearby wells suggested a figure of about 21 thousand gallons per day. Despite frantic efforts to cap the well by pouring 8 thousand barrels of drilling mud and 900 sacks of cement down the well, the flow continued unabated for eleven days. Then it stopped, started again at a lesser rate of flow, and although control efforts are still being made, the area is continuing to discharge some oil. The oil slick covered 800 square miles at its greatest extent (Figure 9.10) and

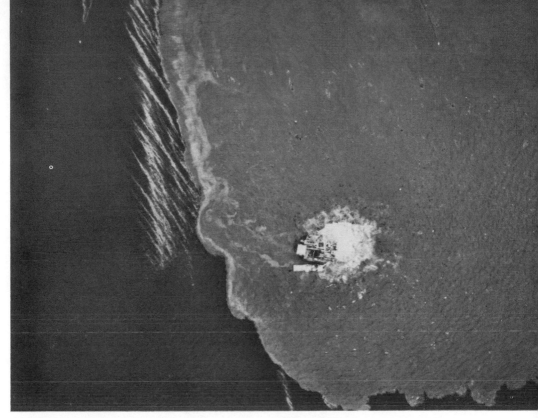

Figure 9.10 Viewed from directly overhead the magnitude of the Santa Barbara oil spill can be better appreciated. The oil slick is the light gray area surrounding the well in the center of the photo. (U.S. Dept. of the Interior, Federal Water Quality Administration)

probably produced an even greater uproar in southern California than the Torrey Canyon wreck produced in Europe.

Before we examine the effects of marine discharges of oil both on oceanic plants and animals and the techniques used to cope with the problems, we must look more carefully at what happens to oil when it is released into the oceanic environment.

Oil and Water

Oil does not float around unchanged until it is washed ashore. All oils, regardless of type, contain some volatiles that evaporate readily. Hence up to 25 percent of the volume of spilled oil is lost through evaporation during the first few days. Then there is a photo-oxidation process in addition to bacterial decomposition that works on the remaining mass. By the end of three months at sea, only 15 percent of the original volume remains—a dense asphaltic substance. It is these black, tarry lumps that most frequently wash up on beaches and stick to one's feet and clothing.

With massive discharges close to shore, however, there is not enough time for much decrease in volume to take place and a thick layer of sticky oil is deposited on anything solid in its path. This was true of both the Torrey Canyon and Santa Barbara spills.

Several approaches were followed in trying to clean up the aftermath of the Torrey Canyon. At sea, the slicks were bombed in an attempt to ignite the oil and burn it off. Because the volatiles that would carry a fire had evaporated, aviation fuel was dumped on the slicks to try to ignite them but without much success. The French used a different technique: they dumped quantities of ground chalk on the slicks in an effort, generally successful, to absorb, then sink the oil and so prevent its being cast up onshore. The technique used on the largest scale, mostly by the British, was to disperse the slick with detergents and emulsify it into small droplets that could be more readily attacked and decomposed by bacteria. Detergents have no effect whatever on oil other than dispersing it. Over 2.5 million gallons of detergents were used in a dispersion attempt both at sea and ashore. Despite the quantity used, the results were disappointing; not only were the slicks not effectively dispersed, but the detergent had a more negative effect on the flora and fauna than that which might have been caused by the oil.

Once the aromatics have evaporated, oil is not especially harmful to organisms, at least not chemically; however, if the layer is thick enough, it may smother creatures unable to move out from under it. Water birds are affected differently. Swimming and diving birds become covered with oil, which mats their feathers, reducing their buoyancy and preventing flight. The insulative value of feathers is also lost and the birds quickly die of exposure in cold water, as would a naked man (Figure 9.11). Oil slicks probably kill thousands of birds, particularly along cold rocky coasts where the birds nest in great numbers.

Figure 9.11 These waterfowl were unlucky enough to mistake an oil slick off the New England coast for a patch of still water and became so thoroughly soaked that they quickly died. (U.S. Dept. of the Interior, Sport Fisheries and Wildlife)

Figure 9.12 Northern murres are one of the most abundant sea birds in the northern hemisphere. (Karl Kenyon from National Audubon Society)

Environmental Repercussions

The murre (*Uria*), a strange little bird weighing about two pounds and vaguely resembling a penguin, is the most abundant sea bird in the northern hemisphere (Figure 9.12). Altogether, there are probably 50 million murres in the world that eat 50 million pounds of fish per week. During the summer they have what may be a critical effect on mineral cycling, and hence productivity, in the Arctic. By feeding on bottom fish, murres, in their excretions, release to the surface water mineral nutrients such as nitrate and phosphate that ordinarily would sink to the bottom of the sea as these bottom fish or their predators died. This fertilizing effect is extremely important in maintaining the fertility of surface Arctic waters. The situation is especially critical off the southeast coast of Newfoundland, where many tankers find it convenient to flush their wastes into the sea. The resulting slicks kill at least 1000 birds a day in this region and perhaps as many as a million a year. Once again man is interfering with a biogeochemical cycle without knowledge of the ultimate effects of his actions.

The effect of detergents used in enormous quantities, however, is obvious to anyone. Although detergent used far away from the shore affects organisms only in the immediate surface layers, treatment of polluted shores has killed large numbers of organisms of varying kinds, as well as producing secondary effects on species not directly affected. Unfortunately, the detergents used in the Torrey Canyon cleanup, unlike those used in household products today, were of the hard type, that is, nonbiodegradable. As with the oil, it is the volatile fraction of the detergent that is most toxic. For this reason, on the open sea, detergent toxicity to organisms declines sharply after a few days.

In addition to barnacles, crabs, and snails, the chief victim of detergent toxicity on rocky coasts is quite ironically the limpet—one of the few organisms observed to remove in its feeding activities the very oil film that generated concern. Another unexpected effect was the great increase in the green alga, sea lettuce or *Enteromorpha,* which grew out of control following the death of the snails that normally keep it in check. This recalled the similar recovery of heavily grazed pastures when the rabbit population of Great Britain was sharply reduced by a disease a few years ago (see Chapter 15).

Several conclusions can be drawn from the Torrey Canyon incident which may prove useful in dealing with accidental spills in the future: (1) aside from its damage to seabirds, oil is an aesthetic pollutant rather than a biological one. Tools used to restore the aesthetic quality of the seashore should be chosen with careful regard for the biological consequences of their use, and biologists should be consulted before any action is taken; (2) natural recovery (browsing by intertidal animals and bacterial decomposition) should be allowed to occur as much as possible, reserving detergents only for those areas of highest recreational use; (3) nontoxic detergents must be developed to disperse oil without destroying organisms.

Nobody's Ocean

The Santa Barbara spill raised other questions of perhaps more basic concern. All accidents cannot be prevented; every oil well has a certain probability of polluting sometime in some way. The circle of potential damage is greatly increased, however, when the oil well is perched on the continental shelf surrounded by very pollutable water. When a drilling site is not only crisscrossed with faults but is highly subject to earthquakes, one wonders how much the oil beneath the surface is really worth. It has been calculated that just two of the many pipelines connecting the offshore drilling platforms with the coast contain as much oil as was recently spilled in the Santa Barbara Channel. To engineer these pipelines adequately against the event of their breaking in an earthquake would be prohibitively expensive. Indeed, if these costs were carried on corporate ledgers, drilling for oil in the Santa Barbara channel would be at best a marginal undertaking. What usually happens is that the costs of environmental protection are either disregarded or passed on to the public at large, and the oil companies continue their highly profitable game of resource exploitation.

The crux of the problem of ocean pollution is that the ocean, unlike Great South Bay or Lake Erie, belongs to no one. Beyond the continental shelf where most nations exert some proprietary interest and perhaps even concern, the ocean is up for grabs. There is no supranational body to pass in judgment on the ocean's future; that alone would seem to seal its fate. But if the ocean belongs to no one, it also belongs to everyone. E. B. White reflected this growing concern for the environment at large when he reacted to a report that the Atomic Energy Commission had authorized the dumping of radioactive wastes into the ocean: "I sometimes wonder about these cool assumptions of authority in areas of sea and sky. The sea doesn't belong to the Atomic Energy Commission, it belongs to me. I am not ready to authorize dumping radioactive waste into it, and I suspect that a lot of other people to whom the sea belongs are not ready to authorize it, either."[2]

FURTHER READING

Dean, D. and H. H. Haskin, 1964. "Benthic repopulation of the Raritan River estuary following pollution abatement." *Limnology & Oceanography*, 9(4), pp. 551–562. Badly polluted aquatic systems *can* recover when attempts are made to check the pollutants.

Foehrenbach, J., 1969. "Pollution and eutrophication problems of Great South Bay, Long Island, New York." *Jour. Water Poll. Control Fed.*, 41(8), pp. 1456–1466. A review of the Great South Bay story.

Hardin, G., 1969. "Finding lemonade in Santa Barbara's oil." *Saturday Review*, 52(19), p. 18. Well-written account of the implications of oil drilling in the Santa Barbara Channel.

Lauf, G. H. (ed.), 1967. "Estuaries." AAAS Publ. 83, Washington, D.C. Many papers in this volume are specialized; but some are general and give valuable insight into the effects of man upon these important geographical features.

Marx, W., 1967. *The frail ocean.* Coward-McCann, New York. A broad survey of man-ocean interactions.

Odum, E. P., 1961. "The role of tidal marshes in estuarine production." *The Conservationist*, 15(6), pp. 12–15. Superbly written rationale for preserving wetlands.

Smith, J. E. (ed.), 1968. "Torrey Canyon." *Pollution and marine life.* Cambridge University Press, Cambridge, England. Carefully documented report by the chief biological investigator at the time.

Teal, J. and M. Teal, 1969. *Life and death of the salt marsh.* Little, Brown and Co., Boston, Massachusetts. A general account of salt marsh ecology and the fate of coastal marshes.

[2] Hardin, G., 1969. "Finding lemonade in Santa Barbara's oil." *Sat. Rev.*, 52(19), p. 18.

CHAPTER 10

THE AIR AROUND US

On Thursday, December 4, 1952, a large high-pressure weather system spread slowly southeast across the British Isles.

The system brought with it light variable winds, dry air, and rather frigid temperatures. . . . Considerable fog began to form late Thursday evening and during the early hours of Friday morning. At first it remained comparatively clean and harmless—but not for long. As the city awoke, tons of smoke from millions of domestic chimneys were hurled upward into the cold motionless, foggy air. Huge power stations added still more tons of coal smoke and sulfur oxides to the atmosphere. Cars, trucks, buses, and a variety of factories and industrial plants all contributed their pollutants. In a short time the fog had become massively contaminated by a mixture of smoke, soot, carbon particles, and gaseous wastes. Now yellow, now amber, now black, the great killer smog held London in its grip, and by early evening, only twelve hours after its onset, the first of the city's inhabitants began to die.[1]

Lest we think air pollution a contemporary phenomenon, John Evelyn in January 1684 wrote: "London, by reason of the excessive coldness of the air hindering the ascent of the smoke, was so filled with the fuliginous steam of the sea coal, that one could hardly see across the streets, and this filling the lungs with its gross particles, exceedingly obstructed the breast so as one could hardly breathe."[2] Edward I (1307–1327) ordered a man put to torture for burning coal and fouling the air, an act sure to be recalled with some relish when one is trapped between two chain smokers on a long overseas flight.

Is it possible to go far enough back in time to encounter a halcyon

[1] Wise, W., 1968. *Killer smog: the world's worst air pollution disaster*. Rand McNally, Chicago.
[2] Ibid.

period when the air was pure? If we think of pure air as we think of pure
water, probably not. Air is always a mixture of many things. In addition
to the three gases, nitrogen, oxygen, and carbon dioxide, air also includes
varying amounts of water vapor, dust, and every conceivable element
and compound that wind or man distributes.

Long before man, however, dust storms, fires, volcanoes, and ocean
storms polluted the air with vast quantities of particles and impurities of
various sorts. The pollution continues: in 1883, a volcanic explosion pul-
verized the island of Krakatoa in the East Indies and threw very fine dust
high into the atmosphere to circle the earth for several years before
finally settling out. This suspension, probably without lasting effects, did
produce a long series of lovely sunsets through the reflection and diffrac-
tion of the sun's rays by the many dust particles.

Man's appearance, of course, compounded the picture. While controlling fire to his ends, one of its natural products, smoke, has been associated with man since that first fire back in the cave. The problem has
been due not only to the increasing number of fires built by ever increasing numbers of people, but to the increasing complexity of the materials
combusted and increasing numbers of uses to which the fires are put.

SULFUR DIOXIDE SMOG

For thousands of years wood, or perhaps dried dung or peat, was the
only fuel used, but at some point it was discovered that the black, shiny,
strangely lightweight stone that we now call coal could make a very hot
flame that lasted longer than a wood fire. Apparently coal was used, if
sparingly, for some time in the Far East prior to its "discovery" by Marco
Polo. Supposedly, Polo introduced coal, or at least the concept of its
combustibility, into Europe. Considering the decimation of forests around
population centers and the growing scarcity and expense of wood and
charcoal, one would think the idea of using coal would have been welcomed with open minds. But this was a Europe where the Church considered almost any new idea heretical.

When coal is burned, sulfur compounds, especially sulfur dioxide,
are released in the smoke. While this was unpleasant enough in itself,
perhaps the fumes suggested to the medieval mind the sulfur and brim-
stone associated with Lucifer. Whatever the reason, coal replaced wood
with surprising slowness. But the gradual enlightenment of the medieval
mind and perhaps the greater effect of the soaring price of wood finally
brought coal into common usage throughout Europe; with that, of course,
came the sulfur dioxide fumes. John Evelyn, in 1661, recognized the
problem and proposed a reasonable solution for his time: banish the
offending industry, although the individual coal hearth pushing its sulfur-
laden smoke through thousands of chimneys all over London was equally

to blame. Evelyn's suggestions were ignored and for almost 300 years nothing was done about the problem. Once accepted and even romanticized by nineteenth-century novelists, the beloved British hearth was not to be easily given up.

But all Englishmen did not ignore the problem. In 1905 a London physician, Dr. Harold Des Voeux, described the combination of smoke and fog as "smog." However, it was not until years later that the word was again used—and misused at that.

The Effects of Sulfur Dioxide Smog

The Meuse Valley in Belgium is heavily industrialized; there are coke ovens, blast furnaces, power plants, glass factories, and steel mills. During the first week of December 1930 the valley was covered with a stagnant air mass that entrapped a smothering blanket of smog. Over a thousand people became ill and sixty died, ten times the normal death rate. In late October 1948, Donora, a small mill town on the Monongahela River in Pennsylvania, similarly surrounded by low hills and having a steel mill and a zinc reducing plant, reported seventeen deaths after three days of similar conditions. An additional 42 percent of the population became ill. The normal daily death rate was two, hence, the increase could only be attributed to the smog. The symptoms in all examples, London, Meuse Valley, and Donora were throat irritation, hoarseness, cough, shortness of breath, nausea, and feeling of chest constriction.

The major cause of each of these disasters was sulfur dioxide. This compound is found in greatest concentration in the air surrounding the major industrial cities of the world. This does not mean, however, that the amount of sulfur dioxide was in itself lethal. Rather, as the concentration rose the sulfur dioxide, in addition to the many other pollutants, produced lethal results.

Sulfur dioxide is not an especially toxic gas, but in a humid atmosphere it is converted into sulfur trioxide or sulfuric acid, and adsorbed onto the fine particles or *fly ash* that also result from combustion. Most of the human body is protected from the environment by thick skin, and much of that in turn is protected by clothing. The major point of vulnerability thus lies in the delicate membrane lining the eyes, nose, and respiratory tract, which is far more sensitive to injury than the skin and much more absorbent. Aerosols or fine particles carrying damaging compounds are inhaled into the lungs.

Particulate matter is usually eliminated by the cells lining the walls of the respiratory system. Each cell has a hairlike cilium, which beats twelve times per second. Beating in waves, these cilia work dust particles and other foreign material into the mouth where it can either be expelled or swallowed. A polluted atmosphere containing sulfur dioxide, ozone, or nitric oxide can inhibit cilial action, allowing particles to remain in the respiratory system and cause damage.

Four major types of respiratory damage from air pollutants are bronchitis, bronchial asthma, emphysema, and lung cancer.

Chronic bronchitis is characterized by permanent damage to the bronchial tubes, resulting in reduction or failure of ciliary action, and overproduction of mucus by gland cells. Because cilial action cannot dislodge the extra mucus, a chronic cough develops. The mucus also constricts the opening of the bronchial system, causing shortness of breath.

Bronchial asthma is usually the result of the allergic reaction of the bronchial membranes to foreign protein or other materials. The membranes swell and make difficult the expulsion of air from the lungs. This explains the characteristic symptoms of wheezing and shortness of breath.

Emphysema follows the constriction of the finer branches of the bronchial tubes, the bronchioles (Figure 10.1). When air is exhaled, more air remains in the tiny air sacs of the lungs (alveoli) than should; when new air is inhaled, the overinflated sacs balloon larger and larger until they explode. This causes two adjacent sacs to unite. The gradual reduction in the number of air sacs destroys the capillaries through which oxygen is taken up by the red blood cells and slowly pushes out the chest, giving a characteristic "pigeon-chested" appearance. The loss of oxygen

Figure 10.1   The lung alveoli are tiny air sacs providing surface area which facilitates uptake of oxygen by the bloodstream. The graph shows that the amount of air held in the lungs of an emphysema patient is far less than that inhaled by a normal person. (Courtesy of the Oregon Tuberculosis and Health Association.)

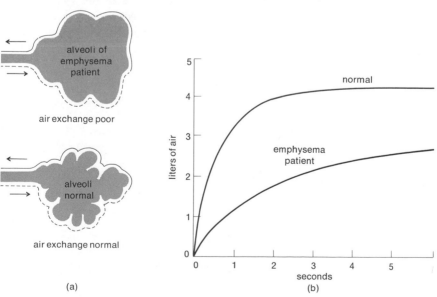

exchange capacity of the lungs leads to slow oxygen starvation of the entire body and chronic shortness of breath.

Lung cancer is stimulated by various substances called *carcinogens*: one, benzpyrene, is characteristic of coal smoke, others like cigarette tars are inhaled deeply by millions of smokers. Both remain in the lungs, in part because of the inhibition of the cilia which might otherwise remove them. Another prime source of benzpyrene is charcoal-broiled steaks. The singed fat drips onto the coals and produces smoke rich in aromatic hydrocarbons, including benzpyrene, as much per steak as is contained in 600 cigarettes. This then coats the meat. But you need not give up barbecued meat; just trim off most of the fat before cooking.

When you consider that over a lifetime, the average person breathes 580 million times, sampling 8 million cubic feet of air, it is not surprising that when air contains pollutant particles various chronic lung conditions may develop. In Great Britain, where a large proportion of the population lives in very smoggy cities, fully 21 percent of men from forty to fifty-nine years old have chronic bronchitis. This condition causes 10 percent of all deaths in Great Britain.

The connection between asthma and air pollution was clearly demonstrated after World War II among servicemen stationed near Yokohama, Japan. Individuals with no previous history of asthma developed the classic symptoms of wheezing, coughing, shortness of breath, and sleeplessness. Even transferral to Tokyo did not improve the condition. Once out of the Kanto Plain where both Tokyo and Yokohama are located, most cases rapidly cleared up. There was even one case where a soldier, racked by asthma in Yokohama, spent six asthma-free years in the United States only to have a recurrence of asthma on a later visit to Yokohama. Even the common cold and other upper respiratory tract diseases are more frequent in cities, showing a distinct correlation with increase of pollution level in the late fall and winter, abetted quite probably by the extremely dry condition of most houses and apartments during the winter season (see Chapter 18).

Despite the strong circumstantial evidence linking air pollution to chronic disease, many are not convinced. A vice-president of a large oil company recently wrote:

> This complex and troubled world we live in is so full of significant, pressing, and perplexing problems that we can ill afford the luxury of wasting effort on imaginary problems or trying to discover problems where none exist. Because human health is of such vital personal concern to each of us that we are naturally inclined to get emotional about it, we should be particularly cautious in ascribing or assuming a cause-and-effect relationship between air pollution and health until scientifically reliable supporting evidence is available.[3]

[3] *Scientist and Citizen*, 7(3), p. 3.

This somewhat cautious view was not shared by the Surgeon-General of the United States.

... much of the speculation and controversy about whether or not air pollution causes disease is irrelevant to the significance of air pollution as a public health hazard. We are accustomed to thinking that a disease state is brought about by a single cause—a carryover from a period in public health history when virtually total emphasis was placed on the bacterial or viral agent which had to be present before a communicable disease could be recognized and dealt with. . . .

New criteria must be employed in assessing the damage of air pollution—criteria which include statistical evidence that a disease condition exists in a population, epidemiological evidence of the association between the disease and the environmental factor of air pollution, reinforced by laboratory demonstration that the air pollutants can produce similar diseases in experimental subjects. . . . But the qualitative message at hand conveys a clear message. There is no longer any doubt that air pollution is a hazard to health.[4]

Sulfur dioxide and plants. Sulfur dioxide affects plants as well as people. In the form of sulfite at concentrations of less than 0.4 ppm, it kills leaf cells, causing large red or brown blotches on leaves between the veins. Since sulfur often accompanies ores of silver, copper, and zinc, smelters which roast these ores to recover the metal have until recently poured huge quantities of sulfur dioxide into the atmosphere with disastrous effects on the surrounding vegetation. In the early 1900s, Ducktown, Tennessee, located near a copper deposit, released forty tons of sulfur

[4] Ibid.

Figure 10.2 Thirty years *before* this photo was taken in 1943 the sulfur dioxide fumes from this copper smelter near Ducktown, Tennessee were eliminated. But the damage to the land (rich forest) had been done. Almost thirty years *after* the photo was taken the landscape for 100 square miles remains much as it appears in this photo. (U.S. Dept. of Agriculture)

dioxide a day into the air of this southern Appalachian valley. Not only were seven thousand acres of the vegetation killed and another seventeen thousand reduced to sparse grass, but most of the topsoil, exposed to rain without protection from vegetation, washed away, leaving a desert in the midst of one of the lushest forests in North America (Figure 10.2). When controls were finally instituted and the sulfur dioxide converted to the useful by-product sulfuric acid, the return on the once-wasted sulfur dioxide proved, ironically, greater than the copper. The desert remains.

Sulfur dioxide smog and materials. Materials do not escape the effect of sulfur dioxide either. A combination of fly ash and sulfur dioxide in a humid atmosphere greatly accelerates the rusting and corroding of metals. Fibers such as wool, cotton, and leather adsorb sulfur dioxide, which oxidizes into sulfuric acid and attacks protein and other organic materials, rotting the fabric. Limestone, certain sandstones, marble, and roof slates—all of which contain carbonates—are converted into soluble sulfates and chlorates which erode and weaken structures composed of these building materials (Figure 10.3).

Figure 10.3 In the more than 300 years since their erection these sandstone figures at Oxford University have been badly damaged by sulfur dioxide smog.

The Control of Sulfur Dioxide Smog

Efforts to control air pollution are usually concerned with the visible smoke, comprised of small particles of fly ash which can be easily controlled. Electrostatic precipitators installed in factory or power plant chimneys can remove up to 99 percent of the fly ash usually scattered into the environment (see Chapter 21). High voltage wires charge the ash particles, attracting them in groups to a set of vertical plates. When these are tapped or vibrated, the ash falls into hoppers below and is removed. Although the market has not been fully exploited, by using fly ash to make concrete, bricks, and paving material, a power plant producing large quantities of fly ash should be able to at least return the installation and operating cost of the precipitators, since it costs at least $2 a ton just to dispose of the fly ash. In the United States, which produces over thirty million tons of fly ash a year, only 18 percent is recovered.

The shift from coal to oil in many homes and industries has resulted in a particulate pollution decline in most cities in the United States. Dust fall in Chicago in 1928 was 395 tons per square mile per year. In 1962 this had dropped to 43 tons per square mile per year.

But few people realize that a smokeless chimney can be just as serious a polluter of the air as one belching clouds of black smoke. Many poisonous substances are colorless—sulfur dioxide for example. At the moment there is no one economical method for removing sulfur dioxide, especially in small quantities, from effluent gases. If the sulfur dioxide content is large enough, *wet scrubbers* such as a sodium carbonate solution can be used to remove sulfur dioxide. Installation costs for scrubbing equipment at one power plant were $10 per kilowatt capacity; subsequent maintenance cost $1.17 per ton of coal used. The system removed up to 91 percent of the sulfur dioxide from the flue gases. If powdered limestone with small amounts of iron oxide as a catalyst is injected into the flame region of power plant boilers, a plant burning 5 thousand tons of coal per day containing 3 percent sulfur can reclaim 300 tons of sulfur dioxide a day at the cost of 465 tons of powdered limestone.

Another approach, adopted by New York and Chicago, is to limit the sulfur content of the fuels, rather than to try to trap the sulfur dioxide as it leaves the chimney. Pretreatment of both coal and oil, although expensive, is certainly practicable.

Utility companies are often reluctant to go into the chemical business, but if you produce 300 tons of sulfur dioxide a day, you *are*, like it or not, in the chemical business. Another problem is that many plants are too small to accommodate a complex sulfur recovery system that may be larger than the original plant itself. In addition, the variable nature of power plant operations, with units starting and stopping in response to power demands, makes it difficult to send a steady supply of by-product chemicals to a market.

In European cities, where the sulfur dioxide problem is aggravated by home fires, a possible solution is conversion to natural gas or oil. From the point of view of sulfur dioxide reduction, electricity would be the ideal solution, but most people cannot afford this type of heating. Natural gas would be the next best choice, and low sulfur fuel oil after that. Until this is done air pollution control can only be partially achieved. The recent discovery of large natural gas deposits in the North Sea may certainly help; however natural fuels are in short supply and the need for a cheaper and more abundant source of fuel is acute.

PHOTOCHEMICAL SMOG

In 1859 Colonel Drake drilled the first oil well near Titusville, Pennsylvania. With the increasing pace of technological development it was only a few decades more before the internal ("infernal" as some pessimists would have it) combustion engine was developed, and soon after that the automobile.

A New Kind of Smog

Because of the relatively late appearance of the automobile, most older cities had somehow to fit it into the scheme of pre-existing roads, urban centers, and suburbs. But Los Angeles grew up with the automobile delaying, perhaps forever, effective mass transit systems and making the city dependent upon the automobile. To accommodate this crush of traffic a freeway system, attracting still more cars which required still more freeways, developed, a transportational illustration of Parkinson's second law—expenses rise to meet income. At the height of this activity a haze began to spread over the city, at first light and occasional. People started to complain of eye irritation and shortness of breath, and certain crops were blighted, particularly leafy vegetables and flowers whose value depended upon their unblemished leaves and petals. Somehow people picked up Des Voeux's descriptive word, smog, to describe the phenomenon. However, the phenomenon in Los Angeles involved neither smoke nor fog.

Believing otherwise, the city cracked down on incinerators and industrial smoke producers. The smog remained. Next the oil refineries were put under strict emission control; the smog got worse. Finally Haagen-Smit, a chemist from California Institute of Technology, identified the problem—automobile exhausts in combination with the unique geography of the Los Angeles Basin (Figure 10.4).

Figure 10.4 Commuters driving to work on a Los Angeles freeway. (Planned Parenthood —World Population)

The geography of photochemical smog. Los Angeles is a beautifully situated city: facing the Pacific in a broad fertile basin, surrounded by mountains up to 10 thousand feet high, and favored with a warm, dry climate. In the early years, people were attracted by the prospect of warm, sun-filled days, groves of orange trees, lovely vistas of soaring mountains. Today the sun is often obscured by a yellowish-brown haze, the orange groves are either subdivided or killed by smog, and clear vistas are limited to occasional days in winter. Apparently the very factors which at one time made the city so attractive have set it up for one of the worst pollution problems in the country.

The city is surrounded on three sides by mountains, the Santa Monicas to the northwest, the San Gabriels to the northeast, and the Santa Anas to the southeast. Although these mountains effectively shield Los Angeles from the hot, dry winds of the Mohave Desert to the east, they also prevent circulation of air. In addition, the weather of most of southern California during most of the late spring, summer, and early fall is dominated by a large high-pressure area capping the basin with dry warm air. Because of the stagnant high pressure and the encircling mountains, the only air movement is a gentle sea breeze from off the cold Pacific Ocean.

The Role of Temperature Inversion

If we were to attach a thermometer to a balloon and have it radio back the temperature as it ascended, we would see a rather unusual profile (Figure 10.5). Generally, the air temperature decreases at a constant rate from the earth's surface well up into the upper atmosphere. But this is not always the case. Sometimes the temperature decreases steadily as the balloon rises only for the first 1500–3000 feet, then it *increases* for a few thousand feet, and finally it resumes a steady decrease. This reversal of the usual temperature profile is called a temperature inversion. The initial decrease in temperature is a result of the cool ocean breeze flowing in at a very low altitude over the basin. The next layer is the warm, dry air of the stationary high which is several thousand feet thick; above this layer, the temperature again declines with altitude. The effect of the warm, dry air of the high-pressure mass is to act as a lid covering the entire basin. Pollutants generated in the basin rise in the cool surface air until they meet the warmer air above at about 1000–1500 feet. Since they cannot penetrate this inversion layer, they spread out laterally and ac-

Figure 10.5 On a clear day in Los Angeles, the temperature *decreases* with increasing altitude. When the area is capped by a mass of warm stable air, the temperature *increases* with altitude for several thousand feet, preventing the dissipation of pollutants generated from below. This increase is shown by the curve at the left of the graph.

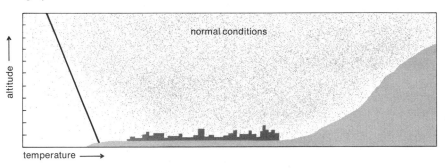

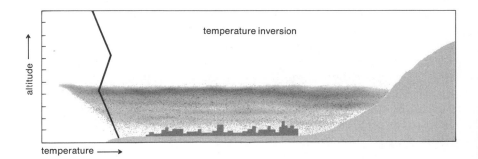

Figure 10.6 Photochemical smog is not limited to Los Angeles. Here, for example, is New York City. Every large city with frequent periods of sunshine and with many gasoline-engine vehicles generates this type of smog. (Planned Parenthood—World Population)

cumulate in the surface air of the basin. The only escape is a very gradual drift southeast where, under influence of sea breezes, the polluted air flows over some of the mountain passes into the Mohave Desert and has been traced as far east as Arizona. Between June and October there is a temperature inversion 80–90 percent of the time over Los Angeles.

There is a third factor. Because the inversion layer is warm and dry, it remains cloudless most of the time. This allows the energy in sunlight to interact with the pollutants generated from below, producing a second generation of new pollutants. For this reason the Los Angeles smog is principally photochemical, not a mixture of smoke and fog as it is in London. There are other differences between the two types of smog as well. Smog in London occurs mostly in late fall or winter in the early morning with temperatures in the 30–40°F range; the humidity is high and fog is usually present. The major effect on people is lung irritation. In Los Angeles, smog reaches its peak during the summer at midday with the temperature ranging from 75–90°F; humidity is low and the sky clear. The effect on people is eye and nose irritation.

Photochemical smog is found in any city that generates a quantity of internal combustion engine exhaust and has a high incidence of sunny days (Figure 10.6).

THE AUTOMOBILE AS A POLLUTION SOURCE

The internal combustion engine was developed with the goals of good performance, smooth operation, and low cost, without any regard for the waste products of the combustion process. In the early years of the automobile it mattered little, because the concentration of cars in proportion to the areal extent of cities was insignificant. But as the number of cities grew their collective combustible wastes grew as well, until pollution became of critical importance. In 1963 daily emission from all sources in Los Angeles totaled 2500 tons of hydrocarbons, 8000 tons of carbon monoxide, 700 tons of nitrogen oxides, 100 tons of particulate matter, and 130 tons of sulfur dioxide. The 3.25 million cars of Los Angeles contributed 70 percent of the hydrocarbons and nitrogen oxides, and 100 percent of the carbon monoxide. So today we are burdened with a highly sophisticated, powerful engine spewing out considerable quantities of pollutants which, once released into the air, are beyond our control.

There are three major sources of pollution from cars: evaporation from gas tanks and carburetors, crankcase blow-by, and tailpipe emission. Although gas caps nominally keep gasoline in the tank, the seal does not prevent the escape of vapors. You can easily demonstrate this by parking your car with a full gas tank in the sun on a hot summer's day. At least a pint, sometimes as much as a quart of fuel is soon dribbling onto the ground. Evaporation also takes place from the carburetor. When the car is decelerated or idles, more gas flows into the carburetor than is burned; much of this excess gas evaporates. Finally, when the engine is turned off, some of the gas remaining in the heated carburetor evaporates. About 20 percent of the hydrocarbons emitted by an automobile is the result of evaporation; altogether, 2.5 percent of all gasoline produced evaporates between the refinery and its combustion.

Because the seating of pistons in the cylinders is not air tight, some combustion products slip past the pistons after the air-gas mixture is ignited by the spark plugs. This exhaust gets into the crankcase where it is released into the air by a breather tube. This source of hydrocarbons is called crankcase blow-by and amounts to 25–35 percent of the exhaust products of the engine.

Most of the combustion products, however, are vented to the atmosphere through the tailpipe. Gasoline burns most efficiently at a 15:1 air-fuel ratio. But a high-compression engine requires a rich mixture of fuel with air: 13 or 12:1. This means that the hydrocarbons in the fuel are not completely burned. This type of engine also requires high octane gasoline, that is, gas with higher proportions of aromatics or volatile components that are also incompletely burned in rich mixtures.

The Composition of Gasoline

If gasoline were a single compound which was completely burned, the end products would be carbon dioxide and water, and there would be no emission problem; but gasoline is not a single compound, nor is it completely burned. A typical gasoline contains three kinds of hydrocarbons: paraffins and olefins, which burn well at low speeds, and aromatics which work well at high speeds. But paraffins often explode spontaneously before ignition, causing engine knock and wasted power. To combat this, tetraethyl lead was added. Lead oxide then formed on the plugs and valves so ethylene dichloride and dibromide were added to clean up the lead deposits. In addition antioxidants, metal deactivators, antirust and anti-icing compounds, detergents, and lubricants have been added. Small wonder that when this incredible mess is burned the combustion products are more than just carbon dioxide and water!

When fuel mixed with air (containing 78 percent nitrogen) is combusted, nitrogen dioxide is formed. As luck would have it, the ideal conditions for nitrogen dioxide generation are high temperature followed by rapid cooling, precisely the conditions provided by an internal combustion engine.

Evaporation contributes volatile hydrocarbons of great variety; combustion adds carbon dioxide, carbon monoxide, and nitrogen dioxide, and of course a large amount of incompletely burned hydrocarbons. These materials, together with sulfur dioxide produced by industry, enter the atmosphere and interact not only with each other, but, more importantly, with sunlight.

Sunlight as a Reactant

Although we are most sensitive to the visible or light portion of the electromagnetic spectrum, it is the shorter wavelengths, or ultraviolet, which have more energy than the longer infrared radiation. It is this ultraviolet radiation that reacts with pollutants to form the secondary products that may cause more problems than the primary ones. Nitrogen dioxide, for example, is split by ultraviolet radiation into nitric oxide and atomic oxygen. Some of this atomic oxygen reacts with the nitric oxide, forming nitrogen dioxide again. This is called an *autocatalytic* reaction because nitrogen dioxide, once formed as a combustion product, can continue to regenerate itself in the atmosphere. In another important reaction, atomic oxygen combines with oxygen to form ozone. Ozone has great oxidative potential and is quite reactive; it attacks rubber, causing it to crack and ultimately to decompose. In fact, one of the clues suggesting the photochemical nature of ozone production in the atmosphere was the observation that rubber cracked only during the day.

The physical form of smog is an *aerosol,* that is, a fine dispersion of either particles or droplets that are less than twenty-five thousandths of an inch in diameter. At least one source of these droplets is the ultraviolet energized reaction of sulfur dioxide and various hydrocarbons, especially the olefins. Other secondary pollutants are the irritants formaldehyde, arolein, and peroxyacetylnitrate (PAN) derived from further reactions of nitrogen dioxide and nitric oxide.

A typical photochemical smog, then, contains the primary pollutants carbon monoxide, sulfur dioxide, nitrogen oxides, hydrocarbon fragments from incompletely combusted fuel, volatile hydrocarbons evaporated from gasoline, and finally materials released from combusted additives, lead, boron, bromine, and so on. Many of these primary smog components react with each other and with sunlight producing secondary and even tertiary pollutants: nitric oxide, PAN, ozone, acrolein, and formaldehyde, to name just a few. How do these materials affect the environment?

EFFECTS OF PHOTOCHEMICAL SMOG

The gross effects of photochemical smog result from its dispersion in the atmosphere as an aerosol. Experiments using aerosols produced from various combinations of pollutants reduced visibility in the atmosphere

Figure 10.7   The upper curve represents the ideal distribution of solar energy on an October day in Pasadena, California. The lower curve is the observed energy. The difference is a measure of the time and density distribution of photochemical smog generated by vehicles having gasoline engines. (Modified from P. A. Leighton, *Photochemistry of Air Pollution,* Academic Press, 1961.)

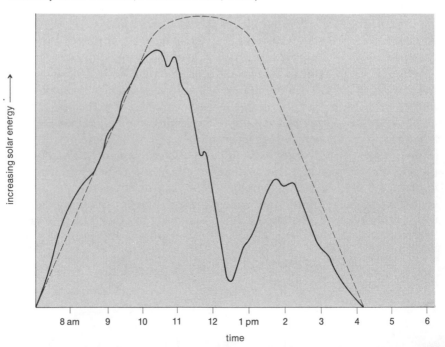

from 26 to 6.5 miles. If the particles were allowed to grow to a point of equilibrium, further reduction of visibility would have ensued. This reduction of visibility is not only aesthetically unpleasant but reduces sunlight received at the earth's surface (Figure 10.7), makes driving and flying more hazardous, and may act together with other pollutants to produce various adverse physiological effects in man.

But photochemical smog has its greatest impact in the differing effects of its major components on man and plants.

Carbon Monoxide

Carbon monoxide is one of the least reactive of the pollutants in photochemical smog, and is tolerated by the body in concentrations up to 10 ppm without noticeable effect. It is potentially quite dangerous because it is odorless, tasteless, and colorless, unlike many other pollutants. In the body it has an especial attraction for hemoglobin, the substance in the red blood cells that normally combines with oxygen, distributing it throughout the body. As the level of carbon monoxide increases in the body, less oxyhemoglobin and more carboxyhemoglobin is formed. Since carboxyhemoglobin does not provide oxygen for the respiratory needs of the body cells, increasing concentrations of this compound endanger survival. A moderate smoker exposes himself to around 30 ppm of carbon monoxide, resulting in a 5 percent level of carboxyhemoglobin. When placed in an urban environment where carbon monoxide concentration in the air may average 20–30 ppm, the additive effect for a heavy smoker may approach the danger level of 100 ppm. Probably 20–30 ppm total should be the maximum allowed, for the physical effects of a decline in sensitivity to environmental stimuli and lack of energy and endurance begin to be felt above this level. Although at a busy intersection in a large · city one may breathe 20 ppm of carbon monoxide, it is possible to get much higher doses, for example, 370 ppm behind a car stopped for a red light.

The Organic Components

The hydrocarbons and nitrogen oxide products such as acrolein, PAN, and formaldehyde irritate the eyes and nose and even attack nylon stockings. Any student who has dissected a frog recalls the tearful fumes of formaldehyde. Typical concentrations of PAN in smog range from 0.05 to 0.6 ppm. PAN enters the leaves of plants through the stomates (Figure 10.8a), and causes the spongy mesophyll cells to collapse (Figure 10.8b). The resulting air space gives leaves a silvered or bronze appearance, ruining their sale as leaf vegetables.

Orange trees, although more resistant than leafy vegetables, are also affected by high concentrations of PAN. The twigs are killed, giving the tree a moth-eaten appearance, and the number of food-producing leaves

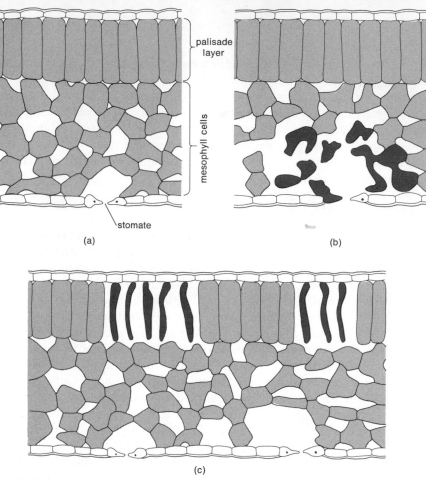

Figure 10.8 (a) Cross section of a normal leaf; (b) Cross section of a leaf whose spongy mesophyll cells have been damaged by smog entering the leaf through the stomates; (c) Cross section of a leaf whose palisade layer has been damaged, causing small flecks to appear on the leaf surface.

is so reduced that the crop of oranges is severely diminished. Because of smog damage, commercial orange production has shifted from the Los Angeles basin to the Sierran foothills on the edge of the San Joaquin Valley.

Ozone

Ozone is also an irritant, but it can impair vision and depress body temperature as well. Its threshold for direct action is around 1 ppm. While the natural concentration of ozone is about 0.02 ppm, it may range as high as 0.5 ppm or more in severe photochemical smog. Ozone attacks the palisade layer in plant leaves (Figure 10.8c), causing brown flecks to appear, especially in tobacco leaves which are very sensitive to ozone. The flecking considerably reduces the value of the leaves, particularly those used as cigar wrappers. Natural vegetation is affected by ozone as well; there has been pronounced dieback of ponderosa pine in the San Bernardino Mountains southeast of Los Angeles. These mountains, which form one rim of the basin, are subject to heavy concentrations of smog drifting out of the Los Angeles area. White pine in the eastern United States develops a similar browning of its needles from ozone. This effect has been noticed in the Cumberland Mountains of Tennessee near Oak Ridge and along parkways in suburban areas such as the Merritt Parkway in Connecticut.

CONTROL OF PHOTOCHEMICAL SMOG

Control the automobile exhaust and you control photochemical smog (Figure 10.9). Like all simple solutions to involved problems, there are hidden complications. As we have seen, the automobile was developed without any particular interest in emission control. Suddenly after eighty years the hue and cry goes up. Control emissions! The first approach has been to control the three sources of exhaust. Crankcase blow-by was handled by recycling emissions back into the carburetor. European cars have long done this; American cars made the change starting with 1968 models. Soon to follow was the requirement of a vapor-tight gas cap. The remaining source, tailpipe emission, is by far the most difficult to control. Until something replaces the internal combustion engine, there are four options: mechanical suppression, alteration in fuel composition, exhaust burners, and engine modification. Mechanical suppression might involve increased average operating temperature, freeing of the cylinders from deposits, and a decrease in the compression rate. Some suppression methods currently being applied are leaner and more precisely controlled

Figure 10.9   Courtesy Oliphant, The Denver Post.

Oliphant in The Denver Post

". . . Then, when the bag is full of hydrocarbons and noxious gases, you simply take it off and throw it away!"

air-fuel mixtures, supplying air to the exhaust gases within the exhaust manifold and more careful design and manufacturing with the view toward emission control.

Alterations in fuel would pose problems, for fuel has evolved into a very complex mixture, as we have seen. But if olefins, which are the most likely hydrocarbons to form aerosols, are reduced to 10–12.5 percent of the gasoline, the olefin content of the exhaust is reduced by 45 percent.

Exhaust burners are of two types, direct flame afterburners, and catalytic converters. The first requires an air supply to support combustion and a spark plug to initiate it. But these devices must have a minimum concentration of exhaust gases to continue to burn, and control only carbon monoxide and unburned hydrocarbons. It would make no sense to have to enrich the carburetor mixture simply to keep the afterburner operating. Catalytic converting works well when new, but becomes less effective as fuel additives coat the catalyst. After twelve thousand miles the catalyst must be replaced. For this reason the major auto manufacturers in the United States are lowering the compression ratio of their engines to eliminate the need for leaded gasoline. But even without lead additives in the gasoline, catalyst lifetimes beyond twenty-five thousand miles are unlikely. Then too, a second catalyst must be present to take care of the nitrogen oxides.

Of the possibilities mentioned, engine modification presents the best possibility for reducing automobile emissions. But this will probably take years of research time and at the moment the theoretical minimum to which emissions from the internal combustion engine can be reduced would still be 25 ppm hydrocarbons, 0.25 percent carbon monoxide, and 100 ppm nitrogen oxides compared with the present levels of 275 ppm hydrocarbons, 1.5 percent carbon monoxide, and 1500 ppm nitrogen oxides.

The major problem in reducing emissions in internal combustion engines by redesign is that the responsibility for keeping engines in good tune, spark plugs unfouled, afterburners adjusted, and catalysts replaced lies with the owner. This is the same owner who finds it difficult at times to fill the gas tank or battery, much less change the oil or air filters. Even though it is in their own interest, the assumption that most people will dutifully follow manufacturers' instructions for maintaining antipollution devices and keep their cars well tuned is unwarranted. Perhaps the only real solution is an alternative to the internal combustion engine.

Alternative Engines

The most reasonable alternatives to internal combustion engines are turbines, steam, and electricity.

A gas turbine consists, basically, of a compressor and a turbine wheel on a common shaft. The combustion chamber lies in the airflow path between them. Since the turbine operates at very high air-fuel ratios, the emission of unburned hydrocarbons and carbon monoxide is quite low, but the production of nitrogen oxides is relatively high because of the high operating temperature of the engine. Reduction in the output of nitrogen oxides would require a catalyst active at less than 2000°F; such a catalyst has not yet been found. Whereas the turbine has the advantages of quick starting, no warmup, low vibration and noise level, fuel consumption at low speeds is excessive, acceleration poor, and the engine cannot be used in braking. But these problems are not insuperable. Once a major manufacturer tools up, turbine cars could be mass produced at approximately the same costs as present cars. But the changeover presents a risk apparently unacceptable to even as large a concern as Chrysler Corporation, whose much touted turbine car quietly disappeared from public view a few years ago.

Steam engines are undergoing a renewal of interest. With modern technology and new materials, compact and low-maintenance reciprocating steam engines are quite feasible, from the standpoint of both performance and emission control. The transmission system required is relatively simple and the long-standing problem of warmup time can be licked by modern boiler technology. Even if steam automobiles are not developed, fleet vehicles such as trucks, buses, and taxis could be converted to steam.

Electric automobiles use energy stored in batteries. Regardless of type, batteries work on the principle of a reversible chemical action between unlike electrodes placed in a conducting solution or electrolyte. The problem with battery-driven cars remains their limited range. The still unattained solution, then, is a battery of long life and high power. In present batteries, as the power increases, the energy level falls off, reducing the range. New battery systems such as sodium-sulfur and lithium-chlorine show promise of combining both power and range, but both require rather high operating temperatures and are in early stages of development.

The most logical solution to the automobile emission problem, short of abandoning cars for mass transport (Chapter 20), would be immediate application of factory-installed mechanical improvements and devices so that pollution levels are at least stabilized as more and more cars come into operation. In the interim, the internal combustion engine should be redesigned to achieve the lowest emission possible. This might even begin to reduce pollution levels. The time gained by these two expedients should be sufficient to develop steam and electric automobiles that would begin to phase out internal combustion engines and ultimately reduce photochemical smog to manageable and perhaps even livable proportions.

CARBON DIOXIDE AND CLIMATE

Because of its natural presence in the atmosphere and its basic role in photosynthesis and respiration, carbon dioxide is not commonly regarded as an air pollutant. But since the beginning of the industrial revolution the use of fossil fuels for combustion has released enormous quantities of carbon dioxide into the atmosphere. At present, the atmosphere contains around 2.3×10^{12} tons of carbon dioxide (0.03 percent) and the ocean another 1.3×10^{14} tons. Exchange between the atmosphere and the ocean amounts to about 200×10^9 tons per year, so the ocean acts as a reservoir or buffer of carbon dioxide in equilibrium with the carbon dioxide content of the air. If amounts of atmospheric carbon dioxide decrease, more is released from the ocean; conversely, if the carbon dioxide content of the air increases, the ocean tends to absorb more. The rate of change, however, is slow. It takes at least 1000 years for 50 percent of the change, up or down, to be accommodated by the ocean.

The Greenhouse Effect

In the atmosphere, carbon dioxide does not affect shortwave radiation. Upon striking the earth, however, shortwave radiation is transformed into and reradiated as longwave radiation or heat. This heat is absorbed by the carbon dioxide molecules and transferred to the atmosphere (Figure 10.10). It is this heat transfer that is ultimately responsible for long-term climate and short-term weather. Because of the balance of incoming light and outgoing heat the mean temperature of the earth remains at about 58°F. This phenomenon of carbon dioxide absorption of heat is called the greenhouse effect. Shortwave radiation or light also passes easily through glass, and is converted to longwave radiation or heat in a greenhouse. Because of the lack of convection and the inability of the heat to pass out through the glass, the greenhouse stays warm. The same thing happens in a closed car on a hot day.

Figure 10.10 Shortwave radiation (light) strikes the earth and is transformed into longwave radiation (heat). Since carbon dioxide absorbs longwave radiation, the more carbon dioxide contained in the atmosphere, the more heat is retained and the warmer the atmosphere becomes.

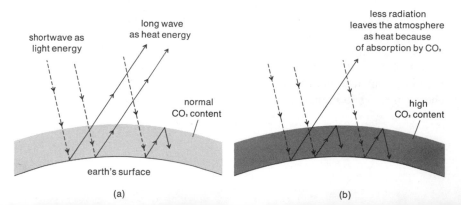

Figure 10.11 The contrails from a jet form patterns that may persist in the form of particles in the atmosphere. The effects of contrails are, as yet, unknown.

A theory has been around for a number of years that as combustion pours more and more carbon dioxide into the atmosphere it will accumulate faster than it can be absorbed by the sea. More carbon dioxide in the atmosphere will absorb more outgoing longwave radiation and the earth's mean temperature will begin to rise. Since a rise of only a few degrees would melt the earth's icecaps and raise the sea level by 400 feet, the problem is not merely academic (see Chapter 9).

Before 1900 the average carbon dioxide concentration in the northern hemisphere was approximately 290 ppm. Today it is 330 ppm, an increase in proportion with the rate of fossil fuel combustion during this time. But at the same time we are dramatically increasing the quantity of dust particles in the atmosphere. This tends to reflect shortwave radiation back into space, thereby cooling the earth. No one is at all sure which of these processes is in the ascendancy or whether there is a balance. As Revelle and Suess said in 1969:

> Human beings are now carrying out a large scale geophysical experiment of a kind that could not have happened in the past nor be reproduced in the future. Within a few centuries we are returning to the atmosphere and oceans the concentrated organic carbon stored in the sedimentary rocks over hundreds of millions of years. This experiment, if adequately documented, may yield a far-reaching insight into the processes determining weather and climate.[5]

If, of course, there is anyone around to record the results!

Another experiment will be initiated by the supersonic transport (SST). The condensation trails or *contrails* of jets are quite conspicuous in the sky (Figure 10.11). Usually they disperse quite rapidly; but some-

[5] Revelle, R. and H. E. Suess, 1957. "Carbon dioxide exchange between atmosphere and ocean and the question of an increase of atmosphere CO_2 during the past decades." *Tellus* 9, pp. 18–27.

times they persist and occasionally they bleed together and cover an otherwise clear sky with a blurry cirrus overcast. The SSTs will be flying at 50–70 thousand feet and carrying fuel loads of 100 tons, two-thirds of which are consumed at flight altitude. Just four flights a day could inject 150 thousand tons of water as well as other compounds from the SST exhaust system into the stratosphere. A well-traveled route might accumulate considerable quantities of such materials. Would these contrails persist in certain latitudes reducing radiant energy received by the earth? Would they affect the radiation balance of the atmosphere? Would they alter precipitation distribution by supplying condensation nuclei? Once again, no one knows.

NOISE

While we are unsure of the ultimate effect of contrails in the stratosphere, there is no doubt of the effect of one of man's activities in the first few hundred feet of the atmosphere—noise. Without question, noise can damage hearing; there is no threshold for ear damage (Figure 10.12). Claims

Figure 10.12 Noise is measured in dynes/cm², watts/cm², or, most commonly, decibels. Some common environmental sounds are listed here in ascending decibel values. (Modified from C. Stark, *Stereo Review*, **23**, 1969.)

decibels

for insurance compensation for loss of hearing have climbed to over a billion dollars a year in the United States alone, prompting industry to take noise reduction programs seriously. But more subtly, noise increases tensions already heightened by the other stresses of urban life. Noise can affect blood pressure and ultimately the heart. Unfortunately we have no earlids to shut out noise at night, hence we can be victimized at all hours.

Urban Noise

Construction, particularly in cities where buildings reflect and even amplify sound, contributes the noise of pile drivers, jackhammers, and compressors to the already high level of city noises. Within the buildings, sound resistance depends on the mass of partitioning walls; the heavier the wall, the greater its inertia and the less sound it transmits by vibration. While the days of lath and plaster walls are gone, contemporary walls can be constructed with cinder blocks and dead air spaces to deaden sound significantly.

Transportation is a close second to construction for noise generation. New York City subways are probably the noisiest in the world, the price paid for being among the first built. Montreal's and Mexico City's subways are models of quiet, efficient operation, the cars running on rubber tires rather than steel wheels. Surface transportation is provided by buses. They are not only noisy but their poorly adjusted and overworked diesel engines emit clouds of smoke and the pungent odor of aldehydes.

One of the greatest sound generators is the jet airplane. Not that noise from piston aircraft was insignificant, but in the 1940s and 1950s before jets were introduced, airports were relatively small and flight volume moderate. Since the introduction of jets at O'Hare International Airport in Chicago in 1958, traffic there has grown to 1200 flights a day; at peak hours jets come and go every forty seconds, moving 23.5 million people through O'Hare every year. In 1960, sixteen airports were served by jets, in 1967, 150, in 1970, 350, and by 1975, 500 will be. The safest way for a jet aircraft to approach a runway is to follow a long, low trajectory and unfortunately this produces the most noise. Even more unfortunately, areas right up to the edges of large new jetports were allowed to be subdivided and are jammed today with residential housing. Why anyone would knowingly and willingly buy a home next to a huge international airport is in itself an interesting question. Still, the problem of jet noise is limited to those living near airports, for once jets reach their cruising altitude, noise levels are not usually a problem.

Faster than sound. The SSTs now being developed generate over twice the noise of standard jets upon takeoff, but the real problem develops when an SST reaches its full speed (1800 mph proposed for the American version). When an airplane exceeds the speed of sound (650 mph at sea level) it generates a shock wave or sonic boom that sounds like a thunder clap. In general, the greater the weight and speed of the plane, the more intense the boom. When the SSTs become operational in the late 1970s close to sixty-five million people in the United States could experience an average of ten sonic booms a day. On the basis of tests over Oklahoma City, St. Louis, and other areas, up to 30 percent of the people felt they could not live with sonic booms; the rest either disliked them, or at best were neutral.

The figure of sixty-five million people being disturbed by sonic booms implies, of course, an overland route. If the SST were used only over water there might be little complaint but then there might be little economic justification for the airlines to fly them, for domestic flights are the bread and butter of most airline operations. Although the danger of physiological damage to hearing resulting from sonic booms is long term, more immediate are the psychological-sociological problems. Damage to property is another matter. Projected property damage claims from broken windows, cracked ceilings and walls may be as much as $85 million annually. More difficult to assess would be the damage to national monuments set up to protect Indian ruins, historic sites, or delicate geological formations; or damage from potential snow avalanches, landslides, or mudslides triggered by the sudden shock waves.

Perhaps the most disturbing aspect about the sonic boom generated by the SST is that so many people will be disturbed or inconvenienced for the benefit of that rather small percentage of the population in any country that will actually fly in the aircraft. As with so many other environmental problems or hazards, we need not tolerate this further degradation of our environment. There is no rational reason why the developers of supersonic aircraft, the Russians, the British-French consortium, and the United States, can not agree to delay production of the SST until the sonic boom problem is mastered by the engineers. Surely a delay of five to ten years would be less dear than the actual and intangible cost of unnecessary shock waves on people beneath the airlanes.

The sky and the sea have long borne the brunt of man's misuse, partly because their very immensity seemed to preclude any long-term effect by man, and partly because both belong to no one. But while it is possible to avoid a polluted sea by living inland, polluted air is in everyone's lungs. Slowly we are coming to realize the subtle cause-and-effect relationship between polluted environments and the increasing incidence of asthma, bronchitis, emphysema, hypertension, and heart deterioration —a relationship so long ignored because these debilities are not glamorous diseases, but the result of the body's inability to cope indefinitely with a broad spectrum of specific stresses.

The significance of all types of air pollution has begun to dawn upon us and the options are clearly focused, perhaps for the first time in man's history. Continued inaction in dealing with these problems is unconscionable for our generation, and suicidal for the next.

FURTHER READING

Burns, W., 1969. *Noise and man.* J. B. Lippincott, Philadelphia, Pennsylvania. Noise viewed as an air pollutant.

Fleagle, R. G. (ed.), 1969. *Weather modification: science and public policy.* University of Washington Press, Seattle, Washington. Examines the scientific basis for weather modification and its implications.

Leighton, P. A., 1961. *Photochemistry of air pollution.* Academic Press, New York. Valuable comparison between different types of urban air pollution.

Lewis, H. R., 1965. *With every breath you take.* Crown, New York. A popular account of air pollution problems, full of fascinating data.

Plass, G. N., 1959. "Carbon dioxide and climate." *Scientific American,* **201** (1), p. 41. Fine review in a popular style of the potential problem of carbon dioxide build-up in the atmosphere.

Stern, A. C. (ed.), 1968. *Air pollution.* 2nd ed., 3 vols. Academic Press, New York. The most comprehensive detailed work available on air pollution.

Wise, W., 1968. *Killer smog: the world's worst air pollution disaster.* Rand McNally, Chicago, Illinois. Graphically dramatizes the London smog episode in 1952. Excellent reading.

Wolozin, H. (ed.), 1966. *The economics of air pollution.* W. W. Norton, New York. A collection of papers dealing with the economic considerations of air pollution.

CHAPTER 11

RADIATION

Radiation is the emission of radiant energy in the form of particles or waves from certain naturally occurring elements. It has always existed to some extent in the environment on earth. But man has, paradoxically, both concentrated and dispersed radioactive materials to an extent unknown in nature. Hundreds of thousands of people have died as a direct result of the concentration of this energy at Hiroshima and Nagasaki and millions in present and future generations are threatened by dispersed radioactive energy in the form of fallout or nuclear wastes and their effects. To understand the danger to life of both concentrated and dispersed forms of radiation it is necessary to know something about the nature of radiation and its manipulation by man.

THE NATURE OF ATOMS

Superficially, objects appear continuous; that is, they have a certain integrity: water is 100 percent water, wood is 100 percent wood, and stone is 100 percent stone. Democritus realized 2200 years ago that beyond the physical subdivision into fog, sawdust, or sand, each of these materials was discontinuous; that their superficial appearance bore little resemblance to the tiny units or atoms of which they were constructed. It wasn't until 1805, however, that John Dalton confirmed the concept of the atom as the basic unit of all matter.

The combining of atoms to form molecules was difficult to understand when atoms were conceptualized as exceedingly small spheres. Atoms actually combine by sharing pairs of electrons. An atom is composed of a nucleus of protons and neutrons surrounded by one or more electrons. The number of bonds that an atom is able to make with other atoms depends upon the number of electrons that it has to share. Hydrogen has but one electron to share. Two hydrogen atoms, each with a free electron to share, combine to form gaseous hydrogen, H_2.

$$H\cdot + \cdot H \longrightarrow H\!:\!H \quad \text{or} \quad H_2$$

Oxygen has two electrons to share, so it can form bonds with two hydrogens, H_2O;

$$H\cdot + H\cdot + O\!: \longrightarrow H\!:\!O\!:\!H \quad \text{or} \quad H_2O$$

nitrogen has three electrons to share forming NH_3, while carbon with four available electrons can form a great number of compounds, the simplest being methane (CH_4). Similarly, all elements have some electrons available for bonding.

The Nature of Radiation

But what is radiation? In 1895 Wilhelm Roentgen, working with an evacuated glass tube with metal electrodes at either end, found that an electric current applied to the electrode at one end of the tube emitted a mysterious ray through the electrode at the other end, which penetrated opaque substances, causing a zinc-sulfide screen to glow or emit light. Roentgen called his discovery the X-ray. Apparently the rays were given off by the metal of the tube plate when stimulated by an electric current. Not long after, Henri Becquerel placed some minerals containing uranium ore on a carefully covered photographic plate and found that they exposed the plate even in the dark by giving off rays similar to those noticed by Roentgen. Becquerel observed also that only minerals containing uranium showed this trait. Pierre and Marie Curie became interested in the phenomenon and found that in addition to uranium, thorium and radium were also radioactive or capable of giving off radiation.

Radium was of particular interest to physicists because it was found to give off three kinds of radiation, called alpha, beta, and gamma. Alpha radiation was later identified as positively charged particles or protons, beta radiation was found to be negatively charged electrons, and gamma radiation was similar to the X-rays discovered by Roentgen.

When radium gives off alpha particles, a gas called radon is formed. The atomic weight (protons plus neutrons) of radium is 226. If an atom of radium loses an alpha particle with an atomic weight of 4, the remaining atom, radon, should have an atomic weight of 222. It does. Later it was found that when an atom of lithium is struck by a proton, it splits into two alpha particles accompanied by an energy release. The splitting of the lithium atom or any other atom into fragments is called *fission*.

Although lithium yields little energy, other atoms like uranium yield tremendous amounts of energy when split.

Isotopes. Many elements were found to have stable forms called isotopes, slightly different in atomic weight but otherwise identical to the regular element. Heavy hydrogen or deuterium, which will someday be a source of fusion power (see Chapter 8), is one example. Irene Curie and her husband, Frédéric Joliot, found that when they bombarded aluminum with alpha particles, neutrons were given off and a strange species of phosphorus appeared. Normal phosphorus has no isotope, but the new species disappeared in fifteen minutes leaving silicon! The Joliot-Curies had produced a new kind of isotope, an unstable one, called a *radioisotope,* which rapidly decayed into another element. Thus began a whole new era of artificially produced radioisotopes, an era that will continue as long as man manipulates atoms.

Today, radioisotopes have been produced for every known element, both intentionally for research and inadvertently in the fallout of atomic bombs. The variety of these radioisotopes is truly fantastic; not only do they vary in the kinds of radiation that are emitted, but in their longevity as well. Longevity of radioisotopes is described in terms of the length of time it takes for one half of their radioactivity to dissipate, the *half-life.* Carbon-14 has a half-life of 5600 years, that is, after 5600 years one half of the radiation will be left, in 11,200 years one half of one half or one fourth is left, and so on. Other radioisotopes have much shorter half-lives than carbon-14: for example, strontium-90, 28 years; cesium-137, 27 years; iodine-131, 8 days. Because of their variety and inherent radioactivity, radioisotopes can be introduced into biological or geological systems and their movement followed by sensitive radiation counters, allowing a better understanding of many processes that involve the movement or cycling of atoms. This is possible because such radioisotope tracers, despite their radioactivity, behave chemically exactly like their normal analogs.

ATOMIC FISSION

The 1930s were heady times for nuclear physicists; discovery piled on discovery, providing new insights into the nucleus and its organization, and also suggesting further manipulations of atomic and subatomic particles. In 1939, Hahn and Strassman, German chemists, who had been looking at the chemistry of uranium when it was bombarded with slow neutrons, were surprised to find barium as one of the by-products. Uranium has an atomic number of 92, barium 56. Usually when uranium or any other element is bombarded with various particles, the atomic numbers of the by-products are rather close to the original. But barium is remote in every way from uranium.

Then Lise Meitner, a German physicist, had the brilliant idea that maybe when uranium absorbs a neutron it splits into two fragments. After all, lithium was known to split into two alpha particles. If so, what was the other fragment? The atomic number of krypton is 36 (92 − 56) and krypton turned out to be the other fragment. But the most exciting part of this disintegration of uranium lay in tallying the neutrons involved. Uranium has 146 neutrons in its nucleus. It absorbs one, making 147; but the two fragments, barium and krypton, have 82 and 47 respectively, totaling only 129. This leaves 18 neutrons, neutrons that are released when uranium undergoes fission. Suppose these neutrons were absorbed by other uranium atoms and more neutrons and energy were released as a result, we would have a chain reaction generating much energy and still more neutrons. The means of unlocking vast stores of atomic energy seemed at hand.

But there are three isotopes of uranium, U-238, U-235, and U-234. Did all three undergo fission, and if not, which one did? Pure uranium contains 99.3 percent U-238, 0.7 percent U-235, and 0.006 percent U-234; of these it was found that only U-235 underwent fission by capturing a slow neutron. Although U-238 captured fast neutrons, fission did not take place. This posed a problem if a chain reaction was to be sustained.

The Chain Reaction

Slow neutrons cause U-235 to undergo fission but in doing so it releases 18 fast neutrons per atom. These are captured by U-238 rather than U-235, stopping the reaction. What of the U-235 which absorbs neutrons? It is unstable and forms a new element, neptunium. This too is unstable and forms still another element, plutonium. Fortunately, plutonium fissions as well as U-235 and can be derived from the abundant U-238 isotope, which is more easily obtained in quantity than U-235. A chain reaction once again seemed possible.

Why the excitement over this fission reaction? If one pound of U-235 or plutonium could be induced to undergo fission, it could release a sudden explosive force equivalent to ten thousand *tons* of TNT, or, if released gradually, power equal to twelve million kilowatt hours!

But *pounds* of U-235 or even plutonium are not easily obtained. Could neutrons somehow be slowed down, favoring their absorption by U-235 instead of U-238? By introducing such light elements as beryllium or carbon, which do not easily absorb neutrons, the velocity of a fast neutron can be slowed as it bounces off these light elements, thereby increasing the opportunity for its absorption by U-235.

Strangely enough, the capturing of neutrons by a fissionable material is related to its volume, while the loss of neutrons is related to its surface area. But as a quantity of matter grows in size its volume increases faster than its surface area. Therefore, the larger the piece of fissionable material, the greater the number of neutrons retained for fissioning the U-235. This brings up the important concept of *critical mass*. The fissionable material must be large enough to retain sufficient neutrons to sustain a chain reaction. Since U-235 is so scarce, and we never use more than is necessary, what is the critical size for U-235? It depends in part on the purity, for impurities usually absorb neutrons as well as U-235. Early calculations indicated a critical mass of 2 to 200 pounds of U-235. This was like asking for a ton of hummingbird tongues! Several possibilities were open: first, processing tons of uranium ore to get pure uranium, then separating U-235 from U-238; second, bombarding U-238 with fast neutrons to get the fissionable plutonium; or, third, introducing purified moderators (beryllium or carbon) into a purified mixture of U-235 and U-238 to produce the desired quantity of slow neutrons.

While all these avenues were explored, the third produced the first sustained chain reaction. In a squash court at the University of Chicago an atomic pile (quite literally) was constructed of bricks of a purified form of carbon (graphite) between which pieces of uranium were placed. To prevent the possibility of too vigorous a reaction, control rods of cadmium, which absorbs neutrons, were inserted into the pile. The graphite blocks slowed the neutrons so that they could be absorbed by the U-235 and enough slow neutrons were available to sustain the reaction. On 2 December 1942 controlled atomic fission was achieved.

Once a sustained chain reaction was obtained, fabrication of an atomic bomb was relatively simple. All that was necessary was to prearrange several noncritical masses of fissionable material, each supplied with a neutron source, fire them together into a critical configuration, and in a fraction of a second enough fission reactions could take place to cause an explosion of incredible dimensions (Figure 11.1).

The hydrogen bomb. Although a large atomic bomb is more than adequate to destroy any military target, a super or hydrogen bomb was developed, an extension perhaps of the "bigger and better" fixation with which modern society seems preoccupied. The hydrogen bomb involved a fusion principle, just the opposite of the fission process in the atomic bomb. Instead of supplying neutrons to split atoms, heat was supplied to fuse atoms of hydrogen into helium, releasing huge amounts of energy. To do this required a seemingly impossible amount of heat—100 million°C! Since this level of heat was possible on earth only in an atomic explosion, it was necessary to trigger the fusion reaction by a fission reaction. So the H-bomb was created, concluding, we hope, the evolution of such grim devices.

Figure 11.1   An atomic bomb test near Bikini Atoll, 1946. (AEC)

A few years ago our attention was riveted to the destructive aspects of atomic energy with Hiroshima, Nagasaki, and a long series of tests in Nevada and the Marshall Islands. More recently, with the adoption by most nations of a test ban treaty, the peaceful uses of atomic energy have come to the fore, first with atomic fission power plants and the longer term potential of power from hydrogen fusion.

RADIOACTIVE FALLOUT

Once a nation has developed atomic bombs there is a strong impetus to continue to play with them, forever testing, much like a baby who has just discovered his thumb. But for the fallout problem the United States and other atomic powers would doubtless still be exploding dozens of test bombs each year.

It was discovered early, however, that nuclear explosive devices create large quantities of radioisotopes in the tremendous pulse of energy liberated in the fission chain. Fifty percent of the bomb energy goes into blast, 35 percent into heat, and 15 percent into radioactivity. The quantity and quality of radioisotopes from an atomic blast vary depending on the materials in the bomb and its supporting tower, the location of the test site, and the composition of soil and water in the test area. Each could contribute an enormously varied list of elements that with an energy input could form radioactive isotopes. Some of these are very short lived, lasting fractions of seconds, minutes, or hours, and decaying before they get very far. Others last years, centuries, even millennia. Regardless, all emit radiation that is potentially harmful in a variety of ways.

Distribution of fallout.   At first it was assumed that fallout particles (dust and debris from the blast) would fall to the earth rather quickly at no great distance from the explosion site and would thereby limit the spread of radioactivity. With small explosions this was possibly true; much of the radioactive material quickly settled out. But with larger and larger explosions more radioactive material was generated and it was thrust higher and higher into the atmosphere.

Generally speaking, the atmosphere has two layers, a lower *troposphere,* the first 30–40 thousand feet, where all of the weather activity takes place. Above this is the more calm, clear, and dry *stratosphere.* Normally any fine materials injected into the troposphere are scrubbed out by rain and returned to the surface. But in the stratosphere, without this scrubbing mechanism, fine particles remain suspended, dispersing and circling the earth several times. Over the equator, the two layers are quite distinct because little circulation occurs between them but toward the poles they overlap a bit forming a slot through which particles can enter the troposphere and be washed out (Figure 11.2). This means that wherever bombs are exploded in the northern hemisphere, fallout will be concentrated in the middle latitudes, just those areas where most of the people live in North America, Europe, and Asia.

This is a classic example of shortsighted experimentation—injecting vast quantities of radioactive materials into the atmosphere without the least notion of the long-term effect on man and the environment.

Figure 11.2   The convergence of trade winds toward the equator results in a zone of quiet air or doldrums, which circles the earth tending to keep separate the weather systems of the northern and the southern hemispheres. (Modified from G. T. Trewartha, *Introduction to Climate,* McGraw-Hill, 1968.)

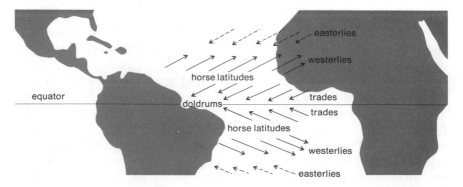

The Fallout Hazard

The hazard presented by a radioisotope depends upon the type of radiation emitted and its energy, its half-life, the rate at which it is absorbed, the time that it is present in the body, and the place in which it is concentrated in the body. Radioisotopes with shortest half-lives dissipate their energy so quickly there is little time for much environmental contamination. Those with very long half-lives, on the other hand, distribute their energy gradually. It is the radioisotopes with intermediate half-lives that are of the greatest concern. Of all the radioisotopes produced by nuclear detonations, the two which we have heard the most about and which have caused the most alarm are strontium-90 and cesium-137. Strontium and cesium fall into this middle category with half-lives of approximately 30 years each. Functionally, each acts like a much more abundant and biologically important element: strontium mimics calcium and cesium mimics potassium. Plants absorb calcium through their roots and use it in calcium pectate, which cements the plant cell walls together, making multicellular plant organization possible. Calcium is also readily absorbed by the human body and incorporated into bone. Since strontium closely resembles calcium, it too is taken up by plants, especially in soils with low calcium content, and then by man in milk and cereals. Strontium, like calcium, is deposited in the bones. Since the marrow of the bones is the principal site of blood cell production, white and red blood cells can be irradiated by the radioactive strontium, adversely affecting their reproduction.

Strontium-90 is transmitted to man in cow's milk and various cereals. Fortunately the cow's gut does distinguish between calcium and strontium even if the plants eaten by the cow do not; given equal amounts, the cow will absorb 100 units of calcium to 10 of strontium-90. Further, when we drink milk our gut also makes the discrimination and selects 100 units of calcium to 17 of strontium-90. So, in a way, despite its strontium-90 content we are better off drinking milk because of this double filtering effect than eating contaminated cereals. The Japanese who have little milk and abundant rice in their diet have a strontium-90 uptake six times that of people in western countries. Although any radiation is potentially harmful, the quantity of strontium-90 in the environment is still rather small. The amount of strontium-90 in the 5 million acres of the Netherlands, for example, is one gram, compared to the 1000 grams of calcium carried around by each of us in our skeletons.

Cesium-137 is immobilized by the clay found in most soils but when as fallout it settles out on the leaves of plants it can be absorbed and translocated to the rest of the plant and even concentrated in some tissues such as those in potato tubers. Since cesium-137 mimics potassium, which is involved with the process of muscle contraction, most radioactive cesium enters the muscle tissue. This means that unlike strontium-90, which is confined to bones, cesium can be readily ingested in meat and in this way concentrated.

Food chain contamination. Food chain concentration has occurred in the arctic tundra. Caribou in the New World and reindeer in the Old World feed extensively on lichens in the winter, as we saw in Chapter 5. Because of the high amount of fallout in the arctic (one-fourth to one-half of that in the temperate zone), and a slow rate of growth, lichens accumulate cesium-137 to a level well above that of other plants. When these radioactive lichens are eaten by caribou or reindeer the cesium-137 is concentrated still more in the meat of these animals. Should the caribou be eaten by a wolf or a man, the level increases again. The cesium-137 concentration doubles at each step in the food chain from lichen to caribou to man. This radiation concentration is especially clear in the arctic ecosystem because of the simplicity of the system and the limited socioeconomic development of the people—in the winter, caribou have only lichens to eat and the Eskimos have little to eat but caribou.

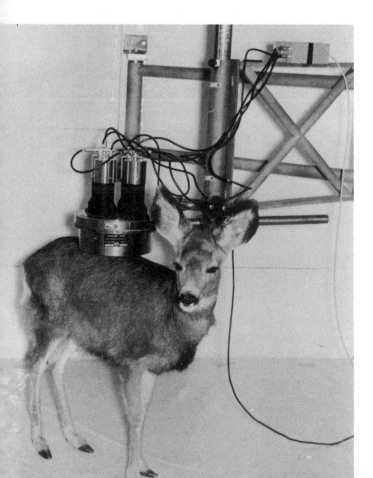

Figure 11.3 The radioactivity this deer has picked up from fallout is being measured by a whole-body counter. (Colorado State University)

There are other examples of radioisotope concentration in eco-systems (Figure 11.3). Zinc-65 is readily absorbed by oysters, clams, and scallops in that order, not only by contact in their gill area but by ingestion of food as well, for diatoms like *Nitzschia,* which we discussed in Chapter 9, absorb zinc-65 quite readily. In late summer of 1956 after the spring H-bomb tests in the Pacific, some of the tuna caught by Japanese fishermen all over the Pacific were disturbingly radioactive. Radioactivity is measured by a device which counts the bursts of radia-tion from a radioisotope. One fish registered 4500 counts per minute (cpm) in its kidney, 1200 cpm in its stomach, 1800 cpm in its intestine, 2500 cpm in its liver, and 1200 cpm in its heart. Although there was little radioactivity in the edible flesh there was concern about the sources of this radioactivity. An expedition sent into the Pacific by the Japanese government discovered that sea water over 1300 miles northwest of the bomb site at Bikini registered over 100 cpm. But that did not explain the excessive radioactivity in the tuna. The answer lay in food chain concen-tration, leading to potentially dangerous contamination of the top con-sumer in the chain, tuna. Over 457 tons of tuna with counts above 100 per minute were destroyed, causing panic in the fish-eating Japanese public, and financial ruin to many in the tuna trade. Picture the conster-nation in the United States if beef was suddenly found to be radioactive!

Not all radioisotopes are produced in so straightforward a manner. In follow-up studies after bomb tests in the Marshall Islands, investi-gators were puzzled to find high levels of cobalt-60 in shellfish. But cobalt-60 was not one of the radioisotopes produced by the blast. Appar-ently some other blast-generated material, a neutron emitter, bombarded the natural isotope, cobalt-59, producing indirectly cobalt-60. The pos-sibilities for harmful contamination are enormous.

RADIATION SICKNESS

All of us come into contact with radiation through fallout from bomb tests. The levels, however, are very low and the effect is probably quite long-term, if it is measurable at all. What about the more spectacular exposure to radiation that leads quickly to sickness and often death?

Ionization

As we saw earlier, there are three major types of radiation, alpha, beta, and gamma. The first two are particles which have relatively little energy; alpha particles are barely able to penetrate the skin and beta particles can penetrate only a millimeter or so. So under normal conditions these two types of radiation cause only skin burns, although when matter emitting either of these radiation types is inhaled or incorporated into the body significant damage may result. Gamma radiation, the third type, is the most energetic and potentially the most dangerous since it can penetrate most substances, certainly the human body, with ease.

Regardless of the form, alpha, beta, or gamma, radiation usually affects living tissue by causing ionizations that, in turn, cause cellular damage. A cell is composed largely of water. When a water molecule is irradiated, an electron is knocked out of orbit. The ejected electron may then become attached to a normal water molecule, making it unstable. These unstable molecules split into hydrogen ions (H^+), hydroxide ions (OH^-), and the free radicals OH· and H·. Free radicals react with various molecules in the cell, which can then no longer function normally, and the cell dies.

Measurement of Radiation and Dose

Although there are many units used to describe radiation in its various aspects we need be concerned only with two types: disintegrations in a radioactive substance producing radiation and ionizations in either air or tissue caused by these radiations. To measure the first type, a *curie* is defined as 3.7×10^{10} disintegrations per second and is roughly equal to the radioactivity of one gram of radium together with its decay products. The radioactivity of materials is usually described in curies, millicuries (10^{-3}Ci)[1], microcuries (10^{-6}Ci), or picocuries (10^{-12}Ci) sometimes called micromicrocuries. But this gives us no information about dose. The roentgen (r), named for the discoverer of X-rays, is a measure of the number of ionizations caused by radiation in air and is defined as the amount of radiation causing 1.6×10^{12} ionizations in one cubic centimeter of air. Since we are more concerned with dose rate in tissues than air, two other units have been devised, the rad, which is 100 ergs (a unit of energy) absorbed by one gram of tissue, and the rem, which is a unit of absorbed dose taking into account the relative biological effect of various types of radiation. But for our purposes a rem or a rad is about one roentgen.

Dose-rate effects. A dose of 1000 r is fatal to all humans. A liver cell receiving 1000 r would be exposed to two million ionizations. Since a liver cell has about two billion protein molecules only one of every thousand protein molecules would be likely to be ionized. Yet this is enough to lead to the death of the cell.

About 50 percent of those people irradiated at 500 r die. This statistical point is called the LD-50 (lethal dose for 50 percent of the population).

[1] This is a shorthand way of writing 1/1000 of a curie. A picocurie would be 1/1,000,000,000,000 of a curie.

At levels of 100–300 r, radiation sickness develops. The first symptoms are nausea, vomiting, diarrhea, and nervous disorders followed by a period of relative well-being. Then the secondary symptoms begin: falloff in red and white blood cell numbers, hemorrhages just below the skin, loss of hair, and ulcerations in the mouth and gut. These symptoms may also disappear, leading to slow recovery. Perhaps the most dangerous phase is during the period of low white blood cell level. Since the white cells defend the body against disease organisms, a person may survive the radiation only to die of a disease like pneumonia or influenza.

The tissues most sensitive to radiation are those which reproduce rapidly under normal conditions: the blood-forming tissues, bone marrow, lymph nodes, spleen, thymus in children, liver in fetuses, gonads which produce either sperm or eggs, epithelium which lines the intestinal tract, and all embryonic tissues which are in a state of flux. The symptoms of nausea, vomiting, and diarrhea are due to disturbances or death of the cells lining the digestive tract; the hair loss is due to injury to the hair follicles. The lens of the eye, which is covered with epithelial cells, is also quite sensitive to radiation and may develop a cataract or opacity.

Survivors of acute radiation in the 100–300 r range often suffer some permanent hair loss, cataracts, low fertility, and an increased tendency toward leukemia. Beyond this dose range, 600 r leads to irreversible destruction of bone marrow, 1000 r to irreversible intestinal damage, and greater than 3000 r, death of the central nervous system. Of three men involved in nuclear accidents, one received 800 r and died in 26 days, a second exposed to 1000 r died in 9 days, and a third received 12,000 r and died in 36 hours. Between 25 and 100 r there are no visible symptoms, just a decrease in the number of white blood cells. Below 25 r there are no measurable changes.

Radiation and aging. One curious question about radiation is whether it induces or initiates aging, or in some way accelerates and enhances it, or whether aging is a time-dependent process that begins and slowly continues with or without radiation. There has been a good deal of research in this area, but because of the subtlety of the problem and the difficulty of extrapolating experimental results from mice to man, no clear-cut answers have emerged. The most plausible hypothesis suggests that background radiation slowly destroys parts of the genetic code contained in the chromosomes, leading to a breakdown of the cell functions controlled by the damaged parts of the chromosomes.

It has been suggested that radiation-induced changes in cells can have an antigenic effect, that is, the irradiated cells will become regarded as foreign protein by the body and be eliminated by the body's immune responses. The accumulation of viruses by cells may enhance this reaction. Also, radiation may decrease the rate of cell production in organs with low regeneration rates to begin with, leading to failure of organ function. From studies with mice there is an indication that radiation may decrease the lifespan by one week per roentgen of total body radiation. But whether this applies to man is not certain, for no one volunteers to be irradiated and those who have been, accidentally, are statistically too few and have not yet completed their expected lifespans.

Radiation and heredity.   So far we have been looking at the effects of radiation on asexual or somatic cells. There is an even greater potential and demonstrated effect of radiation on reproductive cells. By damaging or even slightly altering the chromosomes in the egg or sperm, mutations can be caused that lead to abortions, deformed births, or genetic defects of a less obvious sort. Though only 10 percent of mutations have been calculated to be radiation produced, the mutational rate is proportional to the total amount of radiation absorbed by the sex cells. It makes little difference whether they receive 10 r in one year or 1 r a year for ten years. Most measurements show a linear relationship between dose and damage. So in a sense there is no such thing as a harmless or "safe" dose of radiation. This is why there is such concern in many scientific circles about what seem to be insignificant amounts of radiation, in fallout or in more mundane exposures such as X-ray examinations. James Crow, an eminent human geneticist, stated that even if the maximum fallout of the late 1950s were continued over a thirty-year period and amounted to a total of only 0.1 r in that time, 8000 children in the world's population of the next generation could be born with gross physical or mental defects as a result of that radiation.

Of all the radiation beyond the natural background, X-ray examinations far surpass that received from fallout. The United States population is probably the most X-rayed in the world, yet there is no overt evidence of any increase in mutations, higher incidence of birth defects, or shorter lifespans that can be directly attributed to radiation. Indeed, another human geneticist, Hermann Muller, regards the increased frequency of reproduction by people with genetic defects as a far greater danger to the genetic wellbeing of man than any level of radioactivity currently encountered.

Although we have recently heard much about fallout and artificially produced radioisotopes, radiation has always been with us. At sea level we all pick up at least 0.1 r a year and up to 50 percent more radiation at altitudes over one mile (a result of greater cosmic ray intensity). A quarter of this background level comes from cosmic rays zooming in from outer space, another quarter from various radioactive elements in the body, the rest from soil and rocks. Fallout amounts to 0.001–0.005 r per year for most people, less than 5 percent of the natural background.

Figure 11.4 A radioactive source is located in the mast at the center of the circle of dead and dying trees in an experimental forest. This experiment determines the effect of radiation on a forest area over a defined period of time. (Courtesy of Brookhaven National Laboratory, New York)

In some parts of India and Brazil, radioactive sands emit as much as 17.5 r per year and houses built on these deposits receive 2 r per year, compared with 0.69 r per year in New York City. Long-term studies have been initiated to determine if this higher than normal background radiation has had any measurable effect on the local population.

Radiation and plants. The role of radiation as an environmental variable is somewhat more obvious in plants. A reasonable picture of radiation effects in a natural ecosystem can be seen from an experiment done at Brookhaven National Laboratory by G. M. Woodwell. A piece of relatively homogeneous scrub oak-pitch pine forest was irradiated with a powerful source of gamma rays. The intense radiation killed plants, *but differentially*. It was discovered that pine trees were more sensitive than oak, oaks more sensitive than blueberry, and blueberry more sensitive than a particular sedge (a grass-like plant growing on the forest floor). The radiation was emitted from a point source about fifteen feet from the ground, hence the dieback was circular around the source (Figure 11.4). The trees and shrubs were generally more sensitive to radiation than the sedge, because the sedge buds were at ground level and somewhat protected; the buds of trees and shrubs were fully exposed to the radiation. Variations in radiosensitivity among the trees and shrubs were due, in part, to differences in the number and size of the respective plants' chromosomes. One theory, pioneered by A. H. Sparrow, suggests that cells with a small number of large chromosomes are more likely to be damaged by radiation than cells with a large number of small chromosomes, simply because the former offer a larger target. Moreover, the many small chromosomes of the latter possess more duplicate genetic information so that when one or two are damaged, a number of others can take over their function.

NUCLEAR POWER AND RADIOACTIVE WASTES

Despite the long recital of environmental and health problems elicited by a greatly augmented flow of various kinds of radiation into the environment, there has been a silver lining in the radioactive cloud. Some radioisotopes have been extremely useful in cancer therapy, cobalt-60 and gold-198 for example. Phosphorus-32, carbon-14, and oxygen-18 are used for tracer work in physiology, ecology, and geology. But the most spectacular beating of sword into plowshare has been the use of atomic energy in the production of power.

Figure 11.5 The dome which shields this reactor is a characteristic feature of nuclear fueled power plants, this one at Rowe, Massachusetts. (Yankee Atomic Electric Company)

Figure 11.6 This bundle of uranium oxide fuel elements is being readied for installation in a nuclear reactor. (General Electric)

The Nuclear Reactor

A typical reactor (Figure 11.5) uses as its fuel source a mixture of U-235 and U-238 packed as uranium oxide pellets into stainless steel or zirconium alloy rods about one-half inch in diameter and several feet long (Figure 11.6). These are grouped together in subassemblies far enough apart to prevent a critical mass from being achieved, close enough to sustain a controlled chain reaction. Altogether, a reactor core may contain hundreds of thousands of pounds of uranium oxide. The chain

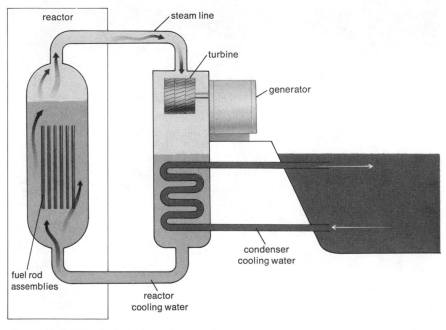

Figure 11.7 The fuel elements in a nuclear reactor are usually cooled by a primary coolant in a closed system. To cool the primary coolant a secondary coolant is necessary. It is the dumping of this secondary coolant into the environment that causes thermal loading problems. (Modified from *Environment,* **II,** 1969.)

reaction is moderated by control rods of cadmium and the core is kept at a temperature of 1000°F by a cooling system usually containing water. As the water, the primary coolant, passes among the fuel elements, it boils, and the resulting steam is bled off to turn a turbine, generating power. Then the water in the system is cooled by a secondary coolant system of water, usually drawn from a large river or lake, and returned to the core (Figure 11.7).

The advantage of nuclear power is that, when the facilities are large enough, it is cheaper to produce than conventional power. In addition, the power plants no longer need to be concerned with fuels or dealing with the pollutants associated with fuels—sulfur dioxide and fly ash. Unfortunately as we saw in Chapter 8, thermal loading from the secondary coolant is about twice that from conventional plants. A pessimist might with some justification claim that technology, while seeking to solve old problems, always manages to find new ones. The new problems, incurred when radiation is used in power plants, are not insoluble, but as more nuclear power plants are built, they become increasingly important.

At the moment, because of their limited number, nuclear power plants probably threaten the environment with radiation only in the form of their low-level radioactive effluents. But, as the number of installations increases, it is more likely that mechanical and human failures will allow the release of high-level radioactivity into the environment. With the present-day design of reactors, an atomic explosion resulting from malfunction is a very small risk even if everything possible goes wrong. However, a nuclear power plant need not go up in a mushroom shaped cloud to release dangerous quantities of radioactivity.

Radioactive pollutants. There are three kinds of radioactive pollutants from nuclear power plants: solid, liquid, and gaseous. Solid wastes may consist of such items as tools, reactor parts, and clothing, which may be quite radioactive depending upon their use and are usually buried in cement drums either in trenches on land or at sea.

Some liquid wastes have low-level radioactivity, resulting from isotopes formed when impurities in the coolant water and corrosion products from the coolant pipes are bombarded with neutrons escaping from the core area. This can be controlled somewhat by demineralizing the coolant water before it enters the heat exchange area but there will always be some radioisotopes generated from this source. A reactor in the 150–300 megawatt range would generate 1–10 curies per year in the form of cobalt-58, chromium-51, manganese-54, iron-59, and molybdenum-99. Because of the low radiation level of these wastes, they are usually discharged into the environment.

The fuel elements themselves are an even more dangerous source of both liquid and gaseous radioactive wastes. Although clad in either stainless steel or zirconium alloy, carefully fabricated to minimize leakage, complete sealing of the fuel elements is apparently impossible to attain or to sustain. Minute cracks allow radioactive fission products to escape into the primary coolant. In some instances, the high radiation level causes the cladding to flake or weaken, allowing still more leakage. When the steam is condensed for return to the pile, the gaseous fission products are separated and vented through a stack. One of these gases, you will remember, is krypton-83 (when U-235 undergoes fission barium-138 and krypton-83 are the two fission products). Others are xenon, argon-41, iodine-131, and nitrogen-13. Not all of these are fission products, of course; some are due to impurities of air and water within the reactor.

Disposal of Radioactive Wastes

The biggest problem lies in disposing of the fission products. Sooner or later, usually in one to three years, fission products accumulate to the point where the chain reaction stops or is *poisoned*, that is, the neutrons are being absorbed at greater rates by the fission products, leaving fewer to sustain the chain reaction. At this point the fuel elements, now extremely radioactive, are removed and shipped in specially cooled and shielded containers to a fuel reprocessing plant. Here they are chopped up and placed in concentrated nitric acid. The fission products and unreacted fuel are dissolved and separated. The unreacted fuel can be used again but the fission products, concentrated somewhat by evaporation, are stored as a liquid in underground stainless steel tanks. With a radioactivity that is usually between 100 and 1000 curies per gallon, enough heat accumulates to keep the tanks boiling like teakettles.

Assuming the present level of efficiency (25 percent), almost one and a half tons of fission products are produced per 1000 megawatts of power. Already hundreds of millions of gallons of high-level wastes are in storage. Since these wastes will remain highly radioactive for hundreds of years while the life of the tank is measured in decades, underground storage tanks represent at best a temporary solution to the waste disposal problem (Figure 11.8).

Figure 11.8 Million-gallon tanks have been constructed at the AEC Hanford Works to contain high-level radioactive wastes. When completed the tanks will be buried in 7–14 feet of earth. (U.S. AEC, Richlands Operations Office)

Though high-level wastes are considered too hazardous to be dumped into the ocean, intermediate- or low-level wastes are disposed of there, often without a clear idea of the hazards involved. One obvious hazard would be containers breaking and releasing their radioactive contents, which might be washed up on crowded beaches. Contamination of food chains, as we have already seen in Pacific tuna, could easily occur also. Concentration of wastes on continental shelves might make the future exploitation of oil, gas, or metal nodules awkward. All too often, wastes are not dumped at either the required distance from shore or the proper depth. At too great a depth the containers may be crushed by sea water pressure, releasing their contents immediately. Also, it is not known how long the containers last on the sea bottom, and how far and in what direction they are moved by underwater currents.

In the future, before the quantity of wastes from increasing numbers of reactors overwhelms the possibilities of storage in liquid form, reduction to a solid state and incorporation into ceramics or glass, both environmentally resistant, may be possible. These stabilized solids could be permanently stored in dry, stable salt mines of which there are large numbers in New York, Kansas, and along the Gulf Coast. By the year 2000, twelve acres a year of such salt mine storage space will be needed to dispose of high-level reactor wastes.

Present disposal of radioactive wastes can best be summed up by citing the three basic approaches: *dilute and disperse,* used for high volume, low-activity wastes; *delay and decay,* used for medium activity wastes where slow movement through soil allows time for radioactive decay; *concentrate and contain,* reserved for high-activity wastes, now stored underground in tanks.

Mine tailings. Radioactive wastes involve more than the reactor and its by-products: uranium must be mined, purified, and processed into fuel elements, and as we have seen, when the elements are exhausted they must be reprocessed. Each of these steps produces radioactive wastes. One example will suffice. Scattered around the Colorado River Basin are huge piles of uranium tailings. Ground to a sandlike consistency to remove the uranium, these tailings contain radium-226 with a half-life of eighty thousand years. Radium and thorium, like strontium-90, are absorbed by the bones. As radioactive dust from the piles blows into rivers and ultimately into Lake Powell and Lake Mead, the radium-226 level has increased in places to twice the maximum permissible level suggested for human consumption. Near Durango, Colorado, a pile of over a million tons of radium-226 sits near the Animas River. As it erodes, this material will certainly increase the radium-226 content of the river. Crops grown on irrigation water from the Animas River already contain twice as much radium-226 as crops growing above the tailings. Since radium is concentrated in hay and alfalfa that is eaten by cattle and in turn by man, the possibility exists of yet another food chain being contaminated by radioactivity. Of course, the tailings should be stabilized by grading and planting, which would greatly reduce erosion into surface drainage. This approach has been followed to some extent by state and federal agencies, but apparently not at a rate that has kept up with continuing production.

Further problems have arisen from the tailing sand being used for children's sand boxes and by contractors as a base for concrete slabs or in the backfilling of basements. As the radium-226 decays, radon, a radioactive gas, is given off. Although radon has a short half-life, it has been established as the prime cause of lung cancer in uranium mine workers. As a result of the use of uranium tailings in construction, many people have been exposed in their homes to levels of radiation many times higher than the maximum dose allowed miners in a uranium mine. In some instances houses have had to be abandoned.

Once again, as with strip mine spoil banks, no one seems interested in the tailings once plants cease operation. The twelve million tons of tailings in the Colorado River Basin may contain as much as 8000 grams of radium-226. When it has been carried by the rivers of the Basin and deposited into the bottom sediments of Lake Powell and Lake Mead, it may well be put out of circulation. But radium-226 may become a much larger problem than it is now as increasing use is made of river water in this arid region.

Radiation levels.   Since radioactivity is so easy to measure and its effects are well-known compared with those of many biocides, standards were set up, somewhat arbitrarily, well in advance of intensified radioactive waste production. Three categories, each with its own permissible level, were established by the International Commission on Radiological Protection. This group recommended a limit to workers in radiation-oriented industries of four rems (roughly 5 r) a year for blood and sex organs, thirty rems a year for the eye, skin, thyroid, or whole body radiation, and fifteen rems a year for other organs. The individual should receive only one-tenth of that permitted the occupational group and the general public one-thirtieth of the occupational level. This assigning of permissible doses evidently assumes a threshold for various types of radiation and places the permissible dose somewhere below this level despite evidence that *any* amount of radiation can cause damage. To be sure, avoidance of all radiation is impossible since background alone provides 0.1–0.5 rems a year. We can and do go beyond this minimum level without grave consequences, but we increase the probability of reproductive cell damage, cancer, shortened lifespan or more subtle effects with every bit of radiation. It is senseless for anyone to expose himself needlessly to radiation. It is even more senseless for whole populations to be unwittingly subjected to any increase in radiation levels that can be prevented; and it is also unnecessary, because the technology exists to remove radioisotopes from effluent wastes, whatever their source. It has simply been cheaper and more expedient to disperse them into the environment. As earth's population continues to grow we can no longer allow the environment to slip from the desirable to the merely tolerable if life is to have any value beyond bare existence.

The Next Generation of Reactors

The next stage of reactor development will be the breeder reactor, so-called because it uses high speed neutrons to convert more U-238 into plutonium than is converted in present reactors, roughly 1.5 lbs of fuel produced for every pound burned. The fuel elements would be cooled by liquid sodium rather than water.

In prospect the breeder sounds good but it will not be without problems. One is the requirement of up to two tons of plutonium as fuel, 0.1 percent of which could become critical in an accident. Edward Teller, who directed development of the hydrogen bomb, was concerned enough about this possibility to say "In an accident involving a plutonium reactor, a couple of tons of plutonium can melt. I don't think anybody can foresee where 1 or 2 or 5% of this plutonium will find itself and how it will get mixed with some of the other material. A small fraction of the original charge can become a great hazard."[2] Even a small explosion generated by hydrogen accumulation could compress various parts of the core and cause a nuclear explosion that would not be containable. This is unlikely, perhaps, but possible. A nuclear reactor that grossly malfunctions only once, however, could cause a disaster of great magnitude. When viewed in this light, confidence in conventional failsafe devices fades quickly.

The AEC and Atomic Energy

Recognizing the threat posed by radiation from atomic bomb explosions during World War II, the United States Congress moved quickly after the war to set up a civilian agency, the Atomic Energy Commission (AEC) to handle both the problems and promise of atomic energy for the nation.

After a period of development, it was decided to open the way for private industry to make use of the new technology under AEC super-

[2] Teller, E., 1967. "Fast reactors: maybe." *Nuclear News.* 10 (8), p. 21.

vision. But if industry was to go into the nuclear reactor business it must own and operate its own reactors. To do this required insurance to protect industry from possible damage claims, standard in any business operation; but no insurance company would underwrite the risk. This seemed to be a coldly logical business response to a poor risk, implying that reactors were unsafe and therefore uninsurable. Here we encounter a basic contradiction inherent in the enabling legislation setting up the AEC. Congress authorized the AEC not only to set safety standards and regulate the uses of atomic energy but to encourage that use in every possible way. This is a bit like asking a fox to guard the chickens, as the first AEC Chairman, David Lilienthal, was aware. Since the contradiction has never been completely resolved, despite a recent executive decision to transfer the authority to set safety standards to a new Environmental Protection Agency, the AEC felt obliged to find a way to put nuclear technology into the mainstream of industry. Rather than taking the hint that the business world did not consider reactors safe, justifiably or not, and rather than delaying commercial use of reactors until private insurance was made available, the Joint Committee on Atomic Energy urged that the government assume the risk. This was done under the terms of the Price-Anderson Act, which guaranteed claims up to $500 million per accident. An insurance cartel was finally set up to cover the first $60 million of this, later increased to $74 million, but that is as far as private capital was willing to go.

It is impossible to judge and perhaps unfair to assume that without responsibility for indemnity, power companies are less concerned about reactor safeguards than they would be otherwise. Yet the internal conflict of interest in the AEC and the apparent private doubts of industry are disquieting.

NUCLEAR WAR: A PROSPECTUS

Most of us are used to living under the Damoclean sword of possible nuclear war. But every day that passes without the ultimate holocaust happening seems to reduce its threat.

Lest we become so numb to the consequences as to be willing to consider nuclear war as an instrument of foreign policy, a quick look at some of the environmental implications is in order. The major hazards of a nuclear explosion are the initial burst of radiation, thermal radiation or heat, shock wave or blast, and residual fallout. The first is usually masked by the second: if you are close enough to the ground target to receive massive radiation, you are more than likely to be burned to a cinder by the thermal pulse. At a distance where heat is tolerable, the blast radiation has been significantly reduced as well, for radiation varies as the inverse square of the distance ($R = 1/d^2$); as the distance, d, increases, the intensity of radiation, R, rapidly decreases.

Effects

The most immediate problem is fire, caused directly through the thermal pulse or secondarily by blast effects such as the short circuiting of power lines and rupturing of gas lines. As one moves away from the target, fire as a direct result of the thermal blast is less important because the suddenness of the pulse does not allow heavy materials to be uniformly heated. Instead they may char or briefly flame, then go out. Light materials, leaves, newspapers, light wood may be ignited and act as kindling, igniting heavier materials. In general anything that can be ignited by a match would probably ignite spontaneously in the thermal pulse. The important aspect of nuclear bomb fires is that many structures are ignited simultaneously. A similar effect was seen in World War II when Hamburg, Dresden, and Tokyo, among other cities, were firebombed. Incendiary bombs scattered all over a city ignited many fires at the same time resulting in incredibly swift destruction by fire. In 1842 a fire spreading from house to house took four days to destroy much of the old part of the city of Hamburg. In a fire raid in 1943, two-thirds of all buildings in a five-square mile area of Hamburg were consumed in twenty minutes. After a few hours the fire died from lack of fuel. Such intense burning causes a huge convectional cell to form over the city as heated air rises, replaced by cooler air rushing in from the suburbs at thirty to fifty miles an hour. The draft increases the intensity and spread of the fire, explaining the rapidity with which a large city can be reduced to ashes. Such a phenomenon is called a fire storm.

Properly timed and placed, the thermal blast of a nuclear bomb could destroy enormous areas of forest. There are seasons when, despite our best control efforts, huge tracts of tinder-dry forests are burned in spectacularly destructive crown fires. It would not be difficult to destroy not only a nation's cities but harvestable crops and timber as well by such means.

Having survived initial radiation, heat and shock waves, one would be faced with fallout. The fallout discussed in a preceding section was low-level because of the dilutant effect of being blown around the world a couple of times. The fallout within a few miles of the blast site is quite another story. In 1954, twenty-three crewmen of a Japanese fishing vessel ironically named the Lucky Dragon were showered with radioactive ash as an aftermath of the Bravo nuclear test on Bikini on 1 March 1954. Although they were over eighty miles from the target, they received between 250 and 330 r and as a result one man later died. Ralph Lapp, an American physicist who visited the crew and examined the ship later, calculated that about fifteen pounds of radioactive ash fell on the Lucky Dragon. Though only a very small portion of this weight was represented by the radioactive atoms, the radiation given off was thought to be equal to that given off by the entire world supply of radium.

Picking Up the Pieces

During the early 1950s at the height of the Cold War, when nuclear war seemed more imminent than it does today, there was much discussion of fallout shelters both public and private. But unlike conventional wars of the past, after the dust settles from a nuclear blast, the problems are just beginning. What would be gained by being protected from a burning house or a collapsing building only to emerge into a world contaminated for weeks with a background radiation level of hundreds of roentgens from local fallout? What kind of survival would be possible where there is no electricity, water, gas, heat, or food? How would you defend yourself in a traumatized world suddenly devoid of the stabilizing social effect of custom, rules, or laws? At one time people were advised to keep a gun in their shelter to protect whatever they had from the have-nots.

In such a context we cannot speak rationally of bomb shelters, protection from fallout, or law and order. Nuclear war is one kind of environmental pollution we absolutely cannot learn to live with. We can only take the greatest pains to avoid atomic war as we should have avoided air pollution, water pollution, and the rest in the past. But, unlike these, with nuclear war, there will be no second chance.

FURTHER READING

Calder, R., 1962. *Living with the atom.* University of Chicago Press, Chicago, Illinois. Easily read discussion of radioactivity in the environment.

Crow, J. F., 1957. *Effects of radiation and fallout.* Public Affairs Committee, New York. This twenty-eight page book treats the genetic effects of fallout on man.

Frye, A., 1962. *The hazards of atomic wastes.* Wash. Public Affairs Press, Washington, D.C. Discusses the origin and disposal problems associated with reactor and industrial wastes.

Hecht, S., 1954. *Explaining the atom.* Rev. ed., Viking Press, New York. Beautifully written account of atoms, atomic energy, atom bombs; easily accessible to the layman.

Lapp, R. E., 1958. *The voyage of the Lucky Dragon.* Harper and Row, New York. Well-written narrative of the fishing trawler showered with radioactive ash from a hydrogen bomb explosion.

Novick, S., 1969. *The careless atom.* Houghton Mifflin, Boston, Massachusetts. Exposé of the dangers inherent in nuclear reactors.

Schultz, V. and A. W. Klement, Jr. (eds.), 1963. *Radioecology.* Reinhold, New York. A good collection of papers dealing with the environmental effects of fallout.

Part III

MAN MAKES
NEW TRAUMAS

CHAPTER 12

BIOCIDES

If it were possible to spray a cotton field with some miraculous compound and kill all the boll weevils without harming any other organism in the community, we would have a pesticide, a material capable of selectively killing a pest. Unfortunately, no compound is known which has this selectivity; if the boll weevils are killed, so are many other species of insects, both destructive and beneficial—spiders, mites, and occasionally fish, amphibians, birds, and even mammals. So the term pesticide is somewhat general, even misleading; insecticide, while more accurate, gives no hint of potential injury beyond insects. The most general term which avoids these overlapping difficulties is biocide. If this has a grim ring to it, like genocide, perhaps it is fitting. For many people still feel that the earth would be better off without insects, mites, ticks, spiders, and all manner of "creepy-crawly" things. Herbicides or plant killers, although much less of a problem in the environment, are also not without their environmental effects; these will be discussed in Chapter 17.

Although odious in extreme usage, biocides *do* have a use and there is no intention here of insisting otherwise. We should, however, have some idea of the role that these materials play in the systems into which we introduce them; we should have a rational basis for determining at what point the cost of biocide use becomes prohibitive in economic and environmental terms. Since most of what we know about the environmental effects of biocides has come from twenty-five years of experience with DDT, many examples will deal with this biocide. But keep in mind that many environmental effects of DDT are potentially possible with *any* biologically active compound thoughtlessly introduced into the environment.

PANDORA'S PANACEA

Almost a hundred years ago, in 1874, a German Ph.D. candidate, Othmar Zeidlar, synthesized an organic compound for his dissertation, as generations of students have since done. Young Zeidlar published his work as a short note in a professional journal, took his degree, and dropped from sight.

The compound Zeidlar synthesized was not unusual, even for 1874; a series of substitutions had transformed the original ethane (C_2H_6) into a *di*chloro-*di*phenyl-*tri*chloro-ethane (Figure 12.1). So dichlorodiphenyl-trichloroethane (DDT) sat on the shelf with a thousand other organic chemicals synthesized over the years until the late 1930s, when Paul Muller, a Swiss entomologist looking at the insect killing properties of various compounds, found that Zeidlar's material was an extremely effective insect killer. DDT was "discovered."

In 1942, some DDT was brought to the United States, where the military quickly appreciated its potential usefulness. As a result of DDT, World War II was the first war where more men died of battle wounds than of typhus and other communicable diseases spread by insects. After the war, DDT was most effectively used to combat a number of insects which were carriers of thirty diseases, including malaria, yellow fever, and plague; within ten years of its first widespread use, DDT had saved at least five million lives. Millions of houses and tens of millions of people were sprayed with DDT in campaigns against fleas, houseflies, and mosquitoes and the diseases these carriers transmit.

Clearly a new age of chemical control of pests was underway. DDT was not the first chemical used against insects, of course. Paris green (copper acetoarsenite), first used in 1867 against the potato beetle, holds that distinction. But DDT represented the first time a totally new synthetic material was introduced into the environment on a large scale.

Figure 12.1 In a series of reactions substituting three atoms of chlorine for three of hydrogen, then two chlorinated rings for two more hydrogens, dichlorodiphenyl-trichloroethane (DDT) can be synthesized from the simple compound ethane.

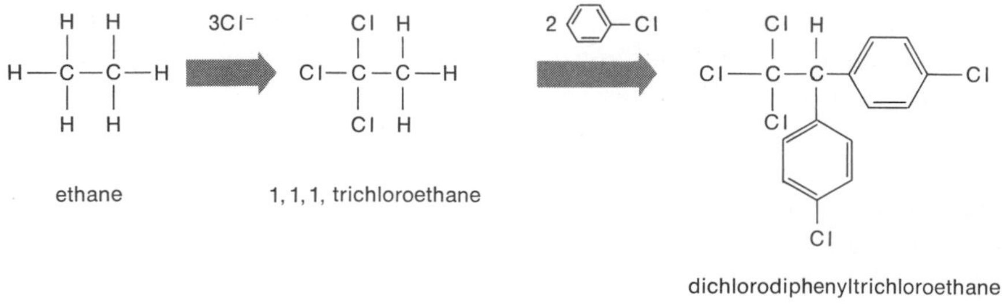

ethane   1, 1, 1, trichloroethane

dichlorodiphenyltrichloroethane

BIOCIDE SPECTRUM

Spurred by the success of DDT, the chemical industry explored the biocide possibilities of other biologically active chemicals.

Today there are six main groups of biocides available to control a
broad variety of animals: arsenicals, botanicals, organophosphates, carba-
mates, organochlorines, and rodenticides. But the two most important and
dangerous from the point of view of environmental contamination are
the organophosphates and organochlorines.

Organophosphates. Among the first organophosphates to be marketed were parathion and malathion, which are still in use today. Their
mode of action on terrestrial organisms is quite specific.

Nerve impulses are conducted across the gap between adjoining
nerve fibers by a compound called *acetylcholine*. As soon as the impulse
has bridged the gap an enzyme, cholinesterase, destroys the acetylcholine
present, preventing further impulse transmission. Organophosphates de-
activate the cholinesterase, allowing a stream of impulses to flow unin-
terruptedly along the nervous system, resulting in spastic uncoordination,
convulsions, paralysis and death in short order. While quite toxic, organophosphates are not persistent and have much less effect on aquatic
than terrestrial organisms.

Organochlorines. Also called chlorinated hydrocarbons, organochlorines include the best known of all the synthetic poisons: endrin,
heptachlor, aldrin, toxaphene, dieldrin, lindane, DDT, chlordane, and
methoxychlor. Unfortunately, there is no clear understanding of just how
the organochlorines work. Apparently the central nervous system is affected, for typical symptoms of acute poisoning are tremors and convulsions. Chronic levels have various effects. In aquatic organisms, which are
especially sensitive to this class of compounds, the uptake of oxygen
through the gills is disrupted and death is associated with suffocation
rather than nervous disorder.

While both classes of biocides were developed to their fullest extent
since World War II and seem to have come upon us rather suddenly,
biocides require a fairly long development time.

Development of Biocides

To develop a commercial biocide, chemical companies must invest a great
deal of time, research, and money. First, a large variety of materials must
be screened for biological activity; toxicity is not always obvious from
a structural formula. Then tests in the laboratory on a wide spectrum

of pests determine toxicity levels of likely compounds. Those that are strongly toxic to one or more pests are tested for the effect of multiple doses and long-term effects, up to a month (a considerable part of the life cycle of many insects). Small plot studies by cooperating state agricultural research stations then work out problems of formulation and field dosage. Next the cooperation of farmers is enlisted for large scale trials on a number of crop plants in different soils, climates, and regions. All of the data from these preliminary steps are then presented to the proper government regulatory agency. If approved, the compound is registered for use and marketed. Long-term toxicity tests continue as the product is used commercially. Ultimately, under this system, the consumer is the final test of any unforeseen hazard. Since the cost of marketing a biocide may run to several million dollars, none of which is returned until the product is sold, chemical companies are understandably anxious after years of development to recover their investment by promoting the use of their product. Fortunately with most large chemical companies, biocides are usually a small part of their total business, which mitigates some of the pressure to rush a product onto the market prematurely.

But despite testing by industry and regulation by government, biocides have caused problems, some resulting from the very characteristics that made them useful pest killers. These problems can be conveniently grouped into four categories: biocides may elicit resistance in the target organism; they may accumulate in food chains; they may persist in the environment; and they are nonselective. Let's look at each of these problem areas in some detail.

Resistance

For a short while, DDT seemed about to push the mosquito and housefly into extinction. But by 1947, trouble developed. Italian researchers reported that the housefly was becoming resistant to DDT; corroboration soon followed from California. Soon other insects became resistant too: the malaria-carrying mosquito, then lice, fleas, and a number of other disease vectors that had formerly been controlled by DDT. Twelve insects were found to be resistant by 1948, 25 in 1954, 76 in 1957, 137 in 1960, and 165 in 1967.

Ironically, resistant pests have appeared everywhere man has sprayed, simply because the development of resistance depends upon killing as many pests as can be exposed to the selected poison. Until recently, this has been the goal of pest control.

When an insect population is exposed to a new biocide, up to 99 percent of the population is eliminated. The remaining 1 percent is by random coincidence immune or resistant to the chemical. A 99 percent wipeout sounds almost as good as extinction. But the few resistant individuals are now able to reproduce almost without competition, a result of both the elimination of their peers and their predators. Breeding at a rate of ten generations a season, a resistant housefly population could become the most abundant strain in about five years.

Simple resistance to biocides is bad enough for the farmer, but in some instances even more serious problems were elicited. After several years' exposure to cyclodiene, cabbage maggot strains were selected that were not only resistant to the biocide, but lived twice as long and produced twice as many eggs as nonresistant forms.

A classic example of unanticipated selection for pest resistance is seen in the lygus bug on California cotton. The lygus bug sucks plant juices from buds, blossoms, and young cotton bolls. Each one of these stages that fails to develop means one less cotton boll. In 1967, a large chemical company marketed azodrin, a new organophosphate; after much advertising, over a million acres were sprayed in California alone. While the new biocide was not persistent, it was toxic to a broad spectrum of insects and eliminated almost all insects in the sprayed fields. Its lack of persistence allowed migration of lygus bugs into the sprayed fields where their natural enemies had been eliminated (predators are always much less numerous than their prey, and therefore recover much more slowly from biocide applications). Then the lygus bug population increased and another spraying was required—then another and another. After a few seasons, it can be expected that the lygus bug will be resistant to azodrin.

The killing of beneficial insects along with the lygus bug increased the population of, and damage from, the boll worm, with the result that application of azodrin may well have decreased the yield of cotton through the increased activities of the boll worm. In addition, studies on the cotton plants showed that many buds do not form bolls and all bolls do not necessarily ripen even under the best of conditions. In fact, many buds and bolls that were attacked by lygus bugs would not have ripened anyway. In an experiment set up by University of California entomologists, the yield of cotton was not significantly different between sprayed and unsprayed plots. Hence it was quite probable that the lygus bug was not even a pest!

This type of situation can easily be repeated, because the grower, often an absentee landholder or corporation, wants someone to sample his fields occasionally, advise on the presence of potentially dangerous pests, and suggest some control. Very often the "someone" is a field salesman for a chemical company. In the cotton growing areas of California, there are a handful of extension entomologists and over 200 salesmen representing 100 companies. Should anyone be surprised that fields are sprayed unnecessarily?

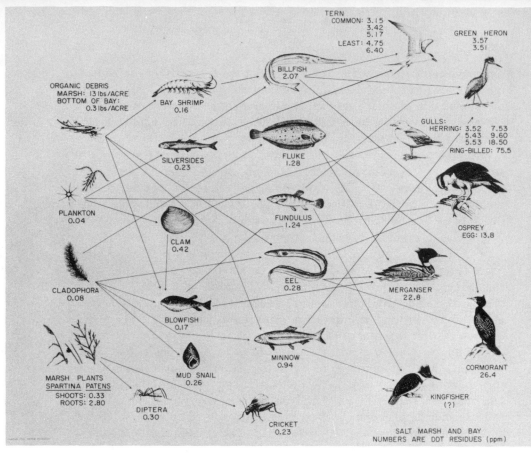

Figure 12.2 Estuarine food chains have been contaminated with DDT. The primary producers—phytoplankton, algae, and marsh plants—absorb DDT from mud or the water and store it in their cellular fat bodies. By passing this DDT on to the consumer, the level may gradually accumulate to large doses. (Brookhaven National Laboratory)

Accumulation

Although DDT is only slightly soluble in water, it is quite soluble in fat. Therefore algae accumulate DDT in their cellular fat bodies on the order of parts per million. An organism that eats algae may accumulate DDT in tens of parts per million. This accumulation and concentration of biocides in a food chain is called *biological magnification*. By means of this process, an extremely small quantity of a persistent biocide in water, in equilibrium with a much larger amount in the bottom mud, has unexpected effects in an ecosystem—far beyond the original intent of those who first introduced these materials into the environment. For example, bottom mud in Green Bay, Wisconsin contained 0.014 ppm DDT. But small crustaceans in the same environment had accumulated 0.41 ppm, fish 3–6 ppm, and herring gulls at the top of the food chain, 99 ppm— enough to interfere with their reproduction (Figure 12.2).

Of course, biological magnification is not limited to aquatic organisms. One early example of this process developed from the attempt to control Dutch elm disease with DDT. The Dutch elm disease is caused by a fungus that plugs the water conducting vessels of the elm, leading to the death of the tree (Figure 12.3a, b, c). Since the fungus is spread by two species of bark beetles, attempts to control the disease have

Figure 12.3 (a) At one time the streets of most northeastern cities and towns were lined with these airy and graceful American elms; (b) but the Dutch elm disease slowly began to kill American elms. (c) Today elms are no longer found in most towns and the stark remains haunt hedgerows and country lanes.

focused, quite naturally, on the insects. The program of spraying elms with DDT began in 1947 and until recently was still being practiced by some communities. While DDT effectively kills the bark beetles and slows the spread of the disease, it seemed in some instances to kill birds too, especially robins. The robin population on the campus of Michigan State University in East Lansing, Michigan, was reported to have dropped from 370 to 4 birds in four years, and almost no nests produced young during these years. In another instance the robin population of Hanover, New Hampshire, a town which regularly sprayed its elms with DDT, fell considerably below that of a nearby town which did not spray. In both Michigan and New Hampshire, large numbers of dead birds were found to contain DDT in excess of the 30 ppm that has been observed to be lethal in robins.

In both instances, spraying seems to have been rather exuberantly carried out, applying DDT in doses much higher than necessary, thus killing more robins than usual. If DDT is applied to trees either in leaf or during dormancy, it drips to the ground beneath the trees and is accumulated by earthworms. The soil beneath a sprayed elm tree may contain 5–10 ppm DDT; an earthworm that ingests this soil for its organic content may contain 30–160 ppm. A robin feeding almost exclusively on earthworms during the early spring, when elms are usually sprayed, might easily receive a lethal dose of DDT through this chain. Many apparently have. In this way a food chain can be poisoned and increasing levels of a noxious substance passed upward through the links.

Birds of prey and DDT. The most serious and best documented example of biological magnification can be seen in the reproductive failure of certain birds of prey, especially the peregrine falcon, bald eagle, and osprey, which have undergone disastrous population decreases since 1945 (Figure 12.4).

The peregrine falcon has completely disappeared from the eastern United States, the bald eagle has become rare, and the osprey uncommon. Charles Broley, who studied the bald eagle in Florida for a number of years, watched the population of eagles in one area fall from 125 nests producing about 150 birds a year in 1940 to 43 nests producing 8 young in 1957. In 1958 only 10 nests and 1 young were found. Similar decreases have been reported in osprey populations in Maine and Connecticut. In all three species eggs fail to hatch or are easily broken and eaten by the brooding bird. DDT concentrations in both the eggs and adult birds were found to be high. Falcons from an area as remote as the Northwest Territories of Canada were reported to contain 369 ppm of DDT in their fat.

How is DDT implicated in the decline of these birds? The high levels of DDT found in the fat of many birds of prey results from the same

Figure 12.4 (a) The peregrine falcon has almost completely disappeared from the United States because DDT has interfered with its reproductive cycle. (b) Ironically, the national symbol, the bald eagle, is also in grave danger. (c) Even the once abundant osprey or fish hawk is rarely seen as persistent chlorinated hydrocarbons are concentrated in the food supply of each of these species. (Photographs from U.S. Dept. of the Interior, Sport Fisheries and Wildlife)

concentration process seen in the earthworm-robin chain. Birds of prey are carnivores; they eat other birds and animals which may have substantial quantities of DDT in *their* fat picked up from *their* habitats. In some birds DDT has been found to inhibit the production of an enzyme, carbonic anhydrase, that controls calcium metabolism; it is, of course, the calcium deposited around the egg which produces a strong shell. This metabolic disturbance results in eggs with much thinner and weaker shells than normal, eggs that are easily broken (Figure 12.5). This was suggested by a study which compared the weight of bird eggs in museum collections before and after the introduction of DDT in 1947. Those birds with declining populations had substantially lighter (hence, thinner) shells after 1947 than those birds with stable populations (Table 1). It should be pointed out that the declining species tend to feed on species which themselves are removed by several steps from the primary consumer. The birds of prey with stable populations usually feed on the primary consumer directly. Hence they are exposed to much lower amounts of DDT.

Table 1 **Relationship of eggshell thickness to population decline of certain birds of prey**[a]

Populations	Weight of eggshell (% change)
Declining	
Bald eagle	−18
Osprey	−25.1
Peregrine falcon	−18.8
Stable	
Red tailed hawk	+ 2.7
Golden eagle	+ 2.9
Great horned owl	+ 2.4

[a] Hickey, J. J. and D. W. Anderson, 1968. "Chlorinated hydrocarbons and eggshell changes in raptorial and fish-eating birds." *Science, 162,* pp. 271–273.

Further evidence pointing to a DDT interference in eggshell characteristics was shown in a study where sparrow hawks, which were fed a mixture of 3 ppm dieldrin and 15 ppm DDT, laid eggs 8–10 percent thinner than control birds. The offspring of these birds laid eggs 15–17 percent thinner than the controls. Clearly DDT interferes with calcium metabolism.

Birds tend to accumulate DDT because they cannot excrete liquids as rapidly as mammals. Some mammals also accumulate DDT, however. A study of mice and shrews in a spruce-fir forest in Maine after one pound of DDT per acre was applied to control an outbreak of spruce budworm

Figure 12.5 DDT interferes with calcium metabolism in many birds, resulting in delicate eggshells which break easily, as in this pelican egg. (Photo by W. Gordon Menzie)

showed that shrews had 10–40 times the amount of DDT residue of mice. This was due to the carnivorous diet of shrews and led to a maximum of 41 ppm the year the forest was sprayed. After nine years, while the DDT residues of mice seemed to be approaching normal, the level in shrews remained unusually high, 2–6 ppm. Even though the level probably had little effect on the shrews themselves, what about the predators feeding on shrews? Reproductive failure has been noted in ospreys feeding on organisms with an equivalent amount of DDT residue to that found in the shrews. One can only wonder what DDT load foxes and bobcats are carrying around in forests, years after the initial introduction of DDT in their environment.

Contamination of the Ocean

It is not surprising that local ecosystems that have been sprayed with various persistent biocides pass these materials on in ever higher concentration through food chains. It is something of a shock, however, to learn that Adelie penguins on the Ross Ice Shelf of Antarctica had traces of DDT in their body fat. Although the amounts were small, 13–115 parts per billion (ppb) in liver tissue and 24–152 ppb in fat, the fact that there was *any* DDT at all, thousands of miles from the nearest possible point of contamination, was alarming. At first it was thought that the presence of several thousand scientific and support personnel nearby might be contaminating the environment in some way through discarded material, clothing, garbage, or fecal wastes, for DDT was not used for any purpose in Antarctica. But subsequent investigation in still more remote areas of

Antarctica indicated that local penguins and cormorants contained 0.001–0.48 ppm DDE (a degradation product of DDT) and 0.011–0.140 ppm DDE respectively. The wide-ranging skua (a gull-like sea bird) contained 0.89–26.0 ppm DDE. Furthermore, pelagic or open ocean birds like the albatross, which touch land only to breed, were found to contain traces of various biocides.

Even more alarming than food chain contamination has been the report that concentrations of DDT as low as a few parts per billion have reduced the rate of photosynthesis in four species of marine algae grown under laboratory conditions. While the present level of DDT in the world ocean is in the parts per trillion range, should it ever increase to parts per billion, photosynthesis by the oceanic algae might be curtailed. Considering that 90 percent of the earth's photosynthetic organisms which produce oxygen are found in the ocean, the oxygen content of the air could conceivably be affected with both predictable and unpredictable results.

The conclusion was unmistakably clear: contamination of food chains was no longer a local affair in direct relation to biocide application at some specific point; the oceanic food chain, remote from land-based biocide usage, was contaminated. How could the enormous world ocean, stretching thousands of miles around the earth, become contaminated, whatever the scale of man's activities? Some hint might be taken from the Bravo series of H-bomb explosions in the Pacific (see Chapter 11), which resulted in the catching of radioactive tuna in Japanese waters thousands of miles from the blast site. Tuna, the last link (before man) in their particular food chain, accumulated fallout to a dangerous degree. Likewise, DDT has collected in atmospheric dust hundreds and thousands of miles from dusting or spraying activities. When washed out of the air by rain, the DDT can enter any ecosystem anywhere in the world. Since oceans occupy 75 percent of the earth's surface, contamination was just a matter of time.

Biocides in Man

If animals can accumulate and concentrate biocides from their food, one wonders how much DDT man has accumulated from his food chain and if there is any evidence of potential harm. Man *has*, along with just about every other organism, accumulated DDT in his body fat. A sample of several hundred people in the United States averaged 8–10 ppm DDT in body fat. Whether this is or will be harmful is the subject of a sharp and continuing debate. Until 1965 there were no medically documented cases of sickness or death in man that could be traced to *proper* use of biocides. Of course any number of children and agricultural workers have died through *careless* handling of biocides, for often people do not read labels or follow instructions. Careless handling of virtually any substance can be disastrous at times, biocides not excepted. In 1962, for example, seven infants in a hospital nursery died when salt was accidentally added to their formula instead of sugar. If, then, there is no evidence that disease or death in humans has resulted *directly* from the proper use of biocides, what are the long-term prospects?

Delayed Effects?

Apparently no one need fear inordinate residues of DDT on fruits and vegetables, for the Food and Drug Administration has set up maximum allowable levels of biocides on food products and keeps careful watch on food shipped through interstate channels. The fear of dying from one poisoned apple is as unreal as the Snow White fairy tale. But, the fact that we are all carrying a supply of DDT, regularly replenished from our environment, leads us to wonder about long-term problems caused by a body burden of DDT. There are three logical categories of DDT effects: no effects; effects that are too subtle to be correlated with DDT levels; and effects that will appear in the future. One recalls with some uneasiness the case involving a group of women who worked in a watch factory in the 1920s painting the dials with a luminescent paste containing radium. At the time, radium was regarded as harmless, so the women habitually pointed the tips of their brushes by wetting them with their tongues. Thirty years later, many of these women began dying of mouth and tongue cancer. Today we can be appalled at the naïveté of allowing people to "eat" radium, but at the time there were no known adverse effects.

In tests where men were fed 3.5 and 35 mg of DDT a day for a period of time, no ill effects were seen. DDT accumulated in proportion to the amount consumed, then reached an equilibrium point between storage and excretion. When the dose was discontinued the body burden began to fall. Apparently the body stores DDT in response to intake rather than independent accumulation, so a person in a high DDT environment will store more DDT than a person in an environment with lower DDT levels. Although no one has died directly because of DDT properly used, many claims have been made and are being made that associate high DDT body burdens with various organ failures and malfunctions. But whether DDT directly or indirectly causes these problems or is just an innocent bystander has not been clearly demonstrated.

As long as there is any doubt about the role of DDT in human metabolism there is risk—an unnecessary risk, for we are not obliged to use DDT or any biocide. It is this doubt that led to the banning of DDT in several foreign countries and in many states of the United States, and will probably force restrictions on the use of other persistent chlorinated hydrocarbons as well. While there has been much obfuscation on both sides of the persistent chlorinated hydrocarbon controversy, anyone who presents an either-or choice, elms or robins, is mistaken because, as we shall see later, we can have both.

PERSISTENCE

We tend to regard the soil as some kind of biological incinerator; whatever we dump is supposed to be decomposed. But some of the organic molecules like DDT are unknown in a natural environment and enzymes that can degrade them are simply not available. The half-life (the time required for half the original quantity of a substance in the environment to decompose) of DDT is around 15 years, toxaphene 11, aldrin 9, dieldrin 7, chlordane 6, heptachlor 2–4, and lindane 2. These are maximum values, but they give some idea of the time it takes for the environment to cope with these materials. Decomposition, fortunately, does not depend solely on microbial attack, or we might never get rid of some persistent organochlorines. Volatilization into the air, photodecomposition by various wavelengths of solar energy, mechanical removal in crops, and leaching are also involved in reducing levels of persistent biocides; but volatilization and microbial degradation are the most important.

The danger of persistence is best seen in the organic biocide with the longest half-life, DDT. Though only slightly soluble in water (ppb), DDT is readily absorbed by the bottom mud of aquatic systems. As DDT in the water is degraded or removed by organisms, fresh supplies from the bottom mud are released. Because of its relatively long half-life, DDT can be released continuously into an aquatic system for years.

Residues

Persistence of biocides is also found in terrestrial environments. When spray schedules are maintained over a long period of time, residue levels build up. The soil beneath some orchards in Oregon was estimated to have retained over 40 percent of the DDT applied over a seventeen-year period. Some well-sprayed orchards accumulate 30–40 pounds of DDT per acre per year; others have totaled, despite degradation processes, as much as 113 pounds per acre after six years of spraying. Persistence is not limited to the newer organochlorines though; arsenic trioxide has accumulated in some parts of the Pacific Northwest up to 1400 pounds per acre—a concentration that often poisons the crops themselves.

Perhaps the most difficult aspect of persistent biocides is the impossibility of their containment. About 15 percent of cultivated land in the United States receives biocides; 3 percent of grasslands and less than 0.3 percent of forests are treated each year. Altogether, 75 percent of the total United States land area remains untreated, and yet virtually all animals, man included, carry traces of DDT and other persistent biocides.

NONSELECTIVITY

Not all of the problems caused by biocides in the environment derive from resistance, biological magnification, or persistence. Killing the wrong organism at the wrong time can be just as great a problem. One example having unexpected effects occurred a few years ago in the Near East. Jackals had been a problem for years in the settlements along the Mediterranean coast of Israel. Attracted by garbage, they stayed on to eat crops and increased greatly in number. In 1965 the government set out bait poisoned with "1080" (sodium fluoroacetate). The bait was widely distributed and no attempt was made to avoid poisoning other animals. As a result of this program, the jackals were eliminated but so were the mongoose, the wild cat, and the fox. With their predators eliminated, hares increased enormously, causing greater damage to crops than the jackals had before. Moreover, Palestine vipers, usually controlled by the mongoose, appeared in large numbers around settlements to the consternation of the inhabitants.

Intensively cultivated land is even more sensitive to the disruptive effects of nonselective biocides. Rice fields which are flooded during the growing season in Louisiana also produce red crayfish, a local delicacy which nets the rice farmer an addition to his income beyond the income from the rice crop. However, to control the rice stinkbug, toxaphene often mixed with DDT or dieldrin was used. Because of an extreme sensitivity to toxaphene, fish were sometimes killed in large numbers. Gallinules and tree ducks were also killed by eating sprayed rice. When the switch was made to malathion and parathion to avoid damage to fish and wildlife, it was found that crayfish were quite sensitive to parathion and substantial crayfish kills took place before a combination of chemicals could be worked out that endangered neither fish, birds, nor crayfish and still protected the rice.

Natural ecosystems can also be strongly affected by the application of broad spectrum biocides which kill insects indiscriminately. The red spider mite normally is kept under good control by a variety of organisms, especially a predatory mite. When spruce forests along the Yellowstone River in Montana were sprayed to control the Englemann spruce beetle, the spider mites, which lived in weblike structures on the undersides of the needles, were protected from the spray, but the predatory mites were killed. The following year, although the Englemann spruce beetle was controlled, there was a huge wave of spider mite damage that was worse than that of the spruce beetle.

Finally, there are examples in the recent past where whole regions, not just fields or forests, were sprayed unnecessarily and indiscriminately. During the early enthusiasm over the innate superiority of chemicals to all other approaches, the concept of mass control arose. By means of various biocides the fire ant was to be eliminated and the Japanese beetle contained.

Figure 12.6 Fire ant mounds in an Alabama pasture seem to be of little concern to the livestock. (U.S. Dept. of Agriculture)

The fire ant.   The fire ant is a minor pest introduced from Argentina into the southeastern United States. It builds mounds one to two feet high (Figure 12.6), inflicts painful stings, and occasionally damages crops and livestock. The USDA with its technological muscle came rushing to the rescue and began spraying large areas of the Southeast with dieldrin or heptachlor at the rate of two pounds per acre. While this was reduced in later years, over 2.5 million acres of land were treated by air in this program. Although the fire ant did not destroy livestock, wildlife, or crops to any important extent, the biocides in some instances did. After $15 million was spent on mass control, the Southeast still had to live with the fire ant—plus heptachlor residues in the environment. Here was one operation whose failure was attested to by the continued life of the pest. Finally it became clear that indiscriminate aerial spraying of millions of acres was not the best way to go about fire ant control. In 1962 a new chemical, mirex, applied on bait near problem ant hills when necessary, began to yield the type of control that should have been used from the first.

The Japanese beetle.   Another example of over enthusiastic mass control was seen in the Japanese beetle eradication program. In the first few decades after its introduction in the United States from Japan, this pest was widely distributed in the Northeast and ate everything in its path. But with the introduction of biological controls (see Chapter 15), particularly a bacterium (milky disease) that attacks the beetle grub, populations have stabilized so that the dense clusters that covered grape

Figure 12.7 During the 1940s the Japanese beetle ranged unchecked. Whole trees
were defoliated by swarms of beetles. (U.S. Dept. of Agriculture)
Figure 12.8 Widespread distribution of a bacterium causing the death of the beetle
grubs has not eradicated the Japanese beetle, but has reduced its number to a tolerable level.

vines, rose bushes, and most garden plants during the 1940s (Figure 12.7)
are no longer a severe problem. Beetles are occasionally seen during the
summer (Figure 12.8) but are nowhere as abundant as formerly.

At first, however, the response by government agencies to the con-
tinual westward spread of the Japanese beetle was a compulsive drive
to eradicate the beetle in the new areas colonized. Consequently, close
to 100 thousand acres in Missouri, Kentucky, Illinois, Indiana, Iowa, and
Michigan were broadly treated with two to three pounds of aldrin, dieldrin, or heptachlor per acre. As might be expected, chlorinated hydrocarbons at this dose rate were dangerous to vertebrates as well as beetles, not
only in initial contact but through stable, persistent residues. A recent
attempt in Michigan by an environmentally oriented pressure group to
prevent mass control techniques from being used against a small Japanese
beetle population was not successful, but the area to be sprayed and the
dose rate were decreased.

The aim in both instances, to eliminate or control potentially damaging pests, was certainly in the public interest, but the technique of mass
indiscriminate aerial spraying was not the best solution. Instead of thinking, "If one aspirin relieves a headache in thirty minutes, thirty aspirin
should relieve it in one minute," we should be thinking, "If aspirin cures
the headache but causes an upset stomach, maybe we ought to try
another remedy."

But the most ominous of all the problems generated by biocides has
been the degree to which farmers have become overdependent on them,
perhaps seen at its most extreme in orchards across the country.

Spray Schedules

A hundred years ago orchards were small, containing perhaps a few dozen or at most a few hundred trees. There were enough adjoining fields and forest to allow a considerable amount of natural control. But as the orchards became larger, introduced pests could easily move from orchard to orchard, and because of the huge concentrations of trees they could inflict considerable damage. With the advent of organic sprays of low selectivity, whatever natural controls existed were reduced or eliminated, letting more pests survive and requiring more sprays. Today a typical commercial apple orchard requires a minimum of nine sprays of various biocides (Figure 12.9):

1. In early spring when the trees are still dormant a spray is applied to eliminate aphid eggs that have overwintered.
2. When the fruit buds begin to show green, trees are sprayed to control mildew, scab, mites, and red bug.
3. As flower buds begin to turn pink, a spray is needed every seven days to attack apple scab. There is no spraying during full flower to allow for the pollinating activities of bees, without which there would be no apples.
4. After petals fall, a spray is applied for codling moth, curculio, red-banded leaf roller, cankerworms, mites, mildew, and scab.
5. A cover spray is added a week after petal fall to get codling moth, leaf rollers, cankerworms, curculio, and scale.

Figure 12.9 These peach trees were being sprayed with parathion at a time (1960) when the potential ill effects of broad spectrum biocides were unrecognized. The tractor driver, without protection, was doubtless being exposed to far more parathion than the peach trees. (U.S. Dept. of Agriculture)

6. After ten more days a second cover spray is added to combat cur-
    culio, codling moth, leaf roller, fruit spot, and scab.
7. In ten to twelve days a third cover spray is added.
8. A fourth cover spray is added twelve to fourteen days later, to
    catch codling moth, sooty blotch, scab, and fruit spot.
9. Two weeks later, a fifth cover spray is added to kill apple maggot,
    leaf roller, fruit spot, sooty blotch, scab, and codling moth.

By mid-July, a second batch of codling moths is likely; then, of course,
there is black rot, white rot, fire blight, and so on endlessly.

The farmer is caught squarely in the middle. If he stops spraying he
is left with either no crop at all or gnarled, wormy crabapples that
no one will buy. If he continues spraying, new pests will appear, old ones
will become resistant, and he will sooner or later be forced out of business, unable to afford spraying twenty-four hours a day.

A similar dilemma was faced by farmers in the Cañete valley in Peru.
Modern organic-synthetic biocides were introduced between 1949 and
1956 to control seven major pests on the chief crop, cotton. At first the
yield of cotton increased, from 406 pounds per acre to 649 pounds per
acre. But by 1965 the yield had dropped to 296 pounds per acre and
major pests increased from 7 to 13, several of them highly resistant to the
biocides. Worst of all, the lowered yield of cotton plus the increasing
frequency and expense of biocide application was driving the farmers
into bankruptcy.

Integrated Control

The solution in Peru was a break with total dependence upon organic,
synthetic biocides and adaptation of a more sensible blend of control
measures called *integrated control*. The heart of this program is decreased
use of biocides and the toleration of a few pests in return for freedom
from biocide-resistant hordes. There are alternatives to the exclusive use
of biocides: development of resistant crops; biological control (see
Chapter 15); and cultural methods like destruction of crop refuse, deep
plowing, timed planting, and rotation of crops, which collectively can
eliminate overwintering stages or upset the often delicate timing that
synchronizes pest and crop.

Knowing the problem and trying to work with it can be more successful in the long run than a blind attack with a broad spectrum biocide. The
California spotted alfalfa aphid becomes a problem when large tracts
of alfalfa are cut. By harvesting the fields in strips, refuges are left in
the unmown strips for the natural enemies of the aphid—fifty-six predators and parasites per square foot in the specially mown fields, versus
fourteen per square foot in fields harvested normally.

The California grape leafhopper developed resistance to the organo-phosphates ordinarily used for control. Workers noted that the wasp *Anagrus* actively attacked the leafhopper but only during the latter part of the growing season, which left the young grape leaves open to leaf-hopper attack early in the season. When wasps were absent from the vine-yards they were attacking another leafhopper on blackberry bushes some distance away. Blackberry bushes had grown in the vineyards, but were regarded as weeds and killed with herbicides. When blackberries were reintroduced to vineyards, the predatory wasp had an alternate prey within easy reach which helped to maintain the wasp population at a high enough level to control effectively the grape leafhopper.

But pest control need not always involve killing the pest. In many parts of the West, reforestation of cut or burned areas by aerial seeding is ineffective because of extensive populations of white footed mice which devour the seeds practically as soon as they hit the ground. If poisoned, new mice from surrounding areas quickly repopulate the area and eat the remainder of the tree seeds. One ingenious approach pioneered by the mammalogist Lloyd Tevis was to condition the mice to avoid all tree seeds by poisoning some with sublethal doses. Experiments indicated that one sickened mouse was worth a dozen dead ones, for once re-covered, the mouse not only avoided all tree seed, poisoned or not, but taught its offspring to do likewise. Unfortunately the variations in individual mouse behavior prevented widescale use of this approach in the situation we have described, but the idea is one that might be studied further.

Integrated control, however, is not just alternatives to organic chemi-cals but moderate use of all controls, including chemicals. A rational, bal-anced approach that considers ecological factors rather than ignoring them might use the guidelines that follow.

Before any control is attempted it must be determined that the sup-posed pest is actually causing the damage attributed to it. Millions of dollars worth of spray chemicals have been wasted on false assumptions. Then a survey should be made of the natural control agents that normally work to limit the population of the pest. If the pest is introduced and has no apparent enemies, locate the predators, parasites, and pathogens from its place of origin (no organism is without these associates!), screen them carefully to avoid making the situation worse by adding more pests, and attempt controlled introduction. Make every effort to preserve and increase the number of natural control organisms, native or introduced. Such biological controls have been much more effective than chemicals in many instances for there is no possibility of the pest developing resis-tance. Do not panic at the first appearance of a pest. Give the natural controls some time to operate. If chemicals prove necessary to achieve control, use the least amount that will be effective, instead of trying to kill every pest in the field. Then spray only when necessary, avoiding routine spray schedules. Remember that overkill and routine spraying quickly lead to pest resistance and elimination of natural controls, requiring still more sprays, which is what we are trying to avoid. Wherever possible, avoid broad spectrum biocides particularly if they are persistent. Persistence tends over the long period to eliminate more beneficial organisms than pests.

If these guidelines are followed, most of the disadvantages of biocides can be avoided. In fact, many of these materials can take their place as useful tools in the continuing war against the great variety of potential pests that man has inadvertently been nurturing through his ecosystem disruption and simplification. If mass indiscriminate spraying continues, problems seen by the late Rachel Carson will seem like bedtime stories compared with the grim realities ahead.

FURTHER READING

Carson, R., 1962. *Silent spring.* Houghton Mifflin, Boston, Massachusetts. The pioneering book on the environmental effects of biocides.

George, J. L., 1958. *The program to eradicate the imported fire ant.* The Conservation Foundation, New York. A well-documented account of the side effects generated by the attempt to eradicate the fire ant.

Graham, F., Jr., 1970. *Since silent spring.* Houghton Mifflin, Boston, Massachusetts. A sequel to Carson's book, updating many of the incidents reported in the earlier book.

Rudd, R. L., 1964. *Pesticides and the living landscape.* University of Wisconsin Press, Madison, Wisconsin. A systematic survey of the various types of chemical pest controls and their environmental effects.

Symposium on Pesticides in the Environment and their Effects on Wildlife, 1965. *Pesticides in the environment and their effects on wildlife.* Blackwell, Oxford, England. A supplement to Vol. 3 of the Journal of Applied Ecology. A wide-ranging collection of papers.

Tatton, J. O'G. and J. H. A. Ruzicka, 1967. "Organochlorine pesticides in Antarctica." *Nature,* **215,** pp. 346–348. Evidence of the wide dispersal of biocides into environments remote from the site of their application.

Woodwell, G. M., C. F. Wurster, Jr., and D. A. Isaacson, 1967. "DDT residues in an east coast estuary: a case of biological concentration of a persistent insecticide." *Science,* **156,** pp. 821–824. Poisoning of a salt marsh food chain by DDT.

CHAPTER 13

INORGANIC POLLUTANTS

Biocides are *assumed* to increase the quantity and quality of crops by suppressing insect attack, but many poisonous chemicals lacking even this redeeming feature are released into balanced ecosystems—some intentionally, some unwittingly. Because our bodies are made of organic matter, we are especially conscious of the effects of organic chemicals on the complex biochemistry of life. But inorganic chemicals often escape the attention they deserve; it is usually long after their use has been initiated that their danger to man's health is recognized.

Of the many inorganic substances that interact negatively with physiological systems, mercury, lead, beryllium, and asbestos are potentially the most dangerous because they are stored in the body and have a cumulative effect.

MERCURY

> The Hatter was the first to break the silence. "What day of the month is it?" he said, turning to Alice: he had taken his watch out of his pocket, and was looking at it uneasily, shaking it every now and then, and holding it to his ear.
>
> Alice considered a little, and then said "The fourth."
>
> "Two days wrong!" sighed the Hatter. "I told you butter wouldn't suit the works!" he added, looking angrily at the March Hare.
>
> "It was the *best* butter," the March Hare meekly replied.
>
> "Yes, but some crumbs must have got in as well," the Hatter grumbled "you shouldn't have put it in with the breadknife."[1]

[1] Carroll, L., 1939. *The complete works of Lewis Carroll.* The Nonesuch Press, London.

In Lewis Carroll's day, hatters *were* often mad, suggesting a condition that was noticeable enough to be enshrined in a once-popular expression, but it was not until recently that the madness was recognized as an occupational hazard of hatters or people who worked with animal skins. At one time fur pelts were treated with mercuric chloride in processing and subsequently anyone who worked with the skins ran the risk of mercury poisoning, especially hatters who, in fashioning men's hats from beaver skins, were constantly exposed to mercury fumes. Today, mercury compounds are no longer used in processing furs, nor are beaver hats wildly popular, so the phrase "mad as a hatter" has passed from common speech and even understanding. Mercury poisoning, however, remains a hazard.

Because of its initially vague symptoms—fatigue, headache, and irritability—mercury poisoning often goes unrecognized at first. At a later stage, arm and leg numbness develops, followed by disruption of balance, blurring of vision, deterioration of muscular coordination, emotional disturbances, and finally wasting away of muscles.

Mercury in naturally occurring stable compounds causes no problems, but its liberation in the form of soluble salts from a group of mercury compounds with industrial or agricultural uses is beginning to cause some concern.

Inorganic Mercury Compounds

Inorganic mercury compounds are used by over eighty industries including plastics, industrial chlorine, and electronics. Until recently most cases of mercury poisoning were derived from occupational exposure: accumulation of inorganic mercury in the kidneys affected readsorption and secretion of sugar, protein, and salts; and accumulation in the brain caused a loss of coordination. But in 1953 a number of people living in the vicinity of Minamata Bay, Japan, became ill with a strange series of disorders. Ultimately over a hundred cases were reported, many ending either in death or permanent disability. Investigations finally linked the disorders with seafood taken from Minamata Bay giving the disease the name, Minamata disease. One case history gives some idea of the nature of Minamata disease:

A 14-year-old boy who had been agile and bright before his illness is said to have eaten a large number of crabs and small fish from a posted area of Minamata Bay during a ten-day period in July, 1958. A few weeks later he noted numbness around his mouth and in his hands and feet. He did not have fever, headache, or a stiff neck. He . . . became clumsy in buttoning his clothes and handling his chopsticks; his family observed that he staggered slightly when walking. His auditory acuity and attention span diminished, and he developed the mannerisms and behavior of a younger child. . . . His memory for most recent events was adequate, but he was unable to perform calculations beyond the eight- or nine-year old level. When observed in 1960, he was still hospitalized in the Minamata City Hospital. . . . His physical disability was considered mild, but his inability to calculate and remember complex Japanese written characters made it impossible for him to continue in school.[2]

Of fifty-two cases originally studied, seventeen died, and twenty-three were permanently disabled.

Other circumstantial evidence linking the disease to the eating of seafood from Minamata Bay was the death of large numbers of cats belonging to afflicted families. Judging from the nature of the symptoms, the illness was a form of mercury poisoning. In Minamata City at the head of the Bay, a chemical plant had greatly increased its production of vinyl chloride; simultaneously the number of cases of Minamata disease increased.

Mercuric chloride is used as a catalyst in the production of vinyl chloride. Considering that sixty grams of mercuric chloride are "lost" per ton of vinyl chloride manufactured, it has been calculated that in one year nearly 2000 kilograms of waste mercury were generated: 1000 kilograms were washed from the vinyl chloride produced and another 1000 kilograms of spent catalyst were discarded. Although the source of the mercury was inorganic and insoluble, when it settled in the bottom mud of aquatic systems, it was converted into water-soluble methylmercury which accumulates in organisms. Although the Minamata plant had settling basins and improved waste treatment facilities, enough inorganic mercury was transformed into organic forms to enter the bay's food chain and poison the bottom mud, shellfish, fish, and finally people. The organic mercury content of mud from the bay had a range of 12–2010 ppm, shellfish had 38–102 ppm, and people who died from Minamata disease, 13–144 ppm in kidney tissues.

These instances of inorganic mercury poisoning are dramatic but they are geographically limited to the vicinity of plants that use inorganic mercury compounds in their processing. Of still greater concern to the general public is the entry of mercury into the environment directly through organic mercury compounds.

Organic Mercury Compounds

There are two types of organic mercury: aryl salts of mercury (those with carbon rings attached) which break down into inorganic mercury in the body, or the alkyl (straight chain hydrocarbon) salts of mercury,

2 Kurland, L. T. et al., 1960. "Minamata disease." *World Neurology, 1*, pp. 370–395.

particularly methylmercury, that are able to diffuse easily through membranes and spread throughout the body. Methylmercury also generates a somewhat different symptom picture than inorganic compounds of mercury.

Mercury in the Environment

Mercury is a *fungicide,* a potent killer of fungi; therefore, organic compounds of mercury, especially methylmercury, have been widely used by pulp mills to keep fungi, bacteria, and algae which thrive on wood pulp from clogging up the machinery. By 1970 methylmercury released into the environment from this source had accumulated in freshwater fish to such an extent (more than 0.5 ppm) that people were advised not to eat fish caught from the St. Lawrence, Oswego, and Niagara Rivers as well as Lakes Erie, Ontario, Champlain, and Onondaga. Another use of mercury-containing fungicides has been the treatment of seeds to prevent or inhibit the growth of molds during storage and after the planting of seed in the spring. In 1970 the tragic poisoning of an entire family who ate pork fed on mercury-treated grains served as an illustration of the danger of such a practice.

Swedish farmers, until recently, regularly dusted up to 80 percent of all grains sown with a fungicide containing methylmercury. The poisoning of birds eating such treated seed led to investigations that showed that eggs produced in Sweden contained four times the mercury of eggs from other European countries, 0.026 ppm contrasted with 0.007 ppm. Although these levels are not considered dangerous, the World Health Organization (WHO) warns against more than 0.05 ppm. When hens were fed treated seeds, the mercury content of their eggs rose to 4 ppm. The question then arose whether grain ultimately developing from treated seeds would retain any of the mercury and pass it along in a food chain.

Experiments in Sweden indicated that seeds of plants grown from mercury treated seeds contained 0.030 ppm mercury, while the plants from untreated seeds produced seeds with only 0.014 ppm mercury. The eggs from hens fed seed from treated seed contained 0.29 ppm mercury; the eggs from hens fed seed from untreated seed contained 0.010 ppm mercury. As a result of such data, changes in the formulation of the fungicide and in the rate of its application were recommended to Swedish farmers. A less toxic form of organic mercury, methoxyethylmercury, replaced the highly toxic methylmercury, and only seeds containing mold were treated before planting, reducing the percentage of treated seeds from 80 percent to 12 percent. It was found that the yield of grain was not reduced by these practices, and the mercury level in eggs decreased from 0.029 ppm in March 1964 to 0.010 ppm in September 1967.

In the United States, methylmercury fungicides are widely used in agriculture, especially in the wheat growing areas. Although preliminary testing by the United States Department of Agriculture has not cor-roborated the Swedish finding that there was an appreciable amount of mercury found in seeds of plants grown from treated seeds, further test-ing might well demonstrate this.

Treated seed is routinely eaten by game; as a result in 1969, pheas-ants in Montana contained between 0.05 ppm and 0.45 ppm of mercury. In California that same year, pheasants with a mercury level of 1.4 to 4.7 ppm were found, and a general level of at least 1 ppm in the pheasant population of Alberta, Canada caused the pheasant hunting season to be canceled in that province.

Although the Food and Drug Administration allowed *no* mercury residue in or on foods until 1970 (in compliance with the much-debated zero tolerance rule or Delaney Amendment), currently the fungicide manufacturer has only to demonstrate that residues are harmless.

At present levels of fungicide use, with or without zero tolerance, the question is not *whether* foods contain mercury (apples often contain 0.1 ppm, tomatoes 0.1 ppm, potatoes 0.05 ppm, eggs and meat 0.1 ppm), but whether these levels are harmful or may be a cumulative threat in the future. Sweden has already banned the mercury fungicide still in wide use in the United States, recommending a less toxic form. Most other European countries have varying tolerance standards for mercury in foodstuffs. Despite the formal zero tolerance level for mercury residues in the United States, few crops are presently being routinely monitored for mercury residues by the United States Department of Agriculture (USDA). It might be prudent, in view of uncertainty about the toxicity of the mercury in food, to substitute less toxic forms of mercury in widely used fungicides. This would probably reduce, in grain directly and in the food chain indirectly, the mercury residue resulting from the ingestion of treated grain accidentally by livestock or birds. The time thus bought might permit the development of fungicides of more selec-tive toxicity, affecting just the target fungus and not all other life as well.

LEAD

Between 1954 and 1964, 128 children died of lead poisoning in the United States, but the lead source was neither air pollution nor careless use or storage of a biocide. Until 1958 most interior paints contained lead pigments; since that time lead has been replaced by titanium. However, some old houses were painted countless times with lead paint; as a result, thick chips of leaded paint fall off as the walls and ceilings peel. Putty also contains lead and is even more likely to be found in substand-ard dwellings. At least 20 percent of all children between the ages of one and five, regardless of social or economic status, ingest nonfood particles of all types, partly out of curiosity and partly because of a behavior syndrome called pica. Although the reason for the syndrome is not completely clear, it seems to be related to a deficiency of iron in the body. Pica is not limited to children; some adults, particularly pregnant women, eat clay or cornstarch in excessive amounts. One out of fifteen children in ghetto areas suffers from lead poisoning resulting from the ingestion of lead in old paint or putty. Although acute lead poisoning has easily recognizable symptoms, chronic or asymptomatic poisoning is either without symptoms or has vague symptoms that could be ascribed to any of a dozen causes. Consequently, chronic lead poisoning may never be diagnosed at all, or it may be recognized only after irreversible damage has been done.

Lead ions are absorbed from paint chips in the stomach and distributed, initially, throughout the soft tissues of the body. While in this soluble ion form, lead is both mobile and toxic. Damage, of course, depends upon the concentration of lead in the body. Because the soluble lead in the soft tissues is in equilibrium with the blood, blood analysis usually gives a fair indication of the amount of soluble lead in the body at any given time. Most of this soluble lead is excreted in the urine, which also provides an indication of the level of body lead. Some lead is deposited as insoluble salts in bone tissue. If it remained there, treatment for lead poisoning would be simple, but under certain conditions, in children especially, the bone lead can be returned in a soluble form to the bloodstream.

Lead most often affects blood, kidneys, and nerves. The symptoms are: anemia from interference with red blood cell production; chronic nephritis, which may contribute to hypertension and, ultimately, kidney failure; effects on the nervous system as seen in behavior problems characterized by convulsions or swelling of the brain. Most of these symptoms, reflecting damage from large doses of lead, are quickly recognizable and it is possible to administer treatment in time to prevent permanent damage. The most frequently used treatment is to bind the soluble lead in the system to special compounds called chelators, which carry the lead out of the body in the urine. But chelators must be used with great care, for if there is a large lead residue in the stomach, chelators may encourage such rapid absorption of lead as to cause death. Chelators may also release lead already stored in the bones, causing further complications. So it is important that the digestive tract of a child with acute lead poisoning be free of all leaded materials before treatment begins. Unfortunately, many children saved from one episode of lead poisoning are returned to the same dangerous environment only to succumb to the same syndrome again.

Asymptomatic Lead Poisoning

It is the asymptomatic aspect of chronic lead poisoning that can be the most dangerous in children. Often several diagnostic tests are required before lead poisoning can be determined and, unless it is looked for, it may go unnoticed. A sample of 1000 children in Chicago ghettos found seventy victims of lead poisoning without symptoms and three with obvious symptoms. It has been calculated that as many as fifty thousand children in the United States suffer from the asymptomatic form of lead poisoning. Even though physical recovery is possible, brain damage often ensues, for the ages one to five are critical years in the growth and development of the brain. In another study of 425 Chicago children suffering from lead poisoning, 39 percent developed various neural injuries leading to mental retardation.

The solution to this problem is in part a medical one. Ghetto children subject to possible lead poisoning should be examined much more frequently to uncover and treat as promptly as possible any cases of lead poisoning that show up. Leaded paints must be removed from substandard housing in a rehabilitation program designed to upgrade this type of housing wherever possible. Where rehabilitation is impossible, urban redevelopment projects should replace substandard housing. This is the only sure way to eliminate thousands of cases, recognized and unsuspected, of lead poisoning. It is bad enough for ghetto dwellers to endure the deaths of lead-poisoned children, but the insidious effect of sublethal doses in causing mental deterioration or retardation is a totally unnecessary result for which the entire community must share the blame and ultimate cost.

Atmospheric Lead

While dramatic and most certainly tragic, poisoning of children from doses of leaded paint does not represent man's sole exposure to lead in the environment. Lead, like most other elements, is present in the natural environment, albeit in very low concentrations. The industrial revolution, however, followed by the proliferation of the automobile, particularly in the northern hemisphere, has gradually increased the background of lead in the environment. Dated samples of snow and ice from the Greenland ice cap near Thule indicate that up to 1750 there were about 20 micrograms (μg) of lead per ton of ice; by 1860 this had increased to 50 μg per ton; in 1940, 80 μg; in 1950, 120 μg; and in 1965, 210 μg. This represents an increase of 400 percent between 1750 and 1940, and a 300 percent increase since 1940 in background levels. The highest samples from Antarctica are equal to the lowest from the arctic. This reflects the great preponderance of land area in the northern hemisphere, the presence there of most of the sources of environmental lead, and the general lack of air circulation between the northern and southern hemispheres, the same phenomenon encountered in the concentration of fallout (see Chapter 11). Although some of this lead comes from industrial sources, the striking correlation of atmospheric lead with other automobile emissions suggests that the internal combustion engine has been the major source of lead in the atmosphere.

Lead has been added to most gasolines in the form of tetraethyl lead mixed with ethylene dibromide and ethylene dichloride plus a marker dye, since 1923. At that time lead from such a source was inconsequential; but in the intervening years, 2.6×10^{12} grams of lead have been combusted and distributed over the entire northern hemisphere. Much of this lead has run off into the oceans, leading to an increase in the lead content of the surface waters.

On a typical day the "autopolis" of Los Angeles generates thirty thousand pounds of lead from the seven million gallons of gasoline that are burned, which helps to explain why the atmospheric lead concentration of Los Angeles is 2.5 μg per cubic meter compared with 1.4 μg in Cincinnati or 1.6 μg in Philadelphia. This urban concentration is about fifty times the concentration of lead in rural air, and 5000 times the natural concentration. At peak traffic periods at congested points, the lead concentration may rise as high as 45 μg per cubic meter. Inhaling city air with a lead concentration of 1 μg per cubic meter, a man in one day would absorb 20 μg of lead. In addition to the lead we breathe in from the air, we ingest about 300 μg of lead a day in foods and liquids on an average.

Since our oral intake of lead every day is fifteen times the amount of lead we breathe, airborne lead is thought by some people not to be a significant hazard. On the other hand, only 5 percent of the lead that passes through the alimentary canal is absorbed, and close to 40 percent of the lead inhaled by the lungs is absorbed. Cigarette smokers, for example, have a slight but consistently higher concentration of lead in their blood than nonsmokers of cigarettes since lead is concentrated in tobacco leaves by lead arsenate residues in the soil from previous spraying or lead aerosols from vehicular traffic.

In either case, little lead is permanently retained by the body, but lead is toxic while it is in its temporary soluble state. When the lead content of the air rises from 1 μg per cubic meter to 15 μg, as it frequently does in large cities, the intake of lead rises from 20 μg to 300 μg a day. This is no longer one-fifteenth of orally consumed lead, but its equal, effectively doubling the intake of lead to 600 μg per day. No city has an *average* of 15 μg per cubic meter of lead in its air today. Nor are any likely to in the future, since starting in 1971 the automobile and oil companies have agreed to eliminate tetraethyl lead from gasoline. Ironically, this is not because of any recognition of the danger from airborne lead in the atmosphere, but rather because leaded gasoline fouls anti-pollution devices required by federal and state governments.

Future Effects

To assess the effect of lead on man from industrial and automotive sources, we must consider whether the rate of environmental accumulation of lead has increased over the last few years, whether such an increase is mirrored by an increase in the body lead, and whether there is a threshold below which atmospheric lead causes no physiological damage.

Observations from the arctic, antarctic, and the world ocean indicate that lead is increasing in the environment. Reports of increasing body burden, however, are conflicting; one survey indicates a significantly higher lead content in American lungs (51 ppm) than in African lungs (26 ppm). But another report analyzing the lead content in the blood of various populations found as much lead in remote tribes in New Guinea as in people living in some large American cities. Unfortunately we have no base line with which to compare contemporary populations. Because of the ubiquity of lead in the environment for hundreds of years, *all* human populations have probably been exposed and contaminated. Since the sea surface has an abnormally high lead content compared with its depths, a result of atmospheric fallout, there is no reason to suspect that any human population, however remote from civilization, can be used as a control. However, a report by workers in California concludes that "further increases in atmospheric lead will result in higher blood lead levels in the population in a predictable relationship."[3]

We know that normal levels of lead in the blood range between 0.05 and 0.4 ppm. The range between 0.4 and 0.6 represents occupational exposure and 0.6 to 0.8 is abnormally high. Above 0.8 ppm recognizable lead poisoning usually results. The average level of lead in the American population is 0.25 ppm. The question is whether this average is "normal" or whether it already represents a serious increase from past levels, thereby reflecting the documented increase in environmental lead. If that is so, then any further increase might well be dangerous. We have the example from the past of chronic lead poisoning in Rome. The use of lead water pipes and, more seriously, lead cooking vessels in ancient Rome by the upper class probably caused the virtual disappearance of this class by increasing the rate of stillbirth. All of which quite probably contributed to the ultimate fall of the Roman Empire.

[3] Goldsmith, J. R. and A. C. Hexter, 1967. "Respiratory exposure to lead: epidemiological and experimental dose-response relationships." *Science, 158,* pp. 132–134.

Whereas we know that 0.8 ppm in the blood can cause lead poisoning, we cannot be sure that 0.7, 0.6, or 0.5 ppm are innocuous. Perhaps we are still too clumsy in our perception of correlations between environmental stress factors and abnormal physiology or psychology to recognize low-level effects when we see them.

BERYLLIUM

Because of its long-standing familiarity and use, the toxicity of lead was recognized quite early. The toxicity of beryllium was much less obvious.

In 1933 some German workers employed in a factory extracting beryllium metal from its ore became ill with severe respiratory symptoms including bronchitis. At the time it was thought that fluorides associated with beryllium were the cause. But in the mid-1940s with the growth of the fluorescent lamp industry, a number of cases of similar description were reported. In 1946, seventeen workers in a fluorescent lamp factory in Salem, Massachusetts became seriously ill and some died. Although the evidence seemed to point to beryllium compounds used to make the lamp phosphor (beryllium carbonate and zinc beryllium nitrate), the United States Public Health Service concluded that beryllium was not toxic. The toxicity of beryllium was finally established a few years later when a similarity of symptoms was observed in foundry, lamp, and neon sign workers, all of whom worked with beryllium metal or its oxide.

Beryllium evokes a variety of symptoms which makes diagnosis difficult at times; victims usually complain of shortness of breath, aches and pains, and a dry cough. This is often followed by severe weight loss and kidney damage. Cortisone or other steroid hormones are sometimes useful in alleviation of symptoms, but because beryllium is difficult to remove from the body, there is no known cure for beryllium disease. Often, by the time of diagnosis, irreversible damage has taken place.

Beryllium Poisoning

Beryllium is indeed quite toxic; whereas 150 μg of lead per cubic meter of air is considered a safe exposure, as little as 5 μg of beryllium per cubic meter of air can cause chronic beryllium poisoning. Further, beryllium behaves in a most peculiar manner. First, there is the neighborhood effect; despite large amounts of toxic arsenic, mercury, and lead released into the atmosphere by industry, no case of poisoning from these metals has been recorded in the neighborhood of the polluting factory. Beryllium poisoning, however, has been recognized in people living as far as three-quarters of a mile from a beryllium processing plant. One such community experienced sixteen cases, including five fatalities. Then there is the oddity that in the neighborhood of the plant, while people living closest to the beryllium source had a greater tendency to develop beryllium disease than those living farther away, within the plant itself this dose relationship did not apply. The beryllium concentration in the air of the neighborhood reporting the beryllium poisoning ranged from 0.01 to 0.1 μg per cubic meter, while that in the plant was 1000 times higher. Yet of 1700 workers only nine developed beryllium disease (0.5 percent) whereas neighborhood people had a beryllium disease rate twice as high. Stranger still, the nine plant workers who became ill were not those who had been exposed longest to the largest amount of beryllium, but people who had been exposed for only a few months or had worked only a short time at the factory. In some instances fifteen years elapsed between exposure and illness, suggesting the delayed response of the cancer-producing radium incident mentioned in Chapter 12. Finally, postmortems on victims showed no relationship between the amount of beryllium concentrated in the lungs and the severity of the disease. Facts like these give many sleepless nights to epidemiologists who study the nature and spread of disease. The best conclusion that can be drawn from these observations is that people vary in their sensitivity to beryllium: some are almost immune and others extremely sensitive to very small amounts.

Control. As a result of these demonstrations of beryllium toxicity, the use of beryllium-containing phosphors in fluorescent lamps was discontinued in mid-1949, and, although cases of beryllium poisoning from broken lamps were described for several years thereafter, once the old tubes had been replaced there were no further complications either in manufacture or use of fluorescent lamps. Most industries that use or process beryllium today exercise elaborate precautions involving high-speed ventilators and shields to eliminate or sharply reduce the amount of beryllium particles or oxide in the air. Standards have been established, although on rather flimsy experimental evidence, that recommend not more than 2 μg per cubic meter of beryllium over an eight-hour day in the atmosphere of a factory. Further, no one should be exposed to more than 25 μg per cubic meter at any time. Finally, neighborhood effects may be eliminated if the atmospheric concentration of beryllium is less than 0.01 μg per cubic meter.

Today the incidence of beryllium poisoning in large plants maintaining responsible programs of industrial hygiene is quite low, certainly comparable with other types of job hazards. No fatalities from beryllium have been recorded since 1950.

But as beryllium finds new uses, the probability of its release into the environment increases. Beryllium has entered the space age with the discovery that suspensions of beryllium fibers provide a magnificent source of power when burned in a rocket engine. When research had progressed to the test firing stage, it was calculated that so much beryllium would be released in the firing of a rocket that someone eighteen miles from the launching site would be exposed to the same dose as a beryllium worker if he were on the job twenty-four hours a day for thirty consecutive days. Should such rockets become operational in the future, yet another toxic pollutant will have been added to the atmosphere without apparent concern for its environmental effect in the excitement over its "usefulness" to man.

ASBESTOS

A few years ago a study was made of the Insulation Workers Union of New York starting with the 632 members enrolled on December 31, 1942. Ten years later the cause of death of union members was compared with a statistical sample of men in comparable circumstances except for their exposure to asbestos. It was found that 255 union men had died compared with 203 men in the comparison group; forty-five workers died of lung cancer or disease, compared with six or seven in the comparison group; twenty-nine died of stomach cancer, compared with nine or ten in the comparison group; and twelve died of asbestos disease, while there were no deaths from this cause in the comparison group.

Some 100 thousand to 125 thousand people work directly with asbestos, and up to 3.5 million or 5 percent of the total work force of the United States work with asbestos at some time in some capacity in the construction and shipbuilding industries. With the widespread use of asbestos, there has been a distinct increase in the number of *mesothelio-mas*. A diffuse cancer that develops on the mesothelium or lining of the chest cavity, mesothelioma was extremely rare until fairly recently. Fourteen of the 120 asbestos workers who died most recently in New York died of mesothelioma. This was to be expected since the connection was made between high concentrations of asbestos fibers in the air and mesothelioma.

Somewhat less expected was the report of forty-seven cases of mesothelioma from a sample in Cape Province in South Africa. Although this area is rich in asbestos deposits, very few of the victims had worked with asbestos directly. Still more surprising was a report in 1960 that of 6312 people X-rayed in a Finnish county that had asbestos mines, 499 had a condition previously associated only with asbestos workers. None of these 499 ever worked in an asbestos mine, but most lived near one. A survey in a nearby county without asbestos mines found no diseased condition in a population of 7101.

Even more remotely circumstantial but still suggestive is the case of

a fourteen-year-old boy with mesothelioma, [who] had no occupational contact with asbestos nor did any of his family. On questioning, the boy's father told the authors that his boy . . . had helped him while he had replaced most of the plaster board during extensive remodeling of his house. Plaster board contains a high percentage of asbestos.[4]

Apparently, when asbestos fibers are inhaled into the lungs they become coated with a material containing iron and an asbestos body is formed (Figure 13.1). While large amounts of asbestos bodies were found in the lungs of asbestos workers, careful autopsies of people supposedly remote from asbestos exposure showed a surprising incidence of asbestos bodies: 26 percent of 500 consecutive autopsies in Capetown, South Africa; 50 percent of 1975 autopsies in New York; 30 percent of housewives; 40 percent of white collar workers; 50 percent of blue collar workers; and 70 percent of construction or shipyard workers. Of forty-two cases of mesothelioma investigated in Pennsylvania, twenty represented people who worked with asbestos, eight lived in the vicinity of an asbestos plant, and three had a familial contact with someone who worked with asbestos. These patterns of neighborhood and familial exposure resemble those of beryllium and suggest that asbestos, like beryllium, may affect people differentially rather than on a linear dose scale.

[4] Lieben, J. and H. Pistawka, 1967. "Mesothelioma and asbestos exposure." *Arch. Env. Health, 14,* pp. 559–563.

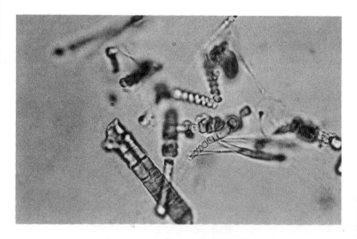

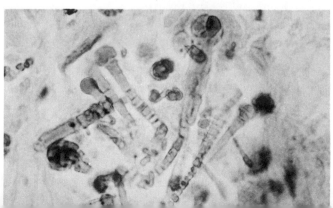

Figure 13.1 The rod-shaped particles in these photomicrographs are asbestos bodies in lung tissue. (Professor I. J. Selikoff, Mt. Sinai School of Medicine, New York)

Fiber Inhalation

The major concern about asbestos is not for the safety of present asbestos workers, for once the danger of asbestos fiber inhalation was recognized several years ago, the way was open for responsible industry to provide, and for responsible unions to demand, a margin of safety by proper ventilation, respirators, and shielding. The chief concern is that asbestos can be extremely damaging to the health of many individuals who are removed from direct contact with the stimulus. Unfortunately, the potential toxicity of asbestos is not at all widely recognized and it is handled quite casually by people who assume it is as harmless as cement or plaster. There are many case histories of people who died some years after such a casual exposure. If this is true of asbestos, what about fiber glass which is being used more in houses, or various other synthetic fibers? We simply do not know what the long-range effects of these materials are.

The basic problem, with all four inorganics mentioned in this chapter, has been the almost total lack of knowledge about either the threshold at which human health is adversely affected or the cumulative effect of low levels of inorganics already present in the body and which are likely to increase as atmospheric pollution worsens. Mercury and lead seem to follow a linear curve of dose-effect; the more lead or mercury, the greater the likelihood of certain symptoms developing in the exposed population. But beryllium and asbestos follow no rules, making them by far the most dangerous to handle and be exposed to. Although you can assign maximum permissible doses of mercury and lead (subject to revision as more information becomes available about low-level effects), the apparent random effect of beryllium and asbestos vitiates the permissible dose approach.

Perhaps inorganic materials are best treated like poison ivy. This is a plant to which people are differentially sensitive. Some people break out with a rash from head to toe by simply walking downwind from an ivy patch, while others can roll in it without ill effect. The only advice botanists or physicians can give to the general public about poison ivy at our present crude level of understanding is to avoid it wherever possible. Maybe this is the best advice for the general public to follow for the present with beryllium or asbestos. But, of course, the public is not always in a position to voluntarily avoid certain potentially dangerous inorganic materials. It then becomes the responsibility of the industry producing these materials and finally the government or agencies controlling the environmental relations of the industries to protect the general public. This can be done by either strictly controlling the emissions of dangerous by-products, dusts, aerosols, fibers, or whatever, to a point well below that demonstrated to have deleterious effects on nonworkers, or by eliminating such emissions altogether.

It is possible that a small amount of these inorganic materials is indeed harmless and that our concern over a few isolated cases represents an alarmist position. But until much more experimental data is available to provide rational scientific guidelines about the quantities of various inorganic materials that can be tolerated by the environment and ourselves, discretion had better play a greater role in our value system than it has previously.

FURTHER READING

Elwyn, D., 1968. "Childhood lead poisoning." *Scientist and Citizen,* **10** (3), pp. 53–57.

Gilfillian, S. C., 1965. "Lead poisoning and the fall of Rome." *Jour. Occup. Med.,* **7,** pp. 53–60. Intriguing account of the effect of lead upon history.

Grant, N., 1969. "Legacy of the mad hatter." *Environment,* 11(4), p. 18.

Kettering Laboratory, University of Cincinnati, 1963. *Symposium on lead.* Archives Env. Health 8. An extremely comprehensive and useful collection of papers dealing with lead from all standpoints.

Lofroth, G. and M. E. Duffy, 1969. "Birds give warning." *Environment,* 11(4), pp. 10–17.

Novick, S., 1969. "A new pollution problem." *Environment,* 11(4), pp. 2–9. Three papers from the same issue of *Environment* give a reasonable picture of the problems caused by mercury in the environment.

Patterson, C. C. and J. D. Salvia, 1968. "Lead in the modern environment." *Scientist and Citizen,* 10(3), pp. 66–79. Elwyn and Patterson give a good idea of the role of lead in the environment.

Selikoff, I. J., 1969. "Asbestos." *Environment,* 11(2), pp. 2–7. An excellent review of the asbestos problem by the world's foremost authority on the medical aspects of asbestos.

Stokinger, H. E. (ed.), 1966. *Beryllium: its industrial hygiene aspects.* Academic Press, New York.

Tepper, L. B., H. L. Hardy and R. I. Chamberlin, 1961. *Toxicity of beryllium compounds.* Elsevier, New York. Stokinger and Tepper present a comprehensive picture of beryllium as an environmental hazard.

CHAPTER **14**

ORGANICS AND THE ENVIRONMENT: EAT, DRINK, BUT BE WARY

It's a very odd thing—
As sad as can be—
That whatever Miss T. eats
Turns into Miss T.

WALTER DE LA MARE

American supermarkets are filled with the greatest abundance and variety of food ever available to man. But the list that sweeps from proletarian potato to gourmet truffle includes more and more foods that have been grown or processed in some way that has included contaminants or additives. *Contaminants* are compounds, organic or inorganic, that are retained accidentally in marketed foods—biocide residues on plants and antibiotics or hormones in meat. *Additives* on the other hand are substances purposely added to foods during their preparation or processing to assure longer shelf life, greater attractiveness, consistency, flavor, or ease of preparation.

THE FEDERAL FOOD AND DRUG ADMINISTRATION (FDA)

Historically, the concern of the public and hence governmental regulating agencies has been with adulterants—the "chalk in milk" or "water in beer" kind of thing that was common until the turn of the century. Public indignation reached a peak in 1906 with the publication of Upton Sinclair's muckraking novel *The Jungle*, exposing conditions in the meat-packing industry. As a result of the widespread demand for standards,

261

the administration of Theodore Roosevelt set up a Food and Drug Administration to control the practices of the food and drug industry and to protect the public. Its first director, Harvey W. Wiley, interpreted the Food and Drug Act quite broadly and within six years had incurred the wrath of his sponsor by attacking Coca-Cola for its caffein content and questioning the safety of saccharin. Wiley resigned under pressure in 1912 and the Food and Drug Administration has since interpreted its mandate to protect the public somewhat more conservatively. However, the scope of the FDA's activities has of necessity broadened as the food industry adapted increasingly modern technology to the preparation and processing of food for the expanding market. Because of the incredible complexity of pest control, as was pointed out in Chapter 12, one of the chief functions of the FDA today is to license the use of biocides, set standards for residues, and monitor food products for their residue content. But of equal importance to the FDA and the public is the increasing flood of additives used by the food industry in processing food for public consumption.

THE ADDITIVE ARRAY

In the "good old days" products were simple, the range limited, and the quality, by today's standards, poor; but processing and packaging were relatively direct (Figure 14.1). Today few foods escape some treatment between farm and market. On your next trip to the supermarket, read some of the labels which, by law, must indicate at least in general terms the contents of all food items. A surprising number of labels (Figure 14.2) fairly bristle with such words as *acid, alkali, anticaking agent, antioxidant, bleach, buffer, disinfectant, drying agent, emulsifier, ex-*

Figure 14.1 This recreation of an old general store is a marked contrast to today's discount emporium, where most of the items come prepackaged in boxes or plastic. (New York State Historical Association)

tender, artificial flavor, fortifier, moistener, neutralizer, preservative, sweetener, thickener—to mention just a few. Table salt, for example, may contain sodium silico aluminate, dextrose, tricalcium phosphate, potassium iodide, and polysorbate, in addition to sodium chloride. Some food additives are common inorganic salts, but many are compounds whose effect may extend well beyond their intended sphere of activity, much like the detergents discussed in Chapter 9 which kept sudsing and sudsing and caused environmental problems well beyond their manufacturer's intent. With additives, this larger sphere of unintended action is the human body.

The Pros and Cons

From the very beginning, the use of additives by the food industry polarized the consuming public. The larger group accepts whatever the industry wishes to include in its processing, assuming that the industry knows best and that if the ultimate product is "bettered" somehow, then it is worth the cost. A smaller but much more vociferous group equates the word chemical with poison and assumes that *any* chemical, organic or inorganic, added to food is poisonous. Unfortunately many faddists have joined the cause, so that the issue is no longer whether certain additives are harmful or not; the desire for *wholesomeness* and the yearning for the *natural* blend into a great crusade against all food additives.

The food-additive protest, however multifarious its background and nature, is by no means as irrational as its critics suggest. Some additives *are* poisonous when taken in large enough doses, although few people have been literally "poisoned" by an additive. Most problems have arisen from the implication through animal testing that large quantities of certain organic compounds may cause cancer in a great variety of body

Figure 14.2   The FDA requires a listing of ingredients and additives on the labels of most edible products. The lists are often appallingly long.

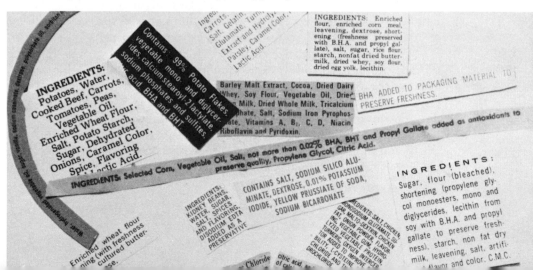

sites and organs. Since the threat of cancer is often brandished as a side-effect of many additives, we should take a closer look at the disease and its relationship to chemical additives.

Additives as carcinogens.   The nature and cause of cancer is one of the great medical puzzles. Something apparently goes wrong in a cell, which causes it to divide without differentiation. That "something" might be a virus, a particle of beryllium oxide, an asbestos body, or some organic compound, perhaps innocuous in itself, that is metabolized by the body into a form that interferes with cell metabolism and causes cancer to develop. The proliferation of cancerous cells ultimately becomes so demanding of the body's resources that organ failure and death ensue. We cannot examine the structure of a prospective additive and predict if, where, and when it may cause cancer. We cannot predict how such a compound will be metabolized by the body, whether the metabolite will be rendered harmless or whether it will be far more destructive than the original material ingested. Today we can only infer from animal testing on a purely trial and error basis that if a compound causes cancer in a rat or monkey then it might do likewise in man. When cancer has been indisputably demonstrated in animals, the offending material is assumed to be a carcinogen in man and is either not used or is withdrawn from use.

It would be extremely convenient if all we had to do to assure our complete safety in using additives was to demonstrate that they did not cause cancer or other debilities in a variety of test animals. But it is not that simple. Some compounds do not cause cancer in all test animals; a dye, 2-naphthylamine, for example, causes cancer in dogs and man, but not rats, rabbits, or cats, while sodium arsenite causes cancer in man but not in experimental animals. In addition there are the problems of deciding on the number of experimental animals that will assure a fair test, the diet on which to place them during testing, the stage in the life cycle to test (young, adult, pregnant, old), the duration of the experiment (some cancers develop in the offspring of the adults exposed to the test material), and whether administration should be oral, by injection, or by application to the skin surface. While many of these problems can be handled by standard experimental procedures, the testing of new drugs or additives is both time consuming and extremely expensive.

Further uncertainties arise from the latent period between the time of incipient damage and its expression in the form of a detectable cancer. It is not the young man of twenty-one smoking five packs of cigarettes a day, inhaling asbestos fibers, or ingesting a coal tar food dye who develops lung cancer, but the prematurely aging man of fifty or sixty. Who can remember what he might have eaten or breathed twenty or thirty years ago that could explain his present stomach, kidney, or lung cancer? Human metabolism varies widely; even in the best of health, no two people are likely to metabolize a given compound in exactly the same way. During illness or organ malfunction a compound may be metabolized differently than during a normal state. Finally, additives may interact; an additive that normally is not absorbed in the intestine may be readily absorbed with ill effect if an emulsifier is present. So the whole subject of the impact of food additives on human health is far more complex than the "purists" and "adulterers" would have us believe.

The position of the food-processing industry is concisely stated by R. Blackwell Smith, Jr. of the Medical College of Virginia:

> substantially every substance, including pure water and table salt, may be harmful if a sufficient quantity is swallowed or otherwise introduced into the body; and conversely, it is a generally accepted fact that there is no substance sure to kill or harm if swallowed or taken otherwise, provided the amount taken be sufficiently small . . . an additive or other chemical is not and cannot be of itself either poisonous or non-poisonous, hazardous or non-hazardous, harmful or safe; but every additive or other chemical may be safe at some level or mode of intake and hazardous at another level or mode of intake. The question, then, is not one of *whether* but essentially one of *how much*.[1]

This argument is typically rebutted by William Longgood:

> Poison is harmful to the human organism. . . . When it is ingested by a human being there is damage. The more poison, the more damage. . . . The fact that the dose may be reduced until damage no longer can be seen or measured by man's instruments does not mean that the damage no longer exists; it merely means that it no longer can be seen.
>
> The vested interests that profit from the sale and use of chemicals in foods are scornful of this attitude. They say it is not scientific. As "scientific" proof of the alleged harmlessness of eating small amounts of poisons in foods, they solemnly point out that it is possible for a person to cram enough salt or water down his throat to kill himself. This is supposed to prove that everything is harmful and even fatal if taken in large enough jolts.
>
> What this strange logic claims is that because a little salt is innocent and a lot is harmful it follows that all other substances that are harmful in large amounts are safe in small amounts.
>
> Of course salt or water and the hundreds of poisons used in food cannot be equated in this way. Salt and water are necessary to life, while virtually none of the food chemicals is necessary or even useful to life; with only rare exceptions, these chemicals are antagonistic to living tissue. The question is not whether they harm those who consume them, but the extent of the harm.[2]

[1] *Science and food: today and tomorrow.* 1961, NAS-NRS Publ., 887. Washington, D.C.

[2] Longgood, W. F., 1960. *The poisons in your food.* Simon & Schuster, New York.

Realizing that we may be faced with a value judgment, perhaps the question to ask is not whether additives harm or how much, but why use them at all. Suppose, because of an atypical quirk of metabolism, one person in a million developed cancer from benzoate of soda used to prevent spoilage of cider. The additive allows this delicious and healthful beverage to be available and enjoyed year round by the 999,999 people unaffected. Would the death of this person be justified by the pleasure or nutrition of the survivors? How about 1 in 100,000, or 1 in 1000? At what point is the cost balanced by the benefit?

CONTAMINANTS

Biocides as contaminants have already been discussed in Chapter 12. But there remain three groups of contaminants which can and have caused problems by not being metabolized as was supposed and so were carried onto the consumer's table.

Antibiotics

The discovery of "wonder drugs" led to their widespread application and use for everything from a minor infection to a major disease. This tendency to over-prescribe led to extreme sensitivity of some people to certain antibiotics, penicillin especially. There was also an increasing resistance of disease organisms to antibiotics, a situation quite parallel to insects and biocides.

When applied to animals, antibiotics gave effective control, often for the first time, of various poultry and stock diseases. But like the overanxious farmers with biocides, stock raisers began to treat their animals frequently, first as a preventive, then as a growth inducer. For it was found that when antibiotics were routinely given to chickens, gut bacteria, which normally slow the rate of growth, were sufficiently inhibited to allow the chicken to be marketed sooner at a greater weight and profit, though perhaps contaminated with the antibiotics. Mastitis, an udder disease in cows, can be treated with penicillin but the penicillin is passed into the milk. In this instance, milk should be discarded for three days following treatment and antibiotic use should be halted for several days before marketing either the animal or the product to prevent contamination. *But this is not always done.*

The danger is twofold: people who are highly sensitive to penicillin or other antibiotics may be exposed without their knowledge or control. In addition, if the population at large is constantly and unknowingly exposed to antibiotics, resistant forms of organisms may develop, making the drugs useless in a time of real need. Furthermore, recent evidence indicates that drug resistance is a genetic factor that can be transferred from one organism to another, even one species to another. Thus, gastrointestinal bacteria such as E. coli found in the large intestine of man *and* animals could acquire and pass on the resistance factor to any of several pathogenic bacteria normally responsive to the effects of antibiotic drugs. Routine feeding of almost 3 million pounds of antibiotics to livestock each year enhances the real possibility of the development and spread of drug resistance not only in livestock and man, but in pathogens as well. Antibiotics are also used to inhibit bacterial growth on dressed chickens in stores. Although most of the antibiotic is destroyed in the cooking process, the danger lies in the false confidence that *all* disease organisms are thereby inhibited. *Salmonella,* a food poisoning organism, is not inhibited and may increase to a dangerous state in a few days. Also, the illusion is created for the consumer that the food is fresher than it is. The consumer can only purchase by appearance, depending on the honesty of the producer or distributor for quality. Often one suspects that this dependence is abused.

Antibiotics, if they are to live up to their name, should never be used routinely, for the short-term profit is never worth the long-term risk to the livestock or the consumer, either during the life of the stock or upon its processing. In their enthusiasm for progressive techniques, food handlers do not always follow precautions, like the English butcher who sprayed his beef with DDT then proudly said that he spent eighteen shillings a week on "modern sprays." One can safely assume that antibiotics are occasionally misused with similar exuberance.

Hormones

With increased interest and research on the development of hormonal contraceptives, a compound called stilbestrol was synthesized that had biological effects similar to those of the female sex hormone, estrogen. It was found that when a pellet of stilbestrol was inserted under the skin of the neck of a chicken, the animal rapidly gained weight and became plump and fat; a further useful effect was its ability to chemically emasculate cocks into capons without castration. Poultry raisers could produce more chickens at lower cost and, of course, higher profit. The pellets, however, did not always dissolve before marketing, which must have led at times to a biologically effective dose of sex hormones for the unlucky consumer of a soup made with chicken necks.

The balance between male and female is a fine one and a very small amount of estrogen-simulating hormone (unaffected by cooking) can prove disastrous. Then too, some chickens were marketed containing as many as twelve pellets, some of which had migrated to other parts of the chicken. Furthermore, it was discovered after a few years of use that although a stilbestrol-treated chicken did indeed look better and weigh more than a normal chicken it was because of increased fat and water content, not increased protein. Even the fat proved worthless, forming a gooey mess when fried. In 1960 the FDA declared stilbestrol off-limits for chickens. Stilbestrol is still being used in beef and lamb production, however. While sixteen cents' worth of stilbestrol implanted behind a heifer's ear supposedly produced $12 of extra beef, or 15 percent faster growth on 12 percent less feed, no such dramatic response was found in pigs, which thus remain untreated. While the amount of synthetic hormone ingested by eating beef treated according to directions is quite below the level necessary to affect the delicate physiology of sex hormones, there are suggestions that synthetic estrogens in men may cause cancer under certain conditions. Certainly, improperly applied amounts with an insufficient period allowed for full absorption and metabolic breakdown present a danger to the beef-eating public. With growing evidence of side-effects from synthetic hormones used as contraceptives (see Chapter 23), it would seem that the risk of contaminating virtually all beef produced in the United States with synthetic hormones for the short-term gain of $12 a head outweighs the benefits.

ADDITIVES

One of the chief problems with food is its relatively short storage life. Since earliest times man has tried to preserve his perishable foods by drying or adding salt, sugar, or spices. Today, nitrates and nitrites are commonly used to preserve meat products, especially cold cuts or frankfurters, while benzoic acid or sodium benzoate is often used to preserve liquids which might otherwise ferment. Sulfur dioxide is generally used to preserve dried fruit, and sorbic acid is used for various other products. At customary levels of usage these preservatives are effective and are probably necessary if we are to enjoy a selection of foods otherwise unobtainable except in areas near the site of production. The danger seems to lie not in the nature of the preservative but in the motivation of the preserver. It is one thing to safely preserve foods to allow broad distribution and reasonable shelf life. It is quite another to embalm food that is too far gone to be acceptable as fresh. Such a practice leads to the adding of artificial colors, then flavors—compounds that might be more dangerous to health than the original preservatives.

Antioxidants

An adjunct to the preserving process is the inhibition of the natural tendency of fatty acids, especially those which are unsaturated to oxidize or become rancid (Figure 14.3). Most prepared foods with unsaturated fatty acids contain an antioxidant to preserve the original fresh flavor of the food and extend shelf or storage life. Some commonly used antioxidants are butylated hydroxyanisole (BHA), propyl gallate, and butylated hydroxytoluene (BHT). The antioxidant properties of these compounds are often enhanced by ascorbic, citric, and phosphoric acid. Once again, as with certain emulsifiers and detergents, the effect of some antioxidants, especially BHT, is suspected of extending beyond the food in which it is ingested, to the body itself, inhibiting the uptake of oxygen by hemoglobin in the red blood cells.

Acids

Most acid additives are perfectly safe and have been used for many years without problems. Baking powder or cream of tartar (tartaric acid) reacts with baking soda and produces carbon dioxide, a leavening agent that makes bread and cake light. Phosphoric, citric, and malic acids are extensively used to counteract the otherwise excessive sweetness of most soft drinks. It is not the acids that are a problem to health, but the overindulgence in pastry and soda to the exclusion of more nutritious food.

Figure 14.3 An unsaturated fatty acid has many double bonds between carbon atoms. When these double bonds are replaced by hydrogen atoms the fatty acid is said to be "saturated." "Polyunsaturated" simply means that there are many double bonds present. These double bonds must be protected from oxidation by additives called antioxidants.

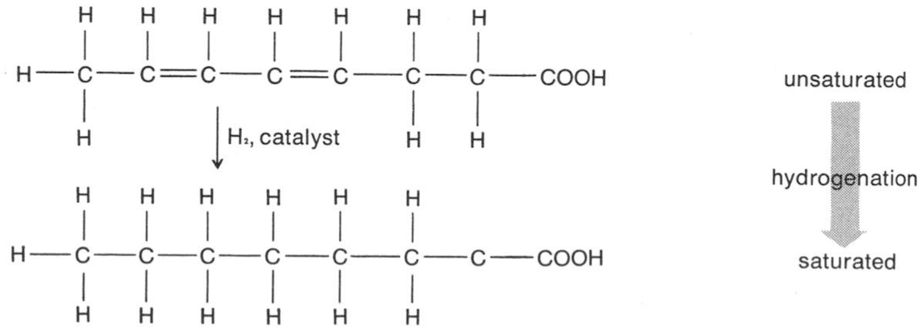

Emulsifiers

In general, an emulsifier breaks up fats and oils into tiny particles; when used in bread it acts as a softening agent. Homemade bread hardens quickly as it grows stale; commercial bread with emulsifiers retains a soft texture for a much longer period. This gives the shopper the illusion of freshness and reinforces the now erroneous conclusion that soft bread is fresh bread. Emulsifiers also allow the water content of bread to be increased, resulting in the characteristic soft spongy texture of commercial white bread. Pound for pound there is likely to be more air and water in commercial loaves than in homemade bread—air and water that the consumer pays for in lieu of flour. When used for these purposes no emulsifier is worth risking its possible synergistic effect. One emulsifier, polyoxyethylene, no longer used in the United States, was found to increase greatly the rate of iron absorption in some animals. Excessive deposition in the liver can lead to cirrhosis and cancer. One effect in man has been an increase in the absorption of vitamin A. Too much of a vitamin can be as much of a problem as too little (see Chapter 1).

Other emulsifiers such as glycerides, lecithin, sorbatin, and alginate are used in baked goods, ice cream, and confections. In baked goods, emulsifiers improve the fineness, uniformity of texture, and softness. Smooth consistency in ice cream depends upon the uniform size of ice crystals, which is controlled by alginate or emulsifiers and stabilizers.

Artificial Sweeteners

Even a food purist would have to agree that natural sugar poses no direct problem to health. But some substitute sweeteners, compounds that taste sweet but have no food value, may have adverse effects upon metabolism. Saccharin, discovered in 1879, has been used for many years as a sugar substitute especially by diabetics who must control their sugar intake. In 1950 a second group of non-nutritive sweeteners was introduced, the cyclamates. While saccharin has over 300 times the sweetening power of sucrose (cane sugar), cyclamates were only thirty times as sweet.

Since saccharin is not metabolized by the body, it is safe within recommended dose limits (0.3 g per day). However, many countries restrict or prohibit its use, not because of ill effects but to avoid misleading people into substituting a nonfood for a food of high caloric value. Cyclamates *are* apparently metabolized, at least by some people; some 20 percent of cyclamate users generated a breakdown product that, when applied to rat cells, caused chromosomes to break up. Further possible danger was suggested by the linking of cyclamates to birth defects in chickens. For this reason cyclamates were banned from foods in 1970.

While saccharin is a safe and useful sugar substitute for diabetics, cyclamates were totally unnecessary. They entered the market in response to the recent fetish for losing weight by every possible means except reduction of food intake. Rather than give up soft drinks or even cut down consumption, the weight-conscious public was only too happy to have its soft drinks without their calories. But deprived of energy-yielding sugar, a soft drink becomes a nonfood, an expensive way to restore body fluids compared with water.

Food Dyes

Of all the food additives, nonvegetable or coal tar dyes are at once the most hazardous, based on past experience, and the most unnecessary. No brief can be made for their nutritional or preservative value, only the rather weak rationale that the consumer demands vivid colors in his food —oranges must be orange, hot dogs pink, and cherries scarlet. Many dyes have been demonstrated to be carcinogenic; therefore, many closely related compounds are suspect. But which to suspect is an apparent source of confusion. England allows thirty dyes, the United States fifteen; but the English ban nine of our dyes and we ban most of their thirty. This reinforces real doubt about the wisdom of ingesting *any* coal tar dyes.

Many of the additives mentioned are, of course, combined in certain products; indeed, a few of these products contain little but additives. Maraschino cherries, for example, are preserved with sodium benzoate, firmed with calcium hydroxide, bleached with sulfur dioxide, artificially flavored, then given their bright green or red color with a coal tar dye.

A few years ago a blueberry pancake mix was marketed by two manufacturers depicting luscious blueberries prominently on the packages. The "blueberries," according to the FDA, were made chiefly of sugars, nonfat dry milk, starch, coconut pulp, artificial coal tar coloring, artificial flavoring, and a very small amount of blueberry pulp. The second firm's "blueberries" were of an equally startling composition: sugar, gum acacia, citric acid, starch, artificial coloring and flavoring, and some blueberry pulp.[3] Since these purple pellets were hardly blueberries, the FDA forced the manufacturers to change the labels.

This is not a unique example of exaggeration in the food processing industry. Another product that has reached the point of processing to the absurd is the nonfood synthetic cream. One especially acerbic housewife reacted as follows:

Dear Sirs:
Under separate cover I am returning the free sample of your product, "Dreammix" which was placed in my mailbox. I am a consumer who has

[3] Longgood, W. F. *The poisons in your food.* Simon & Schuster, N.Y.

learned to read labels of foodstuffs. When I read "Hydrogenated Vegetable Oil, Sugar, Sodium Saseinate, Emulsifiers, Preserving Stabilizer, Artificial Flavour, Colour" as the ingredients of this concoction, I decided that your product is foodless trash, not worthy of being ingested into the human body. However, I was faced with a dilemma. Pouring the contents of the box down the kitchen sink might clog the plumbing. Throwing it into the woodstove might lead to an explosion. Placing it on the compost pile might interfere with proper bacterial action. Tossing it on the town dump might be deleterious to scavenging birds and wildlife. Burying it in the ground might harm the soil. So, after due deliberation, I am returning the stuff to you, its creator. May I suggest that your copy writers might have named this product "Nightmare Mix" rather than "Dreammix". On those occasions when I desire whipped cream, I prefer to go straight to the cow.[4]

While doctored cherries, fake blueberries, and make-believe whipped cream can easily be eliminated from the diet, other more basic foods have a depressingly long list of additives on their label. Assuming that the ideal food is transported from the field to the dinner table with the minimum amount of tampering and maximum retention of plain good taste and food value, bread has fallen farthest from the ideal. The wheat grain as harvested is mostly endosperm, a storage tissue of starch and protein attached to the small embryo or wheat germ. Surrounding the grain are the seed coats. They contain vitamins E and B, certain minerals, and amino acids, while the wheat germ contains unsaturated fatty acids.

Previously in flour processing, the grain was ground into flour including both the germ and the seed coats or bran. Thus all of the vitamins and minerals were preserved and passed on through the baked bread to the consumer. But to many, the bread was coarse and heavy, and the flour quickly spoiled by the activities of insects attracted to the rich grain and by the fatty acids in the wheat germ becoming rancid. In the late nineteenth century a new process was developed to roll the grain instead of grinding it. In this process, the starchy endosperm powdered while the oily germ rolled flat. When shaken through sieves, the germ could be separated from the flour. This allowed the flour to be kept longer without spoiling. But flour must be matured or aged to reach its maximum workability; during this period of storage there is an opportunity for insects to attack the flour.

Then it was discovered that nitrogen trichloride or agene (today dyox or chlorine dioxide is used because of potential danger from agene) could mature the flour instantly. But at the same time it bleached the flour white, decoloring the pale yellow fragments of the seed coats left from the milling process. While this produced a flour of indefinite keeping power, it was practically devoid of its vitamins and minerals—vita-

[4] Grant, D., 1961. *Your bread and your life.* Faber & Faber, London.

mins A, B, E, calcium, and iron. What was left was mostly starch and 7–11 percent protein.

In previous years, when bread was an integral part of every meal and fresh fruit and vegetables were not available year round, many people depended upon bread for vitamins. During World War II, white bread was finally enriched by adding vitamins. Bread producers continue to advertise the fantastic energy and vitamin values of their ultrasoft and moist bread, which must often be toasted to attain sufficient substance to retain the contents of a sandwich. Few appreciate the irony of a situation where the flour processor mills out some twenty vitamins and minerals, puts four back, then touts its enrichment!

Of course this need not be, for flour can be milled to preserve much of the nutrition without sacrificing the storage life of the flour, and matured without resorting to vitamin-destroying bleaches. But the industry seems more concerned with the aesthetics of the product than its nutritional value. This faulty sense of priorities, in all fairness, extends well beyond one industry to the consuming public itself—an overconcern with appearance coupled with a lack of awareness or concern for the real purpose of food. If bread is merely a convenient way to sop up gravy, then we might as well use our spoons. If it is a source of vitamins and minerals necessary for maintaining good health, then it is senseless to mill out these vitamins for whatever reason and replace a few at the consumer's additional expense in the name of good nutrition. The alternative is not a whole wheat bread, which some faddists tout as a cure-all, but a flour milled to retain most or all of its native nutrients, obviating the expense of adding them in the baking process.

Because of the decreased importance of bread in our diet, its nutritional value is perhaps a moot point. But there remains a tremendous list of foods with additives that are regularly prepared and consumed. One estimate claims that additives totaled three of the 1400 pounds of food consumed per capita by the American public per year. As the demand for convenience foods continues to increase, the consumption of an even greater variety of additives will follow. For additives have become indispensable in assuring the palatability of the instant heat 'n' eat, shake 'n' bake, brown 'n' serve, or whip 'n' chill products.

Fortunately much processing with its concomitant additives can be avoided by old-fashioned cooking from scratch, mashing your own potatoes, whipping your own cream. While the time lost is inconsequential, the gain in flavor is readily apparent. But for the majority of consumers who apparently wish for even more highly processed foods, much more careful screening and testing is required. A careful re-examination of the "safe" category maintained by the FDA is certainly due. After all, cyclamates were marketed for fifteen years before their possible harmfulness was announced.

Considering the ubiquitous contamination by biocides, hormones, and the spiraling use of additives, we may never again have the luxury of eating simple, nutritious food, direct from the field, orchard, or pasture. But we have every right to be protected from materials that might prove injurious to our health.

FURTHER READING

Bicknell, F., 1961. *Chemicals in your food, and in farm produce: their harmful effects.* Emerson Books, New York. A fairly low-key presentation by an M.D.

Grant, D., 1961. *Your bread and your life.* Faber & Faber, London, England. Grant gives an interesting, somewhat emotional account of bread and its role in nutrition. There is also a fine recipe for whole wheat bread.

Kohlmeier, L. M., Jr., 1969. *The regulators: watchdog agencies and the public interest.* Harper and Row, New York. The ground rules between government and the consumer.

Longgood, W. F., 1960. *The poisons in your food.* Simon & Schuster, New York. As the title suggests, this book is in the muckraking tradition.

National Research Council Food Protection Committee, 1961. *Science and food, today and tomorrow.* NAS–NRS. Publ. 877, Washington, D.C. A somewhat stiff, but revealing, presentation.

Part IV

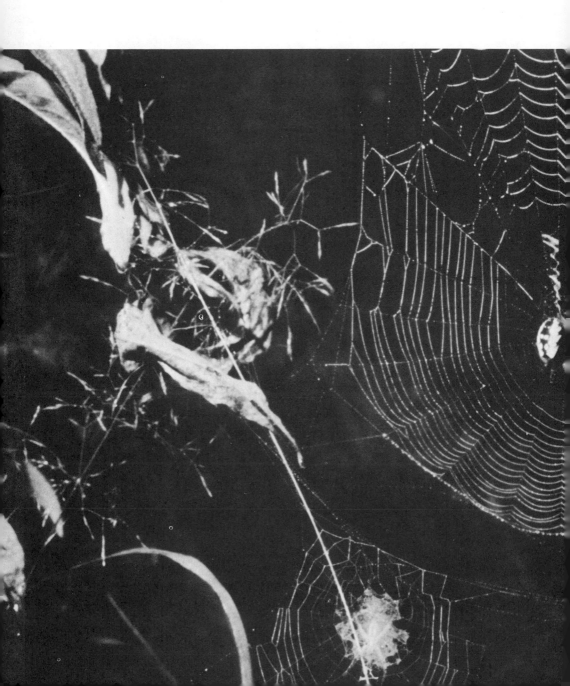

THE BIOTIC WORLD
AND MAN

Photo by J. Paul Kirouac

CHAPTER 15

INTRODUCTION OF EXOTICS INTO OUR ECOSYSTEMS

In the geologic past the continental land masses were much closer together than they are now, allowing a much broader natural distribution of plants and animals than is possible today. At that time there were fewer climatic or geographic barriers to isolate species. Even as late as the Tertiary period, which began sixty-five million years ago, much of the land area of the earth was covered with broad belts of tropic, subtropic, and temperate types of vegetation and inhabited by animals of a more general and wide-ranging sort than we have today (Figure 15.1).

Figure 15.1 If this highly schematized diagram is superimposed over any land mass the pattern of vegetation just before the Tertiary can be seen. Notice the far northern extension of deciduous hardwoods. (Modified from D. Axelrod, *Evolution,* **20,** 1966.)

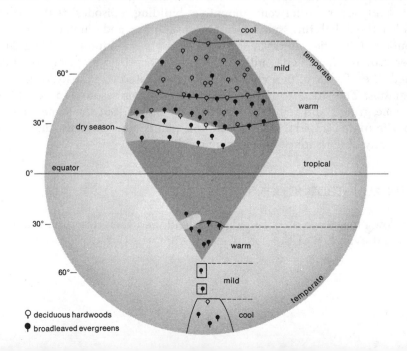

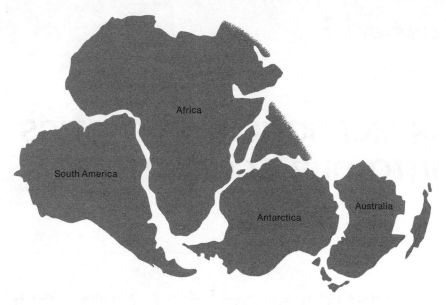

Figure 15.2 A hypothetical land mass called Gondwanaland included the present continents of South America, Africa, Antarctica, and Australia before being broken up by continental drift. (Modified from A. C. Smith and A. Hallam, *Nature*, 1970.)

About 100 million years ago, South America and Africa, Australia and Antarctica, Europe and North America began to be pulled apart by the forces of continental drift—slow moving but enormous convectional cells whose circulation deep within the earth moved the shallowly rooted continents about like the film on scalding milk (Figure 15.2). Continuous land masses began to be broken by straits that became channels, then seas, and finally oceans. Because of continental motion, long quiescent land areas underwent mountain building episodes that further divided the inhabiting species from one another and limited their distribution. Climatic alteration, related both to mountain building and other global phenomena, ultimately led to continental glaciation that ended the Tertiary period and further isolated the distribution of organisms. The stresses produced by these changes resulted in extinction for some groups and reduction in range for others; but the same stresses also stimulated rapid evolution in the plants and animals that were genetically able to adapt to changing conditions.

NATURAL INTRODUCTIONS

The slow emergence of savannah, grassland, and then desert encouraged the rapid evolution of the host of grazing animals that can be seen today in East Africa. These groups reached their greatest geographic isolation towards the middle of the Tertiary period. The later formation of a land connection between North and South America (the Isthmus of Panama) and between Siberia and Alaska (the Bering land bridge) allowed the movement of animals and plants from their isolated centers of specialization and speciation, movements that are still taking place today. Glaciation further mixed the once-isolated stocks of plants and animals.

South America was invaded by tapirs, llamas, peccaries, foxes, dogs, cats, otter, and deer from North America, driving many of the primitive marsupials into either extinction or migration. One marsupial, the opossum, invaded North America from South America and is still extending its range northward. The porcupine and armadillo are also fairly recent invaders of North America. The Bering land bridge allowed elephants and deer to move into North America from Asia, while it also allowed horses, camels, and tapirs to move from North America to Asia.

Usually this process of range extension is extremely slow, but it may be quite rapid. Both the cardinal and mockingbird, formerly limited to the South, were considered rare visitors in Pennsylvania or Connecticut thirty years ago. Today, both birds breed in these northern states and seem to be extending their range farther north. Shortly after World War II, cattle egrets were moved, with the help of storm winds, to South America from Africa. They then worked their way to Florida and now regularly breed along the southern coast from North Carolina to Texas.

By and large, the post-Tertiary plant and animal redistribution had stabilized by the time man arrived on the scene. But once again these relative isolations are being disturbed, not by continental drift, climate, or land bridges, but by the unique mobility of man. The wanderings of man have encouraged the introduction of plants and animals on a scale never before possible. For thousands of years these introductions were limited to crop plants and weeds, domesticated animals and their pests; but more recently, a greater and more varied number of plants and animals has been shifted about, accidentally or purposefully—sometimes with catastrophic results.

MAN THE INTRODUCER

While other large land areas were receiving exchanges of species through land bridges, Australia and New Zealand remained isolated. For over seventy million years New Zealand and Australia have been isolated from the floral and faunal developments that swept most of the rest of the world. These continents have a unique fauna and flora containing many species found nowhere else in the world, while the plants and animals characteristic of the rest of the world, especially Europe, were, until

Figure 15.3 This painting, *The Peaceable Kingdom,* by the American artist Edward Hicks, depicts the peaceful coexistence of man and animals. Unfortunately the introduction of a variety of disparate animals is never peaceable. (Courtesy of the Brooklyn Museum)

recently, altogether lacking. New Zealand, for example, had only two native species of mammals, both bats. The Maoris, early Polynesian settlers of New Zealand, brought in dogs, but it was not until white settlement that widespread introduction of animals began.

Because New Zealand was so far from their home in the British Isles, halfway around the world, the early white settlers were wretchedly homesick. Not only were there no animals to remind them of home, but the shrubs, trees, and flowers bore no resemblance to anything known in England. So great was this feeling of isolation and homesickness that the strange plants and animals were attacked as an evil to be replaced with something familiar. (The same attitude might have taken root in North America had the plants and animals not resembled those of Europe so closely.) Plants and animals began to be introduced at a rapid rate so that New Zealand would look like England as quickly as possible. What followed was an orgy of introduction. In the 120 years since white settlement, 207 species of vertebrates have been introduced. Of these, ninety-one became established, a fairly high rate of success due almost certainly to the lack of competition from native mammals. With such an auspicious start one might picture New Zealand today as a kind of zoo without bars like that famous primitive American painting, *The Peaceable Kingdom* by Edward Hicks (Figure 15.3). But of the ninety-one successful introductions, twenty-nine became problems of one sort or another.

Originally native to Europe, red deer were introduced early to New Zealand. With no serious grazing competition and no predators, there were soon many more deer than could be controlled by hunting. They began to compete with sheep, destroy crops, inhibit forest regeneration, and finally, by overgrazing, to cause serious soil erosion. Today there is no limit or closed season on red deer. In fact, the government offers free ammunition to anyone who will kill them, but the red deer are still flourishing.

A similar story can be told of Australia. Unlike New Zealand, Australia had many vertebrates, but because of Australia's long isolation its vertebrates were mostly the primitive marsupials, possum, kangaroos, wallabies, and the like, which were generally outcompeted when faced with the more advanced mammalian stock introduced by settlers. Like the New Zealanders, new Australians missed the familiar animals of home and soon introduced the rabbit. To their credit, they introduced the fox as well, as a predator. But in Australia foxes and rabbits seemed to go their own ways; the fox never amounted to much as a rabbit predator while the rabbit became one of the greatest pests of all time, as will be discussed later in this chapter.

The Biological Context

What went wrong? Why did perfectly well-behaved European animals suddenly become pests? No animal or plant exists in a world limited solely by physical and chemical factors. With few exceptions, for every plant or animal there is a potential consumer or predator that serves to check its population growth. There are also parasites, and parasites of parasites. A few years ago, John Tilden took a long careful look at a plant that grows on the sand dunes of California and Oregon. "Associated with it he found, by two years of systematic observation, 257 species of arthropods of which 221 were insects. On the latter 65 were parasites and that was not complete. There were 51 species of primary herbivores—leaf-nibblers, leaf miners, stem-borers, leaf-suckers, root feeders, gall-makers; 23 species of predators; 55 species of primary parasites; 9 of secondary parasites; and even one tertiary parasite."[1] All of these organisms interact so that there are rarely too many or too few objects of their predation. This is why in a stable ecosystem there are

[1] Elton, C. S., 1958. *The ecology of invasions by animals and plants.* Methuen, London.

usually relatively constant numbers of plants and animals from year to year. But when you pluck such a plant or animal from its ecological context and fling it into another environment it will either disappear almost immediately or proliferate with unbelievable fecundity. This is a direct measure of the importance of ecological associations in maintaining stable populations.

Even today, we know far too little about the ecological relationships of plants or animals to be able to predict which course the introduced plants or animals will take in a new environment. Obviously elephants would be short-lived in Labrador, but what about zebras in New Mexico or bamboo in Louisiana? Of course not all introductions have been disastrous, nor have they all been intentional; let us examine the record.

Types of Introductions

Few insects have been intentionally introduced except in a desperate attempt to biologically combat proven pests, but accidental introductions have been rife. The Japanese beetle has already been discussed in Chapter 12. Of even greater impact has been the gypsy moth. Leopold Trouvelot, a French astronomer employed by the Harvard Observatory, tried to crossbreed several silk-producing moths with the aim of finding a strain resistant to a disease threatening the French silk industry. One of his subjects was the gypsy moth, notorious in Europe for its destructive effect on the leaves of shade trees. In 1869 Trouvelot accidentally allowed some gypsy moths in his collection to escape near his home in Medford, Massachusetts. Seemingly, they disappeared into the blue. But twenty years later in 1889, Medford was overwhelmed by an incredible swarm of caterpillars.

> The numbers were so enormous that the trees were completely stripped of their leaves, the crawling caterpillars covered the sidewalks, the trunks of the shade trees, the fences and the sides of the houses, entering the houses and getting into the food and into the beds. . . . The numbers were so great that in the still, summer nights the sound of their feeding could plainly be heard, while the pattering of their excremental pellets on the ground sounded like rain. Valuable fruit and shade trees were killed in numbers by their work, and the value of real estate was very considerably reduced. . . . To read the testimony of the older inhabitants of the town, which was collected and published by a committee, reminds one vividly of one of the plagues of Egypt as described in the Bible.[2]

In response to this scourge, the state government was persuaded to employ some control measures which were moderately successful until

[2] Howard, L. O., 1930. *A history of applied entomology (somewhat anecdotal).* Smithson. Misc. Coll., 84.

1901 when Massachusetts decided enough money had been spent killing caterpillars and canceled the appropriation. In 1905 the gypsy moth infestation was not only out of control again but beyond the resources of state action; the federal government stepped in and has been attempting to prevent further spread of the gypsy moth ever since (Figure 15.4). For

Figure 15.4 (a) This young maple was attacked by gypsy moth caterpillars early in the summer. The green leaf blades were soon skeletonized to the tough veins and petioles, which are still attached to the twigs. (Photo by Dan Jacobs) (b) Two months later this same tree has put out a second set of leaves, completely obscuring the earlier attack.

many years, spread of the infestation was checked by the Hudson River. Then a series of hurricanes carried moths to the west, embroiling the federal government in a controversial mass spraying program over much of the northeast. Taxpayer suits finally dissuaded the shotgun approach and today other approaches to control are being explored. These approaches are discussed later in this chapter.

Practically every crop in the United States has at least one introduced pest of economic importance. Introductions have not been one-way, however. Two examples of a United States native causing problems abroad are the Colorado potato beetle and the grape phylloxera. The potato beetle, when originally discovered, was feeding quite contentedly on a wild relative of the potato in the Rocky Mountains. In 1859 potato cultivation came to Colorado. Sometime that year the beetle discovered the domestic potato. By 1874 the beetle had reached the Atlantic coast and a short time later it was inadvertently carried to Europe. In 1930 most of western, central, and southern Europe had become infested.

During the Korean War, North Korea and China accused the United States of dumping canisters of potato beetles on their potato fields. But in view of the expansion of its range in the hundred years since its discovery, the Colorado potato beetle hardly needs the assistance of the CIA to manage its dispersal!

The phylloxera is an insect that sucks juices from the rootstock of grape vines. Native to the grapes of eastern North America, the phylloxera usually does no particular damage. When introduced into France on the rootstocks of American grapes just before the Franco-Prussian war in 1870, the result was catastrophic. At a time when France could least afford further calamity, vines began dying throughout her world-famous wine districts. For a few years grape juice for wine production was limited to that wrung from dried grapes shipped in from uninfected areas abroad, surely a low point for the French spirit. Within ten years the insect was killing vines in Germany and Italy as well. But the wild American grape came to the rescue by providing resistant rootstocks to which the superior European grapes could be grafted. In another few years, production was back to normal with the curious result that today all French grapes have "American" roots.

Introduction of fish. Fish introduction *seems* to have caused few problems. But we have no way of being sure since displacement and even extinction of anything but the most popular game fish go totally unnoticed by most people.

Probably the most widely introduced fish is the mosquito fish, which is native to the southeastern United States. The mosquito fish has been introduced into subtropical and mild climates all over the world because of the avidity with which it feeds on mosquito larvae, hence the effectiveness with which it controls the incidence of malaria. Although its total ecological effect beyond mosquito control has never been assessed in any of the areas into which it has been introduced, the mosquito fish has apparently caused remarkably few problems.

Trout and salmon have also been widely introduced into areas where native sport fish are absent or poorly regarded. Again, the impact of these exotics on native fish has never been investigated, but no one is apt to complain about the abundance of trout or salmon in a stream, and the side-effects have not yet been noticed.

While fears of released piranha fish tearing Florida swimmers to ribbons have not yet been realized, a catfish introduced from Asia has caused some concern in southern Florida. Able to breathe out of water for up to twelve hours and wriggle cross-country, these catfish can live under conditions lethal to other species. Naturally the Florida Fish Commission is concerned that with its voracious appetite the "walking" catfish will outcompete more desirable freshwater fish.

Another fish widely and thoughtlessly introduced by people is the goldfish. Most people have compunctions about flushing unwanted gold-fish down the toilet so they release them in nearby ponds or streams where they often reproduce vigorously to the exclusion of more desirable fish—goldfish may be pretty, but tasty they are not.

Introduction of birds. For centuries, birds as pets have been moved about the earth at the whim of man. During the Victorian age this effort reached an almost frenzied peak. In city after city acclimatization so-cieties were formed to introduce the avian glories of Europe and the East to the urban areas of the northeastern United States. Birds such as the nightingale, skylark, and English robin were introduced again and again, but without success; even the English sparrow (house sparrow), which had long been regarded as a pest in Europe, was repeatedly introduced. Some measure of the extent to which this craze gripped otherwise conservative institutions can be seen in the purchase and re-lease of 1000 sparrows by the city government of Philadelphia in 1869.

Unfortunately, the sparrow succeeded where the nightingale and skylark did not (Figure 15.5). Soon the noisy, dirty, little birds were everywhere, chasing away native birds, nesting in great piles of trash and under eaves and eating large quantities of grain. By 1883, Phila-delphia thought better of its humanitarian action and passed a law making it lawful to kill or destroy in any way possible the bird it had so enthusiastically introduced.

Before long, however, the house sparrow had extended its range all over the United States and Canada. But like the house fly, the Norway rat, and the common pigeon or rock dove, the sparrow was never able to sever its connection with man. It was unwilling or unable to wander too far from the largesse provided by man's disturbance of the environ-ment—the true measure of a pest or weed. In a few decades, house spar-

Figure 15.5 Much of the success of the house sparrow in the urban environment results from the largesse provided by man and eagerly accepted by sparrow. (Fox Photos)

row populations gradually declined, perhaps because of the decline of the horse when the automobile came into widespread use. Horse droppings undoubtedly provided a nutritional margin that helped carry sparrow populations through severe winters. Today house sparrows remain common and seem to have made a secure niche for themselves as an accompaniment of human civilization, but they are nowhere overly abundant.

Another bird associated with man and his cities is the rock dove or city pigeon (Figure 15.6). Kept alive and thriving almost exclusively by what they can glean from abundant city garbage or crumbs generously scattered about by well-meaning urbanites eager for the sight of any bird, pigeons have increased into enormous flocks in many cities. Ledges, windowsills, and the exuberant architectural ornaments on older buildings are used as nest sites and the pigeon's copious excrement has become one of the vicissitudes of life in most cities. Pigeons have persisted despite occasional efforts to reduce their numbers. It has become clear, however, that pigeons are capable of transmitting several diseases to man—parrot fever (psittacosis or ornithosis), cryptococcal meningitis, histoplasmosis, and encephalitis. Today, efforts to control urban pigeon populations are taken more seriously. By generously spreading pigeon feed treated with a contraceptive drug in suitable gathering places, large numbers of pigeons are rendered infertile, allowing the ravages of age,

Figure 15.6 Pigeons in the city provide more than local color: they can carry disease and disfigure valuable property. (Fox Photos)

alley cats, and taxicabs gradually to reduce the pigeon population to manageable terms. There are still pigeons in New York City, but not the huge flocks that there once were.

Land animals. In 1859 Thomas Austin imported a dozen pairs of European rabbits to Australia and released them on his ranch. Six years later, after killing twenty thousand of them, Austin still had ten thousand left. This fecundity is quickly explained by noting that a female rabbit can breed at four months, then average six litters of six young (called kittens) a year thereafter. Theoretically, one pair of rabbits can generate about 13.7 million offspring in three years! Fortunately, mortality is around 80 percent and adults rarely live more than one year. Nonetheless, in a few short years Australia was faced with a national problem because five rabbits ate as much grass as one sheep. Trying to capitalize on the grave mistake of introducing the rabbit, Australia soon became the world's major supplier of rabbit meat and rabbit fur. But the loss of sheep range more than balanced any economic gain from the rabbit. Besides direct grazing competition, the rabbit burrows extensively undermined buildings, initiating soil erosion and further ruining range land.

All means of control were attempted—trapping, shooting, netting, poisoning, fumigation of burrows, and even erecting a 1500-mile fence designed to keep the rabbits from moving into western Australia, but these measures were of little avail. The combination of ideal habitat and minimal natural predation allowed the rabbit population to increase by 1953, despite control measures, to an estimated *one billion rabbits* over an area of a million square miles (Figure 15.7). Some of the economic burden this rabbit population placed on the Australian economy can be seen by the calculation that if there were no rabbits in Australia the land could support another 100 million sheep. Recently the rabbit population was reduced considerably by biological control. This kind of control will be discussed later in this chapter.

Australian experience with the rabbit is perhaps an exception, for although there have been problem introductions of muskrat into Europe and nutria into the United States, the most severe problems have usually arisen when various mammals have been introduced to small islands. Occasionally chains of errors have followed an initial introduction. Many years ago rabbits were introduced as a supplemental food source on some rather barren windswept islands southeast of Australia. When the rabbits began to compete with the sheep, the economic mainstay of the islands, cats were brought in to reduce the rabbit population. But the cats preferred the eggs of sea birds nesting on the islands and used by the natives as food. In desperation, the islanders brought in dogs to control the cats; however, while the dogs merely chased the cats, they attacked and ate seals on beaches, depriving the islanders of still another food source. In view of the wide range of problems stemming from introduced animals it is heartening that most introductions are unsuccessful.

Figure 15.7 Rabbits gathering at a water hole in Australia during a drought. At the time⁕ this photo was taken there were one billion rabbits in Australia. (Australian News and Information Bureau)

Plants. One of the most important forest trees in eastern North America until about 1900 was the chestnut. A singularly useful tree, the chestnut tree grew tall and straight providing good saw timber; the bark and wood had a high tannin content, which was used in the tanning of hides and which preserved the wood from rapid decay in contact with soil, making the chestnut especially useful for fence posts and rails; the nuts were of excellent quality for roasting and provided much food for game and livestock; and finally, it was a fine shade tree, widely planted and as widely admired:

> Under a spreading chestnut-tree
> The village smithy stands;[1]

The village smithy went out with the horse and buggy but unfortunately so did the chestnut. A wealthy New York gentleman imported a number of foreign chestnut trees to plant on his Long Island estate just before the turn of the century. A few years later he noticed that his American chestnuts were dying. In a few more years, chestnuts were dying all over Long Island, then New York, Connecticut, and New Jersey. By 1940 the chestnut no longer existed as a forest tree (Figure 15.8). The cause of its demise was a fungus introduced on trees brought from China. The Chinese species of chestnut apparently evolved with the fungus and developed a tolerance to it, but the American chestnut

[1] Longfellow, H. W., 1903. *The Complete Poetical Works of Henry Wadsworth Longfellow,* Houghton Mifflin, Boston.

Figure 15.8 In many parts of eastern North America the chestnut was a dominant forest tree. When killed by the chestnut blight its loss was strikingly apparent; particularly in the southern Appalachians where this photo was taken some years ago. (U.S. Dept. of Agriculture)

had no such immunity; in fact it proved extremely susceptible. The fungus slowly girdles the trunk of the tree, filling the water-conducting vessels of the wood with its growth and the crown of the tree, deprived of water, slowly dies. Because the roots remain viable the chestnut continues to sprout, but just enough fungus spores remain in the environment to kill the sprouts after a dozen years or so, preventing the regrowth of any sizable tree. Some chestnuts have been vainly sprouting in this manner for over fifty years. Although the loss of such an important tree was staggering, because of the resistance of the wood to decay, dead trees are still being cut in the southern Appalachians, where they remain a continuing source of the split-rail fencing so popular in suburbia today.

CONTROL OF INTRODUCTIONS

As the list of disastrous introductions lengthened, man slowly became aware that although he had the unquestioned ability to transport organisms—moose, mice, or measles—wherever he pleased, he could not pull the strings that would make these organisms behave as he wished.

Quarantine

The most obvious solution to the problem of introductions is to avoid them in the first place. Most countries have strict regulations prohibiting random importation of any plant or animal stock which might cause problems or which may be a carrier of potentially harmful pests or diseases. The United States requires that all plants entering the country from abroad be fumigated or carefully inspected to eliminate plant diseases or insect pests. Imported wild animals must spend a period of time in quarantine in Clifton, New Jersey, to be sure the animals are disease free. Even then under federal law they cannot be released in this country without a permit from the Secretary of Agriculture. Such strictures in force in the United States have sharply decreased the number of new introductions since 1900. But despite these measures, unwanted plants and animals still occasionally get through.

Eradication

If the infestation is discovered quickly enough eradication is usually attempted. Hoof and mouth disease in cattle or the Mediterranean fruit fly on oranges in Florida are examples where the unwanted organism was eliminated by quick action before it was well-established or spread too far. Eradication is far more easily envisioned than accomplished. Too many times eradication is initiated when the chance for success does not merit the potential damage to other species; this was the case in the previously discussed campaign against the fire ant in the southeast, and the Japanese beetle in the Midwest.

Biological Control

Should an organism evade both border quarantine and eradication attempts and become established, there is a third line of defense that does not necessitate an endless series of biocide sprays. Every organism, as we have seen, is surrounded in its natural habitat by a plethora of other organisms with which it has evolved and to which it is related through producer-consumer, predator-prey, or parasite-host bonds. If the new habitat of an introduced species is a reasonable facsimile of the old and if the web of associates is left behind, as it almost always is, there are few population checks and the species may "explode."

Why are organisms new to a habitat not immediately eaten and eliminated or at least assimilated? The bonds mentioned are developed over long periods of time and are quite conservative. Interlopers have no place. A bird is as loath to sample a beetle it has never seen before as a housewife is to buy rattlesnake meat at the supermarket. After many years, however, predator-prey relationships slowly form and the species is gradually "absorbed" by the pre-existing species as the house sparrow and Japanese beetle have been. But the period of "absorption" may take hundreds of years.

The point of biological control is to speed up this process of natural "absorption" by introducing whatever consumers, predators, or parasites are available from the old habitat that will bring an introduced species under control in the shortest possible time. Biological control is not a new approach, but since it is neither quick, easy, nor necessarily successful, biocides were for a time considered the best solution—that is, until their drawbacks became too glaring to ignore. Today there is renewed interest in the possibilities of biological control.

The first step in applying biological pest control is to know the life cycle of the pest intimately to see where the weak points are, and where controls might be successfully applied. Then the original habitat is carefully studied so that the interrelationships between the control species and the target species are fully understood. Before attempting any introduction of control agents, careful screening must be carried out to be sure that a second pest is not introduced. A convenient rule is that an introduced control species should prefer to starve to death rather than attack some species other than the pest to be controlled. The new habitat must also be generally suitable to the intended control species. Finally, the control species must be reared or cultivated in large enough numbers so that when introduced there is a fair chance of success. Then cultural practices should be modified wherever possible to encourage survival and reproduction of the control species. When these steps are carefully carried out with one or more control species, the target species *may* be effectively controlled.

Despite the time, expense, and uncertainties, many pests have been quite effectively controlled, some spectacularly, using biological methods. Just about any organism from virus to sea cow can be used as a biological control.

Viruses. The European spruce sawfly was introduced into eastern Canada in the 1920s, and before long it had destroyed 3000 square miles of timber. When a virus important to its control in Europe was accidentally introduced, the insect was practically eliminated in three years. Even more spectacularly, the Australian rabbit problem was finally but quickly brought under control by a virus disease, *myxomatosis*. Indigenous to rabbits in Brazil where it causes a minor skin disease, myxomatosis was incredibly virulent when introduced to captive Australian rabbits. After determining that the virus affected only the rabbits, infected rabbits were released, and before the end of a year the virus, spread by mosquitoes, was killing rabbits by the hundreds of thousands. After a couple of years the virulence began to decline, eliminating the possibility of extermination, but the measure still provided better control than could have been achieved by any other method.

The appeal of viruses as a biological control lies in their specificity in contrast to biocides. Of the 280 viruses isolated from various insects, the most promising belong to a group called the nuclear polyhedrosis group. Normally, these rod-shaped viruses are encapsulated in a proteinaceous sheath that insulates them from the environment. This coat is dissolved in the insect gut, and the activated virus penetrates the insect cells, ultimately killing them. When properly diluted, evenly distributed, and applied at the right stage of insect development, viruses can be extremely effective.

Viruses are not yet in commercial production because the FDA is quite properly being cautious about possible effects on other organisms. Tests using the corn earworm virus on a variety of animals and preliminary tests on man have shown no ill effects. As soon as additional tests are completed, viruses will probably be licensed and marketed. If proven safe, as well as successful, other viruses capable of attacking other insects may well replace some of the biocides in use today (see Chapter 12).

Bacteria. For a number of years one bacterium (*Bacillus thuringiensis*) has been commercially produced and used as a means of controlling the cabbage worm and alfalfa caterpillar. The bacterium is extremely virulent and spreads rapidly.

Another related bacterium discovered in New Jersey in 1933 attacks the grubs of Japanese beetles. Because the infected grubs turn white, the infection is called the milky disease. The spores, which are resistant to extremes of temperature and moisture, are easily applied to the soil as a dust. Since the spores pass unharmed through the digestive tract of any animal eating the infected grubs, they remain viable and slowly spread throughout the environment. When large areas are treated, very effective, long-lasting control of Japanese beetles can be obtained without the destructive side-effects of mass aerial spraying (Chapter 12). Between 1939 and 1953 the USDA in cooperation with various state and other federal agencies distributed 230 thousand pounds of spore dust at 160 thousand sites in fourteen eastern states. Much of the reduction in the Japanese beetle population over this period may well be due to the slow but effective spread of this bacterial infection.

Insects. The most widespread and certainly the oldest form of biological control involves insects. Successful control of both plant and insect pests by insects has been obtained many times. Two examples illustrate the possibilities.

The prickly pear cactus was introduced into Australia as early as 1788. It was used then as it still is in Mexico and the southwestern United States as a defensive or sheltering hedge around ranch houses. But by 1870 it began to get out of control. In 1900, ten million acres were covered with the cactus, and by 1920, sixty million acres were infested. Both mechanical and chemical controls were too costly because of the huge areas affected. As early as 1912 research was begun to survey the natural enemies of the prickly pear. Of the several possibilities a plain brown moth called *Cactoblastis* was selected and reared in the laboratory. Over a ten-year period nearly three billion eggs were distributed. The eggs hatch into caterpillars which tunnel inside the flat cactus segments, the stems, and even the roots. When this is followed by bacterial rots, the plants are quickly destroyed. By 1933, prickly pear was under effective control in Australia and although the cactus was not eradicated, there is a large enough resident population of *Cactoblastis* to prevent any widespread regrowth of the prickly pear.

Not long after oranges had been introduced to California on a large scale in the late nineteenth century, the cottony-cushion scale, a sucking insect that weakens the tree and reduces the orange crop, was introduced from Australia. A search in Australia by C. V. Riley, an early pioneer of biological control, turned up a ladybug called the vedalia beetle, which when introduced to California proved extraordinarily successful in reducing damage (Figure 15.9). So successfully did the vedalia beetle control the scale, that it began to be taken for granted. When broad

Figure 15.9 These vedalia beetles are feeding on the large white cottony-cushion scale. The populations of this pest are effectively reduced by introducing the vedalia beetle. (University of California, Department of Entomology, Division of Biological Control, Riverside)

spectrum biocides were first used on orange groves in California seventy years later and the vedalia population declined, cottony-cushion scale again became a serious pest almost overnight.

Sex attractants. The female gypsy moth is such a weak flier that males must seek the female out over a great distance. To guide the male to his destiny, the female exudes a powerful scent which can be smelled by a male several miles away. By grinding up female moths, a lure was developed that could attract males from miles around to a trap or a poison target where they could be killed. Today this scent, identified and synthesized, has been successful in helping to control the gypsy moth. The possibilities of finding and using other sex attractants offer another useful approach to the selective control of pest insects. The insect cannot develop resistance to its own sex attractant without destroying its normal reproductive behavior, so the attractant should retain its effectiveness indefinitely.

Sex sterilants. The object of sterilization is to interfere as unobtrusively as possible with the reproductive cycle of a pest organism so that everything progresses normally except that eggs will not be fertilized and the reproductive cycle will be broken. This has proven quite successful with pigeons, as previously mentioned, and has been used with insects as well. A chemical sterilant is currently being developed that is intended to halt sperm reproduction in rats permanently. One dose is calculated to do this effectively without adverse effect on the rat's sex drive or general health. A female mated with a sterile male displays pregnancy symptoms but does not mate again for eleven days, twice the usual nonreceptive period. This measure reduces the number of possible litters by half and promises to be quite effective in lowering the rat population.

Know your organism thoroughly, and then focus on a weak point in its life cycle. This rule was followed successfully by E. F. Knipling, using radiation on the screwworm fly. The adult is a metallic blue fly about three times the size of a house fly that lays eggs in wounds or even in the navels of newborn warm-blooded animals. The feeding maggots cause a sore whose discharge attracts more flies until hundreds or even thousands of maggots may infect the animal. After five days the maggots drop to the ground, pupate, then emerge as adults. In a few days the adults mate and the female lays her eggs completing the life cycle. In warm regions 10–12 generations a year are possible. The screwworm is limited in the United States to southern Florida and the southern parts of Texas, New Mexico, Arizona, and California.

Knipling's idea was to expose the screwworm pupae to a dose of radiation that would sterilize the males without other side effects; 2500 r was found to do this nicely. Since the females mate but once, mating with a sterilized male would result in sterile eggs. It was found that a ratio of 9:1 sterile to fertile males would result in 83 percent sterile matings. After some preliminary tests, an eradication program was begun to eliminate the screwworm fly from Florida. Using assembly line methods in an airplane hangar, fifty million screwworm pupae per week were irradiated, both male and female, then released. Over a period of eighteen months two billion flies were released over seventy thousand square miles of Florida, southern Georgia, and Alabama. The program was begun in January 1958. One year later the screwworm fly was eradicated from the southeastern states. Reinfestation is prevented by the colder Gulf coast between Florida and Texas which eliminates the possibility of reinfection from overwintering flies and by animal inspection stations which check all livestock crossing the Mississippi.

After this outstanding success, follow-up studies are investigating the possibility of using the method against other large-scale pests such as the gypsy moth, codling moth, boll weevil, and others. But the most valuable lesson demonstrated by this type of biological control is that introduction of sterile but otherwise normal species into a population has far greater impact in reducing the population than simply removing or killing an equal number of individuals. If successful with other major pest species, this technique of sexual sterilization will make biocide pest control unnecessary.

If plants and animals were introduced with care to include the complex consumer-predator-prey-parasite organisms associated with the introduced species, populations would stay under control. But introductions have always been random, a species here and a species there which preclude any short-term stabilization. Population regulation requires an enormous input of energy. Too often we have panicked and caused more problems in our clumsy efforts to exterminate the inexterminatable.

The wisest policy is to avoid introduction until enough is known about the species both in its native home and in the new habitat to evaluate the possible environmental repercussions of the introduction. Once a species is introduced we should learn to live with it, but at the same time we should exploit every means of natural control at our disposal for long-term control by biocides is not only ineffective but self-defeating.

Every species has in its life cycle some point where natural control can be effectively exerted. By seeking out these weak points and exploiting them, man can work *with* natural processes rather than expend large amounts of energy on trying to overwhelm them. Someday man may come to the realization that much of his so-called struggle for existence is unnecessary.

FURTHER READING

Debach, P. (ed.), 1964. *Biological control of insect pests and weeds.* Reinhold, New York. The standard text.

Elton, C. S., 1958. *The ecology of invasions by animals and plants.* Methuen, London, England. Beautifully written classic account by one of the most respected ecologists of the twentieth century.

Laycock, G., 1966. *The alien animals.* Natural History Press, New York. Easy reading in a popular style of the problems encountered when exotic plants and animals are casually introduced.

Mashall, A. J. (ed.), 1966. *The great extermination.* Heinemann, London, England. Documents, in a quite personal way, the loss of plants and animals since settlement in Australia.

Rolls, E. C., 1969. *They all ran wild.* Angus and Robertson, Sydney, Australia. Good account of the introduction of the rabbit into Australia and subsequent problems.

CHAPTER 16

EXTINCTION: THE LORD GIVETH AND MAN TAKETH AWAY

Extinction is a natural process. All organisms are replaced sooner or later by better adapted, newly evolved forms. Although some species like the horseshoe crab have remained relatively unchanged for hundreds of millions of years, the unchanged existence of most species is very much shorter. The average species of bird, for example, lasts for about forty thousand years. A species may become extinct and be replaced by another species, or it may gradually evolve into one or more new species and in this way become "extinct." A good example of this latter process is the evolution of the Darwin or Galapagos finches.

Apparently, a single species of finch found its way to the isolated Galapagos Islands 600 miles off the coast of Ecuador, perhaps a million years ago. Competition within the species, combined with an abundance of available unfilled niches, allowed the slight variants comprising the species to evolve into a number of specialized forms—a seed eater, an insect eater, a fruit and berry eater, and, most spectacularly, a woodpecker-type which instead of developing a woodpecker-like tongue and beak learned to use a thorn to probe nooks and crannies for insects. Today the original finch species that first came to the islands has been replaced by its specialized descendants and no longer exists.

Extinction may often be more abrupt than this and a particular genetic line may end and be replaced by functionally similar but genetically different forms. It is probable, for example, that grazing dinosaurs were replaced by grazing marsupials, and these finally by higher mammals; the niche remained the same, but the animals filling it changed radically. An important aspect of natural extinction is that niches remain occupied.

With the entrance of man as an ecological factor there has been a shift from the gradual replacement-type extinction to abrupt niche-emptying. Species that have undergone a long and quite natural decline resulting from their inability to cope genetically with some change in their environment, or that have become overspecialized to the point of inflexibility, have become endangered or quite suddenly wiped out (Table 1).

Table 1 **Some endangered animals in the United States[a]**

Animal	Number in coterminous United States
Timber wolf	400
Grizzly bear	850
Black-footed ferret	Unknown, but rare
Southern sea otter	500
Florida panther	100–300
Guadalope fur seal	600
Florida manatee or sea cow	Unknown, but rare
Key deer	300
California condor	40
Florida everglade kite	10
Southern bald eagle	500
American peregrine falcon	Extinct south of Canada 5000 in Alaska and Canada
Attwater's greater prairie chicken	750
Masked bobwhite	Extremely rare
Whooping crane	70
Eskimo curlew	Extremely rare
Puerto Rican parrot	200
Ivory-billed woodpecker	Probably less than 12
Kirtland's warbler	Less than 1000
Ipswich sparrow	Less than 100
Pine barrens tree frog	Unknown, but rare
Atlantic sturgeon	A few thousand, probably
Black fin cisco	Very rare—last specimen taken in 1955.
Atlantic salmon	Rare in rivers where formerly abundant

[a] Source: *Rare and endangered fish and wildlife of the United States, 1966.* Bureau of Sport Fisheries and Wildlife, Resource Publication No. 34.

At times it seems that man is bent on reducing the thousands of species of birds and mammals to a few dozen that he considers desirable; the rest are viewed as expendable. The way man has attacked the organisms of the world has varied from outright assault to insidious nibblings, both of which have had the same destructive result. In this chapter we will look in greater detail at the negative impact that man has had on his fellow organisms.

EXTINCTION BY DIRECT ASSAULT

Many animals have disappeared simply because they were edible. Others have become extinct because they became fashionable in man's eyes. A few examples of more recent extinctions are discussed here.

The passenger pigeon. The passenger pigeon, a bird half as large again as the mourning dove and at least as savory, was probably the most abundant bird ever encountered by man (Figure 16.1), yet in a few decades he had hunted it to extinction. In 1810 the ornithologist Alexander Wilson saw a flock of pigeons in Kentucky, which he calculated to be 240 miles long and a mile wide. Wilson estimated the flock to contain 2.23 billion birds! Unlike geese or ducks, which flock only during migration and never in numbers approaching this scale, passenger pigeons were gregarious throughout the year. Nesting was erratic, depending upon the availability of food. Usually, when beechnuts were plentiful, the birds nested in Michigan and Pennsylvania; when acorns were abundant, Wisconsin and Minnesota were favored. In 1871 one of the largest nestings ever observed took place in Wisconsin. The pigeons nested in almost every available tree over a strip seventy-five by fifteen miles covering over 850 square miles. Anywhere from five to a hundred

Figure 16.1 A popular sport in the early nineteenth century was shooting wild pigeons, here the passenger pigeon. (Bettmann Archive)

nests were built in each tree. Such a nesting naturally attracted hunters from all over the country. One conservative estimate placed the number of pigeons in that nesting area at about 136 million.

The males and females sat on the nest alternately, the males leaving at daybreak, returning at midmorning to relieve the females, who fed until early afternoon, returning to allow the males a final afternoon feeding. This traffic to and from the nesting area presented a unique opportunity for hunters. One eyewitness gives this account of the proceedings:

> And now arose a roar, compared with which all previous noises ever heard, are but lullabys, and which caused more than one of the expectant and excited party to drop their guns, and seek shelter behind and beneath the nearest trees. . . . Imagine a thousand threshing machines running under full headway, accompanied by as many steamboats groaning off steam, with an equal quota of R.R. trains passing through covered bridges—imagine these massed into a single flock, and you possibly have a faint conception of the terrific roar following the monstrous black cloud of pigeons as they passed in rapid flight in the gray light of morning, a few feet before our faces. . . . The unearthly roar continued, and as flock after flock, in almost endless line, succeeded each other, nearly on a level with the muzzle of our guns, the contents of a score of double barrels was poured into their dense midst. Hundreds, yes thousands, dropped into the open fields below. Not infrequently a hunter would discharge his piece and load and fire the third and fourth time into the same flock. The slaughter was terrible beyond any description. Our guns became so hot by rapid discharges, we were afraid to load them. Then while waiting for them to cool, lying on the damp leaves, we used, those of us who had [them], pistols, while others threw clubs, seldom if ever, failing to bring down some of the passing flock.[1]

Hunting was not limited to birds on the wing; birds were attacked on the nest with sticks, woods were set afire, and trees were chopped down. After a few more years of this kind of hunting, people began to wonder where the pigeons had gone; some said Canada, others, Australia. But there was no escape for the passenger pigeon from the unremitting pressure of the hunter. Why was the passenger pigeon not saved from extinction? Surely a few thousand birds of all those billions could have been preserved.

Here the bird's social habits worked against him. When the large flocks were broken up into hundreds or even thousands, the breeding instinct was somehow inhibited and nesting was sporadic or not even attempted. With continued hunting and cutting of the hardwood forests which supplied most of their food, nesting, and roosting sites, the pigeons

[1] Schorger, A. W., 1937. "The great Wisconsin passenger pigeon nesting of 1871." *Proc. Linn. Soc. N.Y.*, 48, pp. 1–26.

faded into oblivion. Once extinction was apparent and perhaps unavoidable, a great wave of remorse swept the land, but to no avail, and the last passenger pigeon, an aged female called Martha, died on September 1, 1914, in the Cincinnati Zoo. We must conclude that the passenger pigeon could not be saved: when reduced to numbers congruent with the space we were willing to share with it, reproduction was apparently impossible and extinction inevitable.

The buffalo and the great auk. Similar sentiment bathed the last pitiful remnants of the buffalo that once blackened the American prairie. However, we are forced to admit that the buffalo and the passenger pigeon would not be compatible with present day use of either the Great Plains or the eastern deciduous forest. Picture today a herd of even one million, much less the original sixty million buffalo, wandering from Texas to Montana and back every year, or a flock of one million passenger pigeons settling on an Ohio county to feed! Some buffalo still remain, but they are found only in zoos or reservations.

However, we cannot evade responsibility by assuming the inevitable extinction for *all* wild animals predicated on an essential incompatibility of man and beast. Unlike the passenger pigeon, the great auk was never overly abundant. A flightless sea bird resembling the penguin, the great auk once nested on rocky islets around the periphery of the North Atlantic from Maine to Spain. Although a superb swimmer and diver, the great auk was extremely clumsy on land and when ashore was easily killed. Fishermen from northern Europe on extended fishing voyages found the auk a welcome source of otherwise unobtainable fresh meat. By the time Jacques Cartier visited the coast of Newfoundland in 1534, the great auk was already becoming much less widely distributed and locally rare in northern Europe. On Funk Island off Newfoundland, Cartier found huge colonies of auks and soon the slaughter began in the New World as well as the Old. As the Grand Banks off Newfoundland became known to fishermen, the North American population of auks was vigorously reduced by expeditions that herded whole flocks into stone compounds where they were methodically clubbed to death and rendered for their oil or salted for their meat.

By the early eighteen hundreds, the North American great auks were gone. A few lingered on rocky islets off the coast of Iceland but with the butchering of two birds in 1844, the species was clubbed into extinction. Had today's concept of renewable resource been in existence in the nineteenth century the great auk could well have been saved, for even at that time a brief study could have determined how many birds could be taken for their flesh, oil, or feathers without seriously reducing the population. Unfortunately, animals were not then viewed as a val-

Figure 16.2 In 1840, at the height of the golden age of whaling, when this water color was done by a New Bedford whaler, the odds were fairly even between whale and man—many big ones did get away. (The Whaling Museum, New Bedford, Mass.)

uable resource. We assumed when one was gone there was always another, but we have seen our last great auk.

Whales. While it would be reassuring to think that times have changed and that nations are now concerned with maintaining their animal populations as a valuable resource, the recent dramatic decrease in the population of several species of whales indicates that this is not yet the case.

The great and romantic age of whaling flowered between 1825 and 1860 (Figure 16.2), then it withered as kerosene replaced whale oil. While several species of whale produced good oil, only the sperm whale provided ambergris, a cheeselike substance still used as a perfume base, and spermaceti, a fine oil of great value. Unlike the toothed sperm whale, the right, bowhead, and gray whales have a mouth full of horny plates through which they strain sea water containing their shrimplike food. This was the whalebone used for years in corset stays and buggy whips. Although pursued energetically in the nineteenth century, whaling was inefficient and limited to the most easily caught whales. In the twentieth century, as new uses were found for whale oil and whale meat, interest in whaling revived, and using modern techniques, whaling again became a serious business enterprise with huge factory ships capable of locating, killing, and processing *all* useful species of whale with great efficiency (Figure 16.3).

After World War II, an attempt was made by the International Whaling Commission to establish some rules for whaling, but the nations that signed were only obliged to set their own catch limits and regulations. Scientific advice was offered by the Commission but signatory nations were not bound to follow it. As each nation continued to take what it pleased, whales soon became few and far between. Today only Norway, the U.S.S.R., and Japan continue to maintain high-pressure whaling operations and the focus of whaling efforts has shifted from the Antarctic to the northern Pacific. The sperm whale managed to hold its own until recently when scarcity of other whales increased pressure upon it to the extent that twenty-nine thousand were killed in 1964. Some species like the gray, bowhead, and right whales have been protected, and the humpback whale has been partially protected for several years. But populations reduced to a few thousand may take decades to build up their numbers to relatively safe levels.

Figure 16.3 Today the big ones no longer get away. This factory ship is about to process a fin whale (shown belly up). (Popperfoto)

Figure 16.4 This young flapper seems unconcerned about the fate of the ostrich who provided her costume. (McGraw-Hill Films)

The whale in greatest danger of extinction is the blue whale, the largest animal ever to live. Often more than 100 feet long and weighing up to 130 tons, the blue whale has been reduced to a population of around 600, a level so low that even with complete protection it may slowly slip into oblivion. The only answer is to reduce sharply the annual kill of whales and strictly enforce whatever protective regulations may be necessary to maintain and increase the numbers of the most severely reduced whales. But before this can be done, the remaining whaling nations seem determined to extract the last barrel of oil from the last whale.

Numbers of other animals have been reduced to the extinction point because of their attractive fur or feathers. The sea otter was very nearly exterminated before it was completely protected. Although still limited to only a few points in its former range from northern Japan across the north Pacific to California, the sea otter seems to be slowly increasing in numbers and distribution, allowing the Russians to harvest a small number from their otter population.

Until banned by international agreements, a vigorous trade in millinery feathers at the turn of the century threatened the African ostrich (Figure 16.4), several species of egret, and the birds of paradise found in New Guinea. Today the African leopard and Florida alligator are threatened because of the demand for their fashionable skins. Even though changes in fashion have spared species from extinction in the past, time is short and pressure on these species great. Most reputable furriers refuse to handle skins from animals officially listed by the International Union for the Conservation of Nature and Natural Resources, which maintains a *Red Book* of endangered species of the world. Although the alligator is protected in every state over its range but Texas, it was widely poached in Florida, and even in the Everglades National Park. A national ban on interstate shipment of hides has done much to assure the continuance of alligators in the South; and surely technology will find a substitute for the leopard's skin.

NOVEL USES LEAD TO EXTINCTION

Killing wild animals for food or for their skins can perhaps be rationalized, but killing a gnu to make a flyswatter out of its tail or an elephant to convert its feet into waste paper baskets is perverse (Figure 16.5). In the United States, mummified baby alligators or baskets made of armadillo "shells" are as tasteless as a cookie jar made out of a human skull.

While no one claims the imminent extinction of elephants to satisfy the waste paper basket trade, the walrus is definitely threatened. Although there are perhaps thirty thousand Pacific walruses left, as many as 7000 are killed each year in Siberia and Alaska for their ivory tusks, which are worth $2 a pound. Since the tusks of a large male may weigh twenty pounds each and the ivory carved into ornaments or figures may be worth up to $100 a pound, walruses are actively hunted. Both the U.S.S.R. and the United States regulate the hunting of walruses within

Figure 16.5 Tasteless curios such as these elephant foot wastebaskets can exist only when a thoughtless public provides a market. (Grzimek/Okapia, Frankfurt/Main)

their territory, but walruses are citizens of neither country; without an international treaty to guarantee protection of the walrus, this species may also be doomed.

Superstition

The wild mountain goat or ibex was exterminated in the Swiss Alps in the nineteenth century because of its supposed benevolent powers:

> Almost every part of its body was considered to have some healing virtue or other beneficial property, be it aphrodisiac, talisman, or poison detector. A particularly wide range of medicinal effects was attributed to the hair balls occasionally found in the ibex's fourth stomach compartment. These "bezoar balls" were reputed effective against fainting, melancholy, jaundice, hemorrhoids, hemorrhagic diarrhea, pestilence, cancer, and other ills. The ibex's blood was considered a cure for bladder stones; the heel bone helped combat spleen diseases; the heart yielded a strength-giving tonic; and even the droppings were utilized as medicine against anemia and consumption as well as a rejuvenating agent. The many reputed pharmaceutical properties brought the ibex no benefit; on the contrary, its body was so highly sought after that by the seventeenth century it was already extinct in the Swiss Alps.[2]

Of course today's Swiss are somewhat more sophisticated, but many Chinese still believe that rhinoceros horn in powered form is a powerful aphrodisiac. The result of this superstition has been the reduction of the three Asian species of rhinoceros to the point of extinction; in 1964 there were 600 Indian rhinos, 150 Sumatran rhinos, and only twenty-one Javan rhinos. Even the African species of rhino has been affected by this "magic." Although both the black and white African rhinos are protected, over 1000 are poached each year which has reduced the black rhino to around thirteen thousand and the white to 3900 individuals.

Trophy Hunting

The Arabian oryx is a graceful gazellelike animal with long tapering horns. In recent years its extinction in Arabia was assured by oil-rich Arab notables who took to running down the last fragmented herds in jeeps or airplanes, and machine-gunning them to the last individual. In view of the inevitable results of such "sport," an expedition was mounted a few years ago to capture breeding stock of these animals while they still existed. These were shipped to Phoenix, Arizona, where successful efforts were made to establish a breeding herd to preserve the species

[2] Ziswiler, V., 1967. *Extinct and vanishing animals.* Springer-Verlag, New York.

and ultimately allow reintroduction to Arabia when the safety of the oryx could be assured.

It is only fair to note, however, that trophy hunting is much more regulated than this example would suggest and few other animals are in immediate danger today from this sport.

Competition with Man

In the early years of this century, Boer settlers in South Africa, eking out a living by running stock or farming, viewed any competition from natural sources as expendable and felt justified in leveling their guns on any wild animal in range. In a few years the Cape lion, blue buck, and two species of zebra, the quagga and Burchell's zebra, were extinct. The Cape mountain zebra, bontebok, and white-tailed gnu were reduced to specimens in game reserves.

Similarly, the only parrot native to the United States, the Carolina parakeet, was eliminated because of its fondness for man's crops. It was easy prey because these parrots circled in flocks again and again around fallen members until all were destroyed. The last of these green and yellow birds, a lone specimen in a zoo, died in 1914.

Occasionally the numbers of desirable species are reduced in vain. The tsetse fly carries a parasite (trypanosome) which causes a disease commonly known as sleeping sickness, varieties of which affect both man and his cattle. As a result of the presence of the fly and the parasite, large areas of land in Rhodesia have been closed to settlement and the raising of livestock has been curtailed. Acting under the belief that large mammals were the only reservoir for the trypanosome, control was sought by killing large numbers of zebra, gazelle, and antelope—500 thousand during the control attempt. Later it was found that small mammals and birds also acted as hosts for the trypanosome, making this huge loss of large mammals needless, for the tsetse fly remained unchecked.

Live Animal Trade

The desire of people to see a variety of animals with minimum discomfort and danger has led to the establishment of zoological gardens or zoos. Some, like the zoos at San Diego, the Bronx, or Lincoln Park, are quite extensive, others are quite small. Altogether there are perhaps 100 zoos in the world creating a steady demand for wild animals, particularly the rarer ones. A platypus or giant panda will always draw a larger crowd than a lion or zebra and since many public zoos depend on admission fees or attendance figures to maintain or increase their appropriations, there is constant pressure to exhibit the rare and the unusual.

Figure 16.6 Orangutans were eagerly sought by zoos a few years ago, and their population in the wild was dangerously depleted. Today most zoos are careful not to place a premium on the collection of animals in danger of extinction. (Zoological Society of London)

This pressure is transmitted to collectors and ultimately to animal populations throughout the world. A good case in point is the orangutan (Figure 16.6).

Native to Borneo and Sumatra, this large ape is a favorite exhibit in zoos. But as the orangutan became rarer, the demand increased sharply and irresponsible animal collectors redoubled their efforts. The situation was aggravated by placing a premium on the young apes, usually collected only by killing the mother. In an effort to save the orangutan in its native habitat, most zoos, to their credit, no longer purchase this or other animals in danger of extinction. In some cases where extinction seems imminent, however, a special effort is made to establish a breeding stock in the safety of a zoo, as was done with the Arabian oryx.

Another source of potential danger to wild animal populations, especially monkeys, is the rising demand for animals for experimental purposes, especially in medical and testing laboratories. It is perhaps ironic that as more and more products of potential danger to *man* are generated, such as drugs, additives, and pollutants, more and more *animals* are required in testing programs. Many of these animals—rabbits, hamsters, guinea pigs, and rats—are bred especially for this purpose, but the monkey possesses the best analogue of man's metabolism and thus it is eagerly hunted in Central and South America by collectors. Although some effort is being made today to set up breeding colonies of the most desirable species to increase the supply and to assure greater uniformity and higher-quality animals, it will be several years before the demand can be satisfied.

Both zoos and medical research, however, have used a small number of animals compared with the pet market. Until 1967, parrots were not permitted in this country under ban of the Public Health Service, for parrots were known to carry parrot fever (psittacosis) which could be passed on to man. Recently a wide array of other birds has been found to carry this virus, whose name has accordingly been broadened to ornithosis. Since human contact is much more likely with pet birds than wild ones, parrots remained on the banned list. Then it was found that by holding the birds in an isolation center, treating them for several weeks and observing them for another period, the disease could be controlled. The ban was lifted and a different kind of "parrot fever" swept the country. The five-and-dime stores, which handle most of the parrots sold in the United States, are unable to supply the demand.

The parrots are collected by Indians in the rain forests of the Amazon Basin, mostly in Colombia. Considering the amount of handling necessary from tree to household cage, it has been estimated that only one in fifty birds survives the transition. This means that in 1968, 500 thousand parrots died to supply the ten thousand that were sold. Yearly removal of half a million parrots may be well within the reproductive potential of the species involved, or it may be seriously threatening their survival. The problem is that no one knows what the effect of the soaring market for pet parrots is having in the source country, for the Amazon Basin is still *terra incognita* and none but the broadest ecological generalizations can be made. One effect has been noted though; the Indian collectors, attracted to the easy money offered by traders, neglect their traditional means of livelihood in pursuit of the parrots. After the local supply is exhausted, the trader moves on, leaving the Indian collector economically high and dry.

Until the impact of this increased rate of parrot removal can be ascertained, harvesting of these birds should not be left in the hands of irresponsible entrepreneurs, for it has been just this type of unregulated exploitation that has led to extinction of species in the past.

EXTINCTION BY SUBVERSION

Direct attack by man on a species is an obvious form of extermination or extinction, but more subtle processes often have the same fatal results for the organisms involved. Short of destroying the organism itself, the destruction of its habitat is perhaps the most effective indirect means of bringing about extinction.

Habitat Destruction

For many species closely tied to a specific habitat, destruction or even alteration of that habitat may be devastating. At one time, the island of Madagascar was completely clothed in a rich forest. Its long isolation from the rest of Africa resulted in the preservation of an unusual group of primitive primates, most of which were tree dwellers. In recent years, development has destroyed over 80 percent of Madagascar's forest, leaving this unique collection of primates clinging, literally and figuratively, to the scattered remnants. Already one form is extinct, four more are quite rare, and twenty-three are threatened. Unless steps are taken to preserve some tracts of undisturbed forest as refuges, this entire group of primates will disappear.

    Probably the most publicized species courting extinction is the whooping crane. This five-foot-tall wading bird is completely white except for a bright red crown, black moustache, and black wing tips, a spectacular sight on the ground or in the air (Figure 16.7). The whoop-

Figure 16.7 Few birds can equal the seven-foot wingspread of the whooping crane. These four birds are on their wintering range at the Aransas National Wildlife Refuge in Texas. (U.S. Dept. of the Interior, Sport Fisheries and Wildlife)

ing crane originally nested in marshes in the prairie states and wintered on the Gulf Coast. Never especially abundant, the crane was initially reduced by hunting as it made its long migrations from its breeding grounds in Canada and the United States to its wintering ground. The bird was not especially edible, but it presented a striking target to the hunter. Later, as the prairie states came under cultivation, marsh after marsh was drained. Through a combination of habitat destruction and hunting, the whooping crane was reduced to a low of fifteen birds in 1942 despite the establishment in 1937 of the forty-seven-thousand-acre Aransas Wildlife Refuge designed to protect their wintering range.

Since then the whooping crane population has increased to around seventy birds. Some of these birds are in captivity in an effort to increase their numbers artificially and to provide some insurance lest, despite man's best efforts, a catastrophe eliminate the natural population.

Animal Introduction

Despite unusual examples like the dodo, the great auk, and the passenger pigeon, modern man has directly caused the extinction of relatively few animals, particularly when compared with the destruction wrought by animals he has introduced in certain areas. Most striking among these man-introduced animals are rats, goats, pigs, cats, and dogs in that order. Rats attack all ground-breeding animals of appropriate size and have caused the extinction of nine species of rails (a small marsh bird, often flightless on isolated islands). Though goats are vegetarians, they have with their prodigious appetite reduced more vegetated land to near-desert than any other animal known. Because of their grazing at ground or low shrub level, persistent grazing by goats has on many occasions reduced forest to desert, as we saw in Chapter 2. Goats were widely introduced on many islands by shipwrecked sailors who left them as a kind of living larder to provide a supply of fresh meat for the next visitor. The occurrence of "desert islands" is as much due to these walking vacuum cleaners as any accident of nature. With the destruction or gross alteration of vegetation, indigenous species are deprived of breeding sites and shelter from rats as well as their natural predators; hence the faunas of many of these goat-inhabited islands have suffered impoverishment and extinction.

Another calamitous example of the impact of animal introduction is seen in the sugar cane fields of the West Indies, which were plagued with rats whose depredations destroyed much of the crop. The mongoose, a fierce, weasel-like native of Asia, was introduced to reduce the rat population. The mongoose *did* attack rats for a while but soon began to attack every other ground dweller. Amphibians, reptiles, and birds were preyed upon until several species became extremely rare.

Elimination of an introduced animal is next to impossible, particularly if it is the size of a rat or a mongoose, for the smaller the animal the more difficult the eradication. Even if eradication were to be achieved, once the native animals are gone they cannot be revived.

SUSCEPTIBILITY TO EXTINCTION

Although no species is immune to extinction, some are extinction-prone. Any species that has a naturally low population level or is subject to violent fluctuations in its numbers is in danger. The population of the California condor, like that of the whooping crane, hovers between forty and sixty for a condor pair produces only one egg every two years. With such a low reproduction rate, population growth is unlikely and susceptibility to catastrophe is greatly enhanced. Inbreeding has sharply limited the gene pool. This reduces the genetic variability, which in turn reduces the species' flexibility in the face of environmental change. But reduction to low numbers in itself need not doom a species to extinction. A species that has a high rate of reproduction, when there are no pressures against it, may be able to bounce back from the edge of extinction. For example, the great white heron suffered a 40 percent reduction in the Florida Everglades, as a result of Hurricane Donna in 1960, but by 1963 the population had regained its former level of about 1500 individuals.
uals.

Food specialization. Some species feed on only one or a few other species, which automatically limits their distribution and makes them prone to extinction. Everglades kite, for example, feeds on only one species of snail. As land is drained for cultivation or development, the snail is being exterminated and its overspecialized predator with it.

The Australian koala feeds only on eucalyptus leaves (Figure 16.8). If the eucalyptus forests disappear, the koala will too. The ivory-billed

Figure 16.8 This koala is feeding on eucalyptus leaves, the only food it will eat. (Australian News and Information Bureau)

woodpecker is another species with a limited food source—beetle larvae in *recently* dead trees. Most other woodpeckers, including the closely related pileated woodpecker, accept a broad variety of larvae that successively inhabit dead trees for years after the death of the tree. This need of the ivory-billed woodpecker for recently dead trees requires large tracts of mature timber, which have become rare in the South. This need may doom the ivory-billed woodpecker to extinction.

Specialized breeding sites. The wood duck at one time became quite rare. Unlike most ducks, it requires a nesting cavity, and as old hollow trees disappeared, so did the duck. When this need was recognized and suitable nesting boxes were erected in marshes, the wood duck population sprang back.

Kirtland's warbler, described in Chapter 5, is even more specialized in its nesting requirements, which limit it to a small area in Michigan's pine barrens. Some fish have equally specialized nesting sites. Salmon must have streams with gravel bottoms to receive their eggs and silted streams are unsuitable because the eggs become buried in the silt and suffocate.

Island distributions. Many of the world's birds and mammals, now extinct, were restricted to islands. This both limited their numbers and their ability to survive the introduction of predators. Because many small islands originally had few predators, birds sometimes lost their ability to fly. When man introduced predators they eliminated these flightless birds.

Both lakes and isolated mountain ranges form ecological analogues of islands, limiting the range of a species and often imposing special environmental conditions which lead to the evolution of unique forms distinct from related populations. A subspecies of bighorn sheep became extinct because its limited range in the Black Hills allowed it to be hunted without respite. The freshwater shark and porpoise of Lake Nicaragua, and the seal of Lake Ladoga in the U.S.S.R., are all species that, although still extant, may in the future become extinct for the same reason.

THE COST OF EXTINCTION

During the history of mankind, more than 150 birds and mammals have become extinct; at least half of these have been full-fledged species, the rest have been subspecies, races, or geographical variants of still existing forms. The question we must ask is what are the implications of this loss? With millions of species of animals and plants still remaining, some as yet undescribed, why should 150 animals be missed?

Man, of all these hosts of organisms, is the only animal capable of completely exterminating any other form of life. All other species are bound in predator-prey relationships that exercise checks and balances upon excessive increase or decrease in population levels. Although these levels may cycle with time, they do so within certain limits.

Sometime in the last few million years, man became dissociated from these checks and balances and became a superpredator. But with this ability to drive any species into extinction at will comes at first a need, then a responsibility for restraint. The need develops from the growing awareness that no species exists in an environmental vacuum. Organisms and their environments are inextricably interrelated. Scores, perhaps hundreds, of species may be dependent upon a key plant or animal such as a redwood, buffalo, or sand dune shrub (Chapter 15). But why are all these associates so important that we should guard against the loss of any one of them? The answer lies in the relationship of man to this species diversity.

Man has traditionally tried to concentrate the energy from sunlight into a few species which are easily harvested, the monoculture system that was mentioned in Chapter 8. While this practice accomplishes its aim reasonably well, the price paid in crop protection is high—biocides to control weeds, insects, fungi—hormones to control fruit set, preharvest drop and so on. Despite this tremendous input of energy, things still go wrong. Insects and fungi develop resistance and new pests are encouraged by the destruction of their predators. At times, as we saw in the spray schedule for an apple orchard (Chapter 12) it would seem that we have a tiger by the tail in our attempt to maintain simplified ecosystems.

When species diversity is reduced in natural ecosystems, either willfully or inadvertently, the same kinds of problems seen in monoculture begin to arise. But a natural ecosystem cannot be treated like a cornfield and routinely sprayed to control some beetle or caterpillar. For a biocide that might be tolerated in a cornfield could so scramble the food web in a woodlot that the original problem would be lost in the ensuing chaos. Moreover, while we can afford the energy input to maintain a field of corn because of the value of its yield, we cannot possibly afford the economic burden of maintaining natural ecosystems that have maintained themselves until disturbed by man. Consequently, until we know exactly the role of *every* organism in an ecosystem, how it controls or is controlled by others, we cannot abandon any species as superfluous, not even that pitiful flock of seventy whooping cranes. If we do we run the risk of having to cope with future population explosions of fungi, insects, rodents, or whatever, that might well dwarf any problems seen to date.

However logical our need for all extant species may be, to place the care for continued survival of endangered species solely upon our present or future need begs the issue of responsibility. If man were a passive bystander he could rationalize extinction as a purely natural phenomenon. But it is certainly fair to say that since the emergence of man *no* extinction has been a purely natural one. Just as we bear a certain responsibility toward domesticated species which have become dependent upon us, we have an obligation toward wild species which through no fault of their own have become dependent upon man for their continued existence. Each of these species, gnat or gnu, represents a pool of genetic information which uniquely adapts it to its environment. When this information is lost through the process of extinction a valuable key to survival with all of its potential economic, scientific, and esthetic worth is irretrievably lost as well. It is time that we, as self-aware members of most of the earth's ecosystems, developed the ecological sensitivity to refrain from invoking the enormous power of extinction that we have inherited from our ancestors. We must do this for the sake of enlightened self-interest, to be sure, but also because we have come to feel that all other organisms on earth have a right to exist equal to our own and that our traditional dominion over beast and fowl is not a mandate for mindless exploitation but a charge of responsibility.

THE ROAD BACK

Most extinctions need not have happened. With the exception of a very few species like the passenger pigeon, which seemed unable to adjust its life style to accommodate man in any degree, most species presently extinct might be living now had there been a little care and foresight at the right time. Controlled harvesting, for example, might have saved the great auk, the dodo, Steller's sea cow, or the Black Hills bighorn sheep. Thus it is imperative that we profit from the mistakes of the past and accept the fact that there is no inherent reason why man should hurry any organism presently living into extinction. Large-scale efforts have already saved some organisms from imminent extinction, and for a few others the future seems hopeful. Techniques of wildlife management, breeding, reintroduction, and even resynthesis are becoming more widely used and offer hope for saving species now threatened.

Management

Perhaps the greatest challenge is to institute proper management of a wildlife resource before its numbers become so low that emergency measures are necessary. The concept of management is not new but many mistakes were made before it came to be properly applied. In the early 1900s the mule deer herd on the Kaibab Plateau of Arizona was reduced in numbers; sportsmen blamed the traditional enemy, the predator. The government responded with a predator control campaign, which practically exterminated mountain lions from the plateau. The deer quickly responded, but without natural control, they increased so greatly that their browse could no longer support them, and thousands starved to death. The survivors so severely overgrazed what remained of the vegetation that the effects were noticeable for decades afterward. Although the mountain lion population was allowed to recover somewhat after this fiasco, the damage had been done.

Management is now viewed as going beyond mere protection of the "good guys" from the "bad guys." Important management tools are removal of surplus stock by controlled shooting, manipulation of vegetation, and control over multiple use of summer *and* winter range of wildlife. Habitat management using controlled burning is employed to create nesting sites for Kirtland's warbler in Michigan, and to retain a mosaic of open spaces for the use of the California condor in Los Padres National Forest in California (see Chapter 5).

However, efforts to increase the population of bighorn sheep in the Sierra Nevada between California and Nevada have foundered because of inadequate control of the winter range. The sheep have adequate summer grazing in the meadows of Sequoia and King's Canyon National Parks, but when the sheep move down into the National Forests below to graze during the winter, much of the range is already occupied by domestic sheep or mule deer. Few National Parks or game refuges are large enough to be self-regulating. The same problem has occurred in the Serengeti Game Reserve in Tanzania: animals migrate out of the protected area during part of the year for some favored graze that is not included in the reserve. In such situations arrangements must be made to accommodate the year-round needs and activities of the animals, if they are to be maintained in a truly wild state.

Breeding

A few species have survived only in captivity. Both the gingko and dawn redwood (*Metasequoia*) trees were rediscovered growing in temple gardens in China. Both had been thought extinct but today they are planted throughout the world. Another Chinese species, Père David's deer, was discovered in Peking's Imperial Garden in 1861, long after all wild individuals had become extinct. Today there are several hundred in zoos and private collections in Europe.

The best known breeding program was initiated in 1967 to save the whooping crane by captive breeding. Experiments with the closely related and more abundant sandhill crane indicated that the whooping cranes would probably breed in captivity and would replace eggs taken from the nest. Accordingly, each season several eggs are taken from whooping crane nests in Wood Buffalo Park in Canada and flown in special incubators to the Federal Endangered Wildlife Research Station at Patuxent, Maryland, where they are hatched. The resulting birds are later bred to build a reserve of captive cranes as a stock for possible reintroduction.

A successful breeding program was carried out with the Hawaiian goose or nene, which had been reduced to less than fifty birds when a few pairs were bred in captivity in England. In a few years there were enough nene to restock their original habitat in Hawaii. Today there are over 500 nene, captive and wild.

Reintroduction

By 1900 the Russian antelope or saiga was almost exterminated. A careful research program by Soviet scientists managed to increase the breeding stock to about 1000 individuals. When more information about their habits, food, and nutrition had been gathered, the saiga was reintroduced to its former range. With over 2.5 million saiga in 1960 a controlled harvest was allowed for the first time. Each year since, over a quarter of a million saiga are harvested, producing 6600 tons of meat, 240 thousand square yards of hides, and large amounts of industrial fat. Thus a species in danger of extinction was restored as a resource and sustained productivity was achieved on land unsuitable for cattle grazing.

Similarly, the trumpeter swan of North America was reduced to a few dozen birds in Yellowstone National Park and an adjacent hot spring area at Red Rock Lakes, Montana. This latter area was designated as a wildlife refuge in 1935, and by 1958 enough swans had bred to allow introduction of breeding pairs into refuges in Oregon, Washington, Nevada, Wyoming, and South Dakota where they nested successfully. In 1968 the total population of trumpeter swans was around 5000, with over 800 of them in the United States.

Perhaps of greater interest to hunters was the revival of the wild turkey. This large game animal, which was preferred by Benjamin Franklin over the bald eagle as the national bird to be represented on the Great Seal, was so abundant in colonial days that it was sold for a penny a pound. But by 1920 overhunting, deforestation, and loss of the chestnut, its prime food source, had eliminated the turkey from three-fourths of its natural range. A program initiated by the state game department of Pennsylvania employed the capture and release method to reintroduce wild turkey stock to appropriate gamelands in the state, thereby increasing the number of turkeys in Pennsylvania to nearly fifty thousand. From this stock, turkeys have been reintroduced to New York, Connecticut, and Massachusetts. As the number of wild turkeys increases in these states they may offer hunters an open season as has become possible in Pennsylvania.

Resynthesis

Certainly the most unusual approach to species restoration has been an almost literal raising of the dead—the recreation of the images of at least two extinct European mammals, the wild ox or aurochs, and the wild horse or tarpan. Two German zoologists, Heinz and Lutz Heck, worked independently on the thesis that if domesticated breeds were selected from an original stock, then the genes of that original stock would still be scattered through the selected modern breeds. As Lutz Heck put it:

> No creature is extinct if the elements of its heritable constitution are still to be found in living descendants. All that needs to be done is to apply the experience of the breeder to the assembling of the inherited elements scattered among these descendants. To this end a sure eye is needed to detect the primitive qualities of the various races of cattle: the nearest approach to the horn of the aurochs, or its combativeness, or its long legs, or the colour of its hide, or the small udder of the wild animal. The next step is to devise a well-considered plan of breeding, various races being crossed in order to combine all the qualities similar to those of the aurochs, and so its heritable constitution, in a single animal. If this is successfully done, the aurochs of ancient times must eventually reappear from its tame descendants.[3]

Heinz Heck assembled primitive breeds of cattle from all over Europe and began selection for characteristics that resembled those associated with the extinct aurochs. In eleven years the first aurochs facsimiles were born (Figure 16.9), one of each sex. It is remarkable that when paired they bred true; no throwbacks appeared from any of the many offspring over several generations.

Encouraged, Heck repeated the process seeking to recreate the tarpan, extinct since 1876. When there was some difficulty restoring the stiff erect mane of the wild horse, breeding with the Przewalski wild horse produced the desired effect. Then the unwanted Przewalski genes had to be bred out, for the heavy European cart horse is descended from

[3] Street, P., 1964. "Recreating the aurochs and tarpan." *Animals,* London, 5, pp. 250–254.

Figure 16.9 A small herd of aurochs bred by Lutz Heck at the Munich Zoo from many breeds of cattle. Although these animals breed true they can only be considered a facsimile of the original aurochs, which became extinct long ago. (Photo by Lothar Schlawe, courtesy Tierpark Hellabrunn, Munich)

Przewalski stock while the tarpan apparently gave rise to the graceful Arabian riding horse.

Obviously, neither the recreated aurochs nor the recreated tarpan can be considered the exact genetic equivalent of the extinct animals, for the only genes that have been reassembled are those which the breeder has selected with his "sure eye" from the domestic breeds. While enough of these genes have been put together to breed true as a unit, the breeder is a different selecting agent from the environment that selected the original aurochs or tarpan features. The result must therefore be viewed as somewhere between a true reconstruction and a facsimile.

Of course this technique can work only where domestic breeds preserve some of the genes of the original prototype. Since no domesticated forms were selected from the passenger pigeon, for example, most of its genes disappeared with Martha. Perhaps a few are shared by related species, the mourning, zenaida, and spotted doves, but not enough to synthesize another passenger pigeon even if these related species could be crossed.

Fortunately there seems to be enough public concern to continue federal support for programs of the Department of the Interior that are making strenuous efforts to prevent further extinctions of native American species. This support comes somewhat belatedly, though it is nonetheless welcome, considering that the National Audubon Society has been, since the early 1900s, protecting populations of rare birds. The Boone and Crockett Club and the Izaak Walton League have also been concerned for many years with wildlife values. Internationally the World Wildlife Fund has supplied money for protection of endangered species all over the world (see Appendix 2 for a more complete list of organizations concerned with the preservation of wildlife and its environment). With the concerted efforts of all interested parties there is no reason for further unnatural extinctions to occur.

FURTHER READING

Branch, E. D., 1962. *The hunting of the buffalo*. University of Nebraska Press. Fascinating description of a tragic saga.

Bureau of Sport Fisheries and Wildlife, 1966. *Rare and endangered fish and wildlife of the U.S.* Resource Publ. 34, Sup. Doc., Washington, D.C. This is the best and most current source book available on the problems of American wildlife.

Fisher, J., N. Simon and J. Vincent, 1969. *Wildlife in danger*. Viking, New York. A fine survey on a world basis.

Greenway, J. C., Jr., 1967. *Extinct and vanishing birds of the world*. 2nd rev. ed. Dover, New York. A very thorough, updated account of bird extinction.

Laycock, G., 1969. *America's endangered wildlife*. W. W. Norton, New York.

McClung, R. M., 1969. *Lost wild America*. Wm. Morrow, New York.

Ziswiler, V., 1967. *Extinct and vanishing animals*. Springer-Verlag, New York. These three popular books give a good overview of the problem.

Kroeber, T., 1962. *Ishi in two worlds*. University of California Press, Berkeley, California. A poignant example of man's deliberate extinction of man.

Matthews, L. H., 1968. *The whale*. Simon & Schuster, New York. Sumptuously illustrated presentation of the whale in time and space.

Schorger, A. W., 1955. *The passenger pigeon, its natural history and extinction*. University of Wisconsin Press, Madison, Wisconsin. *The* account of the extinction of the passenger pigeon.

CHAPTER 17

CHEMICAL AND BIOLOGICAL WARFARE

On 22 April 1915, French troops on the front line near Ypres saw

> the vast cloud of greenish-yellow gas spring out of the ground and slowly move down wind toward them, the vapour clinging to the earth, seeking out every hole and hollow and filling the trenches and shell holes as it came. First wonder, then fear; then, as the first fringes of the cloud enveloped them and left them choking and agonized in the fight for breath—panic. Those who could move broke and ran, trying, generally in vain, to outstrip the cloud which followed inexorably after them.[1]

The gas was chlorine, released from thousands of canisters by the Germans, who had patiently waited for just the right wind speed and direction. The tactic was brilliantly successful: four miles of the Allied front collapsed into a swarm of panicked troops fighting their way to the rear. The gas caused fifteen thousand casualties that day, 5000 of them fatal. Had the German army fully realized the impact of their innovation they could have plunged through the gaping hole in the French lines and swept to the channel ports of Dunkirk and Calais, perhaps writing a different conclusion to World War I. But the Germans seemed as amazed as the French and sat gaping at their handiwork. The Allies bitterly protested this uncivilized perversion of honest killing, and then quickly used gas too. Chlorine was soon replaced by mustard gas, then phosgene. By the war's end, some twelve tear gases, fifteen choking agents, four blister agents, three blood poisons, and four vomit gases had been released into the European air (Figure 17.1).

[1] Brown, F. J., 1968. *Chemical warfare, a study in restraints.* Princeton University Press, Princeton.

Figure 17.1 A doughboy and his horse prepare for possible gas attack during World War I. (U.S. Signal Corps)

THE GENEVA PROTOCOL

This was not the first time that chemicals were used in warfare, but the variety, scale, and impact were without precedent. The novelty of gas warfare and long lists of its victims provided fine grist for the propaganda mills and, throughout the war, moral outrage was eagerly encouraged by the Allied governments. At the war's end, people were so overwrought that they demanded that such a horrible use of technology be outlawed by international agreement.

After several false starts, the representatives of a number of countries met in Geneva, Switzerland and hammered out an agreement or Protocol, which reads in part:

> Whereas the use in war of asphyxiating, poisonous, or other gases, and of all analogous liquids, materials or devices has been justly condemned by the general opinion of the civilized world; and,

> Whereas the prohibition of such use has been declared in Treaties to which the majority of Powers of the world are Parties; and,

> To the end that this prohibition shall be universally accepted as a part of

International Law, binding alike the conscience and the practice of nations;

Declare:

That the High Contracting Parties, so far as they are not already Parties to Treaties prohibiting such use, accept this prohibition, agree to extend this prohibition to the use of bacteriological methods of warfare and agree to be bound as between themselves according to the terms of this declaration.[2]

Although the United States signed the Protocol, it was not, at the time, ratified by the United States Senate.

Nerve Gases

But once opened, the Pandora's box of gas warfare was not to be so easily closed. In 1936, mustard gas and phosgene were used against the Ethiopians by Mussolini's troops; despite some international moralizing, no concerted action was taken by world powers against Italy. Then in the late 1930s, Gerhard Schrader, a German chemist, discovered the potency of organophosphates (see Chapter 12). Their effectiveness as insecticides suggested that they might kill more than insects. By the early 1940s, the Nazis had developed a deadly nerve gas, called tabun, from organophosphates, then two more, sarin and soman. Sarin was rated as thirty times more toxic than phosgene. Of the three, tabun, though less toxic, was the most easily produced; after the war the Allies found a German stockpile of twelve thousand tons of tabun.

Considering their desperate straits and maniacal leadership in 1944 and 1945, why didn't the Germans use the deadly new nerve gases? For all their eager acceptance of modern strategy and weapons, such as Panzer tanks and divebombers, the German General Staff was strangely reluctant to use gas. Some military historians attribute the decision to the influence of senior staff officers whose careers embraced an earlier military tradition and who perhaps remembered incidents involving the unpredictability of gas in World War I. Moreover, the Germans assumed that since *they* had discovered the organophosphorous nerve gases, the Allies had probably done the same. Fearing massive retaliation on the increasingly helpless cities of the German heartland, the General Staff did not use its tabun.

Ironically, the Allies had *not* discovered the new nerve gases, and learned about them for the first time when a plant manufacturing tabun was captured by the Russians. The Allied attitude toward the use of

---

[2] Romero, R. and M. Leitenberg, 1967. "Chemical and biological warfare, history of international control and U.S. policy." *Scientist and Citizen*, 9(7), pp. 131–140.

poison gases was unambiguously stated by President Roosevelt early in the war: "This country has not used them, and I hope that we never will be compelled to use them. I state categorically that we shall under no circumstances resort to the use of such weapons unless they are first used by our enemies."[3]

The Japanese used some poison gases against the Chinese during the period 1937–1943, but remarkable restraint was shown by both sides throughout World War II. Despite Roosevelt's feeling about using poison gases, however, much research has been carried out on these weapons both during and since World War II, ostensibly in self-defense. But no defense is realistically pursued without a firm grip on offense. In the late 1950s the Chemical Corps, which had the responsibility of developing and, when called upon, using chemical and biological weapons, began an intensive publicity campaign to counter continuing public revulsion against the use of chemical and biological warfare (CBW). Pushing the theme that certain tools in their arsenal were more humane than many weapons in acceptable use, like napalm (a kind of jellied gasoline), the Chemical Corps announced the development of tear gases and hallucinogenic agents that would disarm and confuse the enemy, obviating the need to kill him. Public repugnance subsequently weakened to apathy and with its new humane image, CBW did not again become controversial until the Vietnam conflict.

While this briefly highlights the use of chemical weapons in modern times, what of the weapons themselves, their nature, mode of action, and toxicity?

THE CHEMICAL ARSENAL

Basically the only difference between chemical warfare and conventional warfare is that the toxicity of chemicals is exploited rather than the energy released by their explosion.

Most chemical warfare gases fall into one of six categories: blood poisoning gases, choking gases, harassing agents, nerve gases, tear gases, and incapacitating agents.

Blood poisoning gases. Hydrogen cyanide and cyanogen chloride interfere with oxygen uptake in the body and are quick acting. But they are needed in large quantities for maximum toxicity and have the disadvantage of a strong odor.

[3] Hersh, S. M., 1968. *Chemical and biological warfare: America's hidden arsenal.* Bobbs-Merrill, New York.

Choking gases. Both chlorine and phosgene (CG),[4] used first in World War I, attack delicate membranes in the lungs. The destruction of these membranes exposes the body to the entrance of microorganisms and interferes with the uptake of oxygen by the bloodstream. Chlorine has a very penetrating and irritating odor, while phosgene smells like new-mown hay. A company of German soldiers recruited from a large city was stationed in a small French village during World War I. Well indoctrinated about the danger of phosgene, they panicked when in early summer the real scent of new-mown hay filled the air. Their concern was well-founded, for 80 percent of the gas fatalities of World War I were caused by phosgene.

Harassing agents. Mustard gas (HD) and adamsite (DM) were introduced during World War I and because of their effectiveness are still useful today. Mustard gas because of its low volatility persists for many days, burning or blistering the skin surface or the eyes. The skin blisters develop into deep ulcerations that are subject to infection, and are slow to heal. Temporary blindness may result when mustard gas comes in contact with the eyes. Because of its very faint odor (suggestive of garlic), the four to six hour delay in the onset of symptoms, and the duration of its persistence in the air, mustard gas is still considered a useful chemical weapon. It can, however, be lethal or cause long-term damage to the lungs. Although mustard gas was not intentionally used in Europe during World War II, an accident in Italy in late 1943 demonstrates its effectiveness. German aircraft had bombed the harbor at Bari sinking some sixteen Allied ships including the SS John Harvey which contained 100 tons of 100-pound mustard bombs. As a result of the persistence of mustard gas floating on the water, eighty-three sailors were killed and 534 seriously injured.

Adamsite (DM), while it is also almost scentless, takes effect much more rapidly than mustard gas and elicits a completely different set of symptoms: headache, chest pains, vomiting, nausea, and sneezing. Because adamsite is rarely fatal it has been widely used in Vietnam as a harassing agent to flush the enemy from tunnels into the open.

Nerve gases. Of the three organophosphate gases, tabun (GA), sarin (GB), and soman (GD), sarin is considered the most effective weapon; that is, it kills the fastest and with the least warning. One milligram (1/40 of a drop) applied to the skin is lethal. The mode of action of the nerve gases, like the organophosphate biocides, is the inhibition or destruction of cholinesterase. This causes nerve impulses to discharge continually along the nerve fibers and results in the contraction of all

[4] The United States classifies chemical weapons according to a letter code rather than generic or common names.

muscles. Since movement depends on simultaneous relaxation and contraction of paired muscle groups, the anticholinesterase nerve gases result
in ineffective vibration of muscles, voluntary or involuntary. Uncontrolled
muscle spasms are followed in quick order by convulsions and death.

Tear gases. There are at least three tear gases that have been used
in Vietnam: CA, CN, and CS. When used in air and in prescribed concentrations, the tear gases are not lethal but harassing, causing copious
production of tears, headaches, sneezing, and coughing. When used
in underground tunnels, the concentration of tear gas may be increased
to the point where the tear gases can sometimes be lethal. Tear gases are
also used in the United States and other countries for riot control. Again,
when improperly used, such gas *can* be dangerous.

Incapacitating agents. Development went on in this area for many
years although only one agent, BZ, was ever standardized or approved for
potential use. Initially it was felt that introduction of such an agent into
an enemy command post or staff headquarters, for example, would slow
physical and mental activity, disorient, and confuse, thus permitting
the attacker to take advantage of the ensuing disruption. However,
BZ may also encourage maniacal or violent irrational behavior and
elicit hallucinations in someone controlling guns, rockets, or nuclear
warheads. Consequently such incapacitating agents have been officially
renounced as weapons except if first used by opposing forces. But many
problems remain.

Nerve gas canisters were designed as offensive weapons; short-
sightedly, no provisions were made for their ultimate deactivation or
disassembly. Consequently years after their manufacture, canisters began
to leak, containers corroded, and explosive charges became unstable.
Clearly, dumping poison gas canisters into the ocean as if they were
simply old mortar shells will not do for we know far too little about the
effects of extremely toxic materials on oceanic food chains to take the risk.

Any chemical, however toxic, can be broken down again into its
basic, innocuous components. We must find simple and convenient means
of doing this with outmoded stocks of chemical weapons at their storage
depots obviating the need for their being shipped across country with
attendant danger for all those along the route as well as to organisms in
the traditional disposal areas.

BIOLOGICAL WARFARE

Although the first chemical warfare was waged by the Greeks, it was not,
as we have seen, until World War I and the development of modern

chemical technology that chemical warfare came to be regarded as a major weapon. On the other hand, biological weapons were consistently used in the past but have rarely been employed in modern times.

In 600 B.C. the Greek general, Solon, took advantage of a besieged garrison by liberally dousing their water supply—a small stream—with the herb helleborus. The ensuing diarrhea proved sufficiently incapacitating to the defenders to lead to their defeat. In 200 B.C. a party of Carthaginians, under attack, abandoned their camp to the enemy, leaving behind a generous supply of wine doctored with a narcotic (mandragora). The drugged attackers were slain by the returning troops.

Perhaps competition with Rome sharpened Carthaginian wits for Hannibal himself, waging war against Eumenes II of Pergamum in 184 B.C., had the ingenious idea of packing snakes into clay pots and hurling these upon the enemy's ships, which retired in confusion.

The quite fortuitous appearance of the Great Plague in 1348 soon suggested strategic possibilities to the warring states and princes of the day; by flinging the corpses of plague victims over the walls into fortified towns or castles it was hoped to introduce and spread both plague and panic among the besieged defenders. But the favor was more often than not returned, for medieval cities under siege usually had a much more ready supply of corpses than an attacking army.

During the French and Indian Wars in 1763, Sir Jeffrey Amherst, Commander-in-Chief of the English troops in the American colonies, suggested that blankets of smallpox victims be distributed to unfriendly tribes of Indians to temper their hostility. This proved so effective that the technique was passed along and, as late as the 1840s, Indians were still being eliminated at no cost to the white man, who though not immune had a much greater resistance to smallpox than the Indian.

The formal advent of chemical warfare during World War I suggested the use of microbiological agents as an intentional adjunct to war rather than a haphazard result. But biological warfare was not used in either World War because methods of preparing sufficient quantities of an agent and then effectively spreading it were not available. Unhappily, subsequent research has filled both these gaps.

Disease as an Instrument of War

Biological warfare (BW) has one obvious major advantage over chemical warfare: lethal disease organisms, once introduced, can be contagiously spread from a single individual through an entire army.

Traditionally, the aim of medical research is to control infections, find drugs and vaccines, and overcome resistant forms. But BW research turns this inside out. Its purpose is to develop effective agents of disease that are drug resistant. Fortunately, and this is almost without precedent, medical skills were developed before BW technology, leaving us in a somewhat better position to cope with the potential widescale use of this weapon.

Considering the hundreds of bacterial and viral diseases that attack man, it is surprising that so few have BW possibilities. If we examine the criteria for the ideal BW agent we can see why this is so. First the agent must be *infectious;* a few cells must be capable of rapid and efficient infection. If millions of organisms are necessary to inoculate an individual, the infectiousness would probably be too low to be useful. Once inoculated, the disease must be *effective,* that is, it must disable; athlete's foot is infectious but hardly disabling. Other diseases which may take weeks, months, or even years to be effective, like syphilis, are useless as well. The ideal disease agent must be *available,* easily and quickly cultured in adequate amounts. But it must also be *stable,* able to withstand drying, high or low temperatures, and ultraviolet radiation to which it is exposed in the environment. It must be *virulent* over a reasonable time and in an environment differing from that in which it was cultured.

Transmission by aerial dispersion is the most desirable means of spreading a BW agent. The transmission of otherwise virulent and effective diseases is often encumbered by complicated routes of infection, various animal vectors or carriers, alternate hosts, or reservoirs: A vector requirement complicates the transmission process enormously for it becomes necessary to deal with two organisms, the disease and its carrier.

The ideal BW agent is *difficult to detect* by the population under attack. Many diseases start out with vague, generalized symptoms similar to the common cold which make their identification difficult and may delay effective treatment until an infection of epidemic proportions has resulted. Finally, the home population must be protected either by antidote, vaccine, or antibiotic against the potential backfiring of a BW agent. So there must be a *treatment,* preferably known only to the user, that will protect the population of the attacker.

There is great difficulty in handling biological warfare agents and large uncertainty of their effect; consequently the United States has officially renounced their use, confining its biological research to whatever defensive measures are necessary to protect from possible attack. In addition, existing stocks of biological weapons were recently ordered destroyed.

While most modern biological agents were microorganisms because of the ease with which they could be cultured, distributed, and dispersed, occasionally the use of higher animals was contemplated. The Army once awarded a contract to a university to explore the possibilities of using birds as BW agents, perhaps forgetting a previous World War II experience with bats. The idea then was to equip bats with tiny incendiary bombs. Theoretically, the bats were then to be air-dropped over Japan where the bombs—with delayed fuses—would trigger hundreds of fires. [The researchers] developed a satisfactory bomb weighing less than one ounce, but the bats never cooperated. The bombs were to be attached by surgery and a piece of string to the chests of would-be bat bombers. As the Army envisioned it, the bats would be dropped over large Japanese cities, quickly find hiding places, chew their strings, and leave the bombs. After two years of research, a trial run was made in Carlsbad Caverns, New Mexico. On the first day, some bats escaped and set off fires that completely demolished a general's auto and a $2 million hangar. The Army project was abruptly canceled. The Navy then took over with a new approach; it theorized that if the bats could be artificially cooled and forced into hibernation they would stop gnawing and thus would not immediately chew through the string. . . . The hibernating bats were packed like eggs in crates, flown to New Mexico and dropped. Theoretically, the bats would tumble out at a certain altitude and begin to awaken in the warm lower air over the desert. But, as one report less than adequately put it, "most slept on"—and fell to their death. By this time it was August, 1944, and the macabre project was finally scrapped.[5]

Such comic relief was rarely provided by BW.

HERBICIDES

Plant killers are not new. Indeed, some arsenicals have been used for years; but shortly after World War II a new group of herbicides was developed that involve plant hormones or *auxins*. Auxins normally are produced by the leaves of a plant and in the proper concentration serve to keep the leaves attached to the stalk or stem. In the fall, auxin levels normally drop and a layer of large, thin-walled cells forms where the leaf connects to the plant. When these cells rupture, the leaf falls. By applying a compound which lowers the auxin content of a leaf it is possible to defoliate a plant prematurely at will. This is commonly done before harvesting cotton to avoid plugging the mechanical cotton harvester with cotton leaves.

Conversely, by adding auxin at the right concentration, leaf fall and fruit drop can be inhibited, thereby decreasing loss from preharvest drop of fruit. If an excess of auxin is applied, however, plants respond by increasing their respiratory activities considerably beyond their ability to produce food, causing the plants literally to grow themselves to death. Strangely, the commonly used herbicides with high auxin activity, 2,4-D, and 2,4,5-T, affect only broad-leafed plants. This makes them useful tools for keeping lawns free of dandelions and other broad-leafed weeds

[5] Hersh, S. M., 1968. *Chemical and biological warfare, America's hidden arsenal.* Bobbs-Merrill, New York.

without damaging the grass. Evidence is accumulating, however, that 2,4,5-T, in high concentrations, can elicit in mice birth defects similar to those caused by thalidomide, and as a result 2,4,5-T is no longer recommended for areas near human habitation in the United States.

Because of their defoliating action, various herbicides have been used in Vietnam to defoliate trees (Figure 17.2), thereby depriving the enemy of hiding places and protecting United States troops from ambush, and to destroy crops upon which the enemy is dependent. Three major herbicides have been used to these ends: a mixture of 2,4-D and 2,4,5-T used for general defoliation work and destruction of broad-leafed crops; picloram for forest defoliation; and cacodylic acid for control of grassy plants, such as elephant grass, and destruction of rice.

Introduced early in the war in 1961 on an experimental basis, the use of herbicides grew until the demands of 2,4-D were so great that the entire domestic production in 1967 was bought by the Department of Defense. By early 1967, 150 thousand acres of cropland had been sprayed and 500 thousand acres of forest defoliated. Rational assessment of the results is difficult from both the military and scientific point of view. Herbicides *have* probably saved American lives, which is why they

Figure 17.2 Two C-123 spray planes dumping herbicide on forested land in Vietnam. Over 500 thousand acres were similarly treated, causing widespread defoliation. (U.S. Air Force Photo)

Figure 17.3 Mangroves are especially sensitive to herbicides. All strategic waterways in Vietnam are now lined with dead mangroves such as these. (Society for Social Responsibility in Science)

were introduced in the first place. But they have not been the unqualified success that their users might wish. Defoliants serve two masters when used to prevent ambushes: they reduce hiding places for the ambushers but open up lines of fire for them as well. They also deprive the troops who use defoliants of the safety of cover. General defoliation has had mixed success as well. Mangroves lining waterways have been efficiently controlled because of their sensitivity to herbicides (Figure 17.3), but the evergreen species characteristic of some of the Vietnam forests are more resistant and even repeated treatments are not 100 percent effective. Furthermore, the whole concept of destroying crops to starve the enemy into submission seems to be spurious. Originally based on the experience of the British in Malaya a decade before, the idea was to deprive guerrilla bands of food supplies and drive them into open clearings where they could be destroyed. This worked in Malaya because there was no refuge; application of this policy in Vietnam seemed only to injure the noncombatants. In fact, captured Vietnamese soldiers were reported to weigh up to ten pounds more than comparable civilians.

Environmental Effects

Although the military effectiveness is relatively easy to assess, the ecological ramifications of herbicide application on such an enormous scale are almost totally unknown. The herbicide 2,4-D is quickly metabolized and 2,4,5-T lingers only a few weeks, but picloram is quite persistent. Coupled with its biological activity (100 times 2,4-D) and solubility, the persistence of picloram means that its percolation into the water table and transfer to unsprayed areas is quite likely. This persistence and potency was strikingly demonstrated in a tobacco field with an unusual distribution of stunted plants. With much effort, this pattern was traced to droppings left by mules used to plow the field. The mules had been pastured in a field treated with picloram, which had passed unchanged through the mule's gut and leached out of the droppings, affecting the tobacco.

If the picloram is completely effective in defoliating forests, there is a possibility that in some areas of Vietnam, laterization may take place as the soil is exposed to the effects of rain and sun (see Chapter 3). Almost 30 percent of Vietnam soils are lateritic and are potentially open to this irreversible process. But since defoliation is rarely complete, and many species of grass are resistant to the action of picloram, the soil is not often exposed, limiting the danger from widescale laterization. Of greater potential import is the invasion of repeatedly defoliated forests by bamboo, which effectively retards or inhibits regeneration of the disturbed forest.

Another concern about the effects of persistent defoliation of forests is the destruction of habitats of various insects, birds, and mammals. Many species occupy quite specialized niches which, when the forest is destroyed, disappear, leading to extinction of the species which occupied them.

A recent government document posed some ecological questions to which there are few answers. Research must be undertaken to determine

toxicity to man and animals, synergistic toxicity of several herbicides in combination, or in combination with other environmental factors, or in interaction with the soil, or other chemical compounds resulting from decomposition; concentration of herbicides along food chains, in surface water, in underground reservoirs, in water plants or organisms, in ground water, or in the soil; effects of herbicides on timber crop and rubber production, or on food supply generally; losses of domesticated and soil animals, fish and birds; threatened extinction of rare species; genetic impairments of animals or plants; encroachment of unwanted species on bared areas; mass destruction of sensitive vegetation (e.g., mangrove) requiring decades to recover; and possibility of laterization of exposed jungle soils.[6]

The only studies so far have been superficially done in the dry season when trees usually lose their leaves anyway. Under present conditions no detailed ecological studies will be possible until the fighting ends.

[6] Report to the Subcommittee on Science, Research, and Development, 1969. "A technology assessment of the Vietnam defoliant matter, a case history." U.S. Govt. Printing Office, Washington, D.C.

CBW IN PERSPECTIVE

Criticism of the use of tear gases and herbicides in Vietnam was loud and shrill, matched only by the general furor touched off so long ago at Ypres over the use of chemical and biological weapons. Military apologists would have us believe that they are humane weapons. Citing World War I statistics the apologists state that of 27 percent American casualties due to gas, only 2 percent died compared with 26 percent fatalities in non-gas activities, and that only 4 percent of the gas casualties were permanently disabled in contrast to 25 percent of those wounded by high explosives.

The outraged cry is that high explosives are bad enough but that the use of poison gas or biological agents is a barbaric and immoral prostitution of human technology. Thus the decision to use CBW ultimately is a matter of conscience. But there are a few points that should be considered while searching for that resolution.

Relative Morality

The crossbow so disturbed the moralists of the thirteenth century that a Lateran Council called by the Pope formally outlawed this heinous weapon. Two centuries later the flower of French knighthood, the Chevalier Bayard, accepted crossbows but was appalled at the use of gunpowder. While he treated captured knights and archers with courtesy, bearers of muskets were summarily executed. Professional soldiers have often resisted new weapons as a corruption of their honor.

Some of the reaction against CBW during and after World War I reflected the natural uncertainty about poison gases. But much of it was the result of carefully cultivated propaganda generated by the Allies against the Central Powers to serve the war effort by encouraging anti-German sentiment. Thus anything new used by the Germans, particularly if it were successful, would be branded as inhuman. "By the time the Armistice was declared, gas propaganda had run the policy gamut—the illegal and inhumane act of a murderous aggressor in 1915; just and humane retaliation in 1916; blackout in 1917; and a triumph of Allied industry in 1918."[7] However, much to the chagrin of a chemical industry hoping to develop a market for gases and a few military planners who had glimpsed a brave new world, the atrocity charges stuck in the public mind. CBW was outlawed and efforts to expend public monies on further research were blocked.

[7] Brown, F. J., 1968. *Chemical warfare, a study in restraints.* Princeton University Press, Princeton.

Despite the passage of fifty years, CBW remains stigmatized while napalm is accepted by many as the most useful weapon employed in the Korean conflict. Thus, there seems to be a strong thread of hypocrisy running through both sides of the CBW controversy. If we accept the premise that killing people for whatever reason is immoral, then means are irrelevant. If we accept the inevitability of war it would seem, at least in the abstract, infinitely preferable to die in a few seconds from nerve gas than to bleed slowly to death. Why is it humane to tear gas men from tunnels into a saturation pattern of fragmentation bombs that blow them to pieces while it is inhumane to kill them in their tunnels with sarin nerve gas?

The issue of humane killing now goes beyond the death of soldiers, however, because civilian populations increasingly bear the brunt of modern war. The death rate of civilians in World War I was 5 percent, World War II 48 percent, the Korean War 84 percent, and perhaps an even greater percentage of Vietnamese civilians in the past decade. What of the environment in which these people must try to live when the ideological profundities have been settled?

No matter how many people are killed by modern warfare, their number is insignificant when compared with the number of people yet to come who must live with the destruction and irreparable alteration of the land itself, land which must sustain life after the last corpse has been buried. The extent to which CBW brings about this environmental destruction is a deeper measure of its immorality than the death of combatants or even the increasing death rate of noncombatants.

Although President Nixon's recent disavowal of BW and reaffirmation of Roosevelt's use of CW only in retaliation is reassuring, the continuing use of tear gases and herbicides with their possible mutagenic effects remains profoundly disturbing. The continuing concern about CBW has been less with its immediate effect than its potential role in escalating war from the use of relatively innocuous tear gas and herbicides to lethal gases and diseases. The nuclear standoff has encouraged the feeling that war is unthinkable, that perhaps for the first time reasonable alternatives to war must be sought. If, however, chemical and biological weapons become accepted as humane alternatives to napalm and fragmentation bombs and war once again becomes an accepted way of settling international disputes, *that* ironic result would be the ultimate inhumanity of chemical and biological warfare.

FURTHER READING

Brown, F. J., 1968. *Chemical warfare: a study in restraints*. Princeton University Press, Princeton, New Jersey. CBW from a political and historical perspective.

Clarke, R., 1968. *The silent weapons*. McKay, New York.

Hersh, S. M., 1969. *Chemical and biological warfare; America's hidden arsenal*. Doubleday, New York.

Rosebury, J., 1949. *Peace or pestilence; biological warfare and how to avoid it*. Whittlesey House, New York.

Rothschild, J. H., 1964. *Tomorrow's weapons, chemical and biological*. McGraw-Hill, New York. Clarke, Hersh and Rosebury produce convincing arguments against CBW, while Rothschild, former head of the Army Chemical Corps, thinks otherwise.

Part V

MAN'S URBAN ENVIRONMENT

CHAPTER 18

THE HOUSE AS AN ENVIRONMENT

And if after the hunt goes past
And the double-barrelled blast
(Like war and pestilence
and the loss of common sense),

If I can with confidence say
That still for another day,
Or even another year,
I will be there for you, my dear,

It will be because, though small
As measured again the All,
I have been so instinctively thorough
About my crevice and burrow.
— ROBERT FROST, 1946.

Woodchucks have had thousands, perhaps hundreds of thousands of years to adapt to their environment by constructing burrows that provide protection from enemies and shelter from a long cold winter. Although man's original shelter was a ready-made tree or cave, migration into areas without convenient natural shelter together with culturation that provided new tools and materials, both required and allowed a variety of artificial structures to be constructed. For the thousands of years that man has occupied earth, these structures slowly evolved by trial and error into some starkly functional and often elegantly simple buildings. Once disdained by architects, this native architecture is being studied today as a superb marriage of form and function. A few examples from greatly contrasting environments will give some idea of the relationship between form, function, and environment.

DIVERSITY OF ENVIRONMENT, DIVERSITY OF HOUSE

One of the best adaptations to an especially hostile environment was the development of the igloo by the Eskimo. Dome-shaped and constructed of blocks of hard packed snow, the igloo offered minimal resistance to the sweep of the wind during the winter when tents used in summer were unsuitable. Internally, the dome shape allowed the largest volume possible with the least supportive structure and made the most effective use of point sources of heat, a simple oil lamp and the people themselves. The igloo was made by layering snow blocks in an arching spiral (Figure 18.1). By carefully packing the chinks with snow and by burning an oil lamp inside, the interior of the igloo was glazed into a shiny, smooth surface, that not only sealed the igloo but reflected light and heat, much as aluminum foil might. When the interior was covered with furs and skins, the igloo was quite comfortable, at least compared with the conditions outside: at midnight, air temperature inside an igloo was 35°F while outside air temperature was minus 30°F. During the day inside air temperature rose as high as 40°F compared with minus 10°F outside.

The other environmental extreme, the desert, has also had a profound effect on the shelters of native peoples. The desert, unlike the arctic, poses problems of daily rather than seasonal extremes. The temperature ranges from 110°F to 60°F over a twenty-four-hour period, a range that puts an enormous strain on any cooling system. These extremes are modified, however, by the use of local materials, stone and clay (adobe) for houses with thick walls and roofs that insulate the interior of the building from the blistering heat outside (Figure 18.2). Temperature in a typical adobe dwelling at 2 p.m. was 85°F inside and 105°F outside, while at night the temperature was 80°F inside and 65°F outside. During the day the thick walls and roof acted as insulators, heating slowly as they absorbed the sun's rays. At night the process was reversed; the desert air cooled quickly, but the warm walls and roof radiated heat throughout the night protecting the occupants from the outside chill.

In the tropics, the problem is not temperature variation, either daily or seasonal, but the need for adequate shade, proper ventilation, and protection from heavy downpours. The native peoples achieve this by reducing or eliminating side walls and erecting a steeply pitched roof, thickly thatched. Thin, coarsely woven side walls take advantage of every breeze, and the thick thatch both insulates the shelter beneath from the heat of the sun almost overhead, and it allows the daily rain shower to be shed quickly and almost silently (Figure 18.3). In contrast, the wooden, tin-roofed shacks that often replace the thatched structures are cramped, stuffy, and noisy.

Figure 18.1 Although small, the dome-shaped igloo is remarkably snug and comfortable when properly constructed. (McGraw-Hill Films)

Figure 18.2 These buildings at Taos Pueblo in New Mexico are built of soft clay bricks or adobe. Their thick walls and roofs effectively insulate the inhabitants from the sharp temperature extremes of the New Mexican desert. (American Museum of Natural History, New York)

Figure 18.3 The thatch roof of a house in Thailand allows rain to roll off quickly and quietly while the loosely woven sidewalls allow circulation of air. (Photo by Alice Mairs)

Figure 18.4 This rather substantial townhouse in Thailand has a tile roof with a broad overhang, wide porch, and louvered shutters to shield from the sun and allow maximum circulation of air. (Photo by Alice Mairs)

Until fairly recently, areas with similar climates exhibited similar architecture no matter where the geographical location. A townhouse in Thailand (Figure 18.4) strongly resembled a house in New Orleans; a stone-adobe house in hot, dry Afghanistan (Figure 18.5) looked very much like a New Mexico pueblo; and a fishing village in Singapore Harbor (Figure 18.6) might be located along the Amazon River in Brazil.

The widespread migration of people into a New World with climates different from those they had known required adaptation to the local environment even when this meant a change in traditional building methods. For example, people from Plymouth, England, where winters are mild and summers cool, were obliged to adapt the type of dwelling with

Figure 18.5 Hot dry climates everywhere elicit the use of thick-walled structures that insulate well, as in these houses in Istalik, Afghanistan. (Photo by Alice Mairs)

Figure 18.6 An ideal marriage of form and function in a small fishing village built on stilts in Singapore Harbor. (Photo by Alice Mairs)

which they were familiar to the much colder winters and hotter summers of Plymouth, Massachusetts. In consequence, the rooms and windows became smaller, the ceilings lower, and the fireplaces massive (Figure 18.7). In contrast, the summers in Virginia and the deep South states of Alabama and Mississippi necessitated large rooms with high ceilings and wide porches or verandas often sheltered by a portico or overhang—all of which allowed maximum insulation, shade, and circulation of air.

As people continued to move, following the frontier to Ohio, the plains states of Illinois and Iowa, then Oregon or California, their houses reflected the changing climate. Pioneers from New England recreated New England villages in northern Ohio, which has a climate similar to New England. Today some small Ohio towns strongly resemble

Figure 18.7 The New England winters are long and cold. Consequently this house built in 1684 in Salem, Massachusetts has small windows and huge fireplaces in every room. (Essex Institute)

parts of New England, even in their names. Farther west this flavor is lost as the farm and ranch houses built from the 1840s to the 1860s were adapted to the influence of local climate.

The basic reason for the wedding of form and function in native architecture lies in the sensitivity to and acceptance of the environment by most native cultures. But times are changing in most countries. The new and old coexist for a time, then the old way disappears. Unfortunately the new usually means a break with the past—the Eskimo gives up his igloo or tent for a wooden-frame house, the African his thatched hut for a tin-roofed shack (Figure 18.8). But much that is good may be discarded with the apparently outmoded materials, techniques, and life styles. With so much of the world in a state of cultural transition between traditional tribal culture and a modern industrialized society, technology and ecologically sensitive architecture could help enormously to bridge the gap between the past and the future.

The Eskimo and his igloo demonstrate this point quite well. No one would want to confine an Eskimo to an igloo and deny him more advanced housing. But the igloo is basically a dome, an architectural form that has been ingeniously developed by American engineer-architect Buckminster Fuller. Using modern methods and materials the igloo could be transformed into a functional and comfortable house in sympathy with the local environment and preserving at the same time a traditional form that could help Eskimos bridge the gap between their cultural past and future. The same approach could be used in desert or tropical regions. Of course, architects have not been insensitive to these possibilities. Much has been learned from traditional dwellings and applied to regional design.

In the United States the gap between function and form widened more gradually. With the rise of mass communication—first national magazines and mail order catalogs, then movies, radio, and television— cultural homogenization began to erode local and regional differences in house types. The growing affluence of farmers and middle-class tradesmen in the nineteenth-century United States allowed a shift from a functional type of dwelling to a dwelling preoccupied with form. The simple farmhouse assumed Greek revival characteristics, and later, at the height of the Victorian era, townhouses became swathed in excessive wooden ornamentation (Figure 18.9).

Today you can drive around any town in the United States and see bungalows built in the 1920s side-by-side with Tudor, Spanish, and French provincial styles from the 1930s, Cape Cod houses from the 1940s, and apartment houses incorporating a variety of styles.

The gap between form and function thus opened has been bridged to some extent by technology, which made possible the thousands of

Figure 18.8 The traditional huts of these Kikuyu tribesmen are being replaced by the ubiquitous tin-roofed shack which may or may not be an improvement. (Photo by Alice Mairs)

Figure 18.9 A redwood baron built this house in Eureka, California as a testament to the versatility of wood. Its Victorian exuberance is unmatched. (Greater Eureka Chamber of Commerce)

square miles of tract housing in the world, struck from a common mold and planted in the ground like some monster crop. There is, certainly, some regional differentiation—shake roofs in California, glass walls in Florida, small windows and full basements in Wisconsin. But one could easily switch houses between Fullerton, California and Oshkosh, Wisconsin and make them work in their new environment. What, then, is the price of making a house work in an environment to which it is ill-adapted—a thousand dollars a year in heating or cooling expenses? Before we can examine the cost, literal or figurative, of housing that is out of equilibrium with its environment, we must first look at what constitutes a suitable external environment for man's dwellings.

The Outer Environment

There have been two tendencies at work in the design of dwellings as packaged environments. One is toward the elimination of the connection between the inner and outer environments by building a windowless, possibly underground house, thereby solving most of the environmental problems that bedevil architects: maintenance of external surfaces; control of flies or mosquitoes in summer and drafts in winter; expense of heating and cooling; glare and shadows, with illumination of the exact intensity and brightness desired. With outside dust eliminated and inside dust filtered out by the obligatory air conditioning system, such a house would be extremely easy to keep clean. Occupancy would, of course, depend on a never-failing power source to heat, cool, ventilate, and illuminate; but auxiliary generators are available. Such an underground house would allow total landscaping of the lot above, particularly desirable where lots are small and expensive. When you pay several thousand dollars for a lot it seems somehow wasteful to cover most of it with your house.

The other approach runs in the opposite direction. The contact between inside and outside environments is broadened to the point where it is difficult to separate the two. This approach is achieved most successfully in mild climates where houses may literally enclose a piece of the outside environment in the form of a garden, such as was done in the atrium of a Roman villa. Large areas of glass wall can be used to remove the feeling of separation between inner and outer environments, or multiple stories and cantilevers can thrust the inner environment into the larger outer environment of treetops or sky.

This approach to environmental integration has been used in cold climates as well as warm climates, but greater dexterity of design is required in the former, since the inner and outer environments are so disparate that their integration must remain illusory. Only in a mild climate, such as that in Florida, along the Mediterranean, in Southern California, or the warm desert areas, can the two be physically integrated.

Figure 18.10 This housing was placed on its site with depressing regularity, ignoring the interest inherent in a hilly site.

The setting. Since most dwellings are very much a part of the surface environment, the orientation on a lot is of critical importance to the distribution of natural light within the dwelling, the ease or difficulty of heat control, and maintenance of a dry cellar or garage. Far too often placement of streets on a tract by the developer reflects the most mathematically efficient way of subdividing property into the maximum number of lots. Houses are built on these lots either parallel to or at some standard angle to the street (Figure 18.10), giving the house buyer little choice about the exposure of his house to sunlight or to wind.

The house exists both in a seasonal environment in which the sun appears from different angles and for varying lengths of time throughout the year and in a daily environment of constantly changing patterns of light and shadow. On September 21 and March 21 in the Northern Hemisphere the sun rises at 6 a.m. and sets at 6 p.m.; on June 21 the day runs from 4:30 a.m. to 7:30 p.m. and on December 21 7:30 a.m. to 4:30 p.m. Fortunately one needs neither a mathematician nor a computer to predict the results of any particular dwelling orientation if one knows the aspect of the site and the shape of the building. There is a machine called a heliogram available to most architects that can reproduce the various seasonal changes in the lighting as the earth circles the sun. Using such a device a small model of a house can be orientated, taking into consideration light, temperature, and exposure factors.

Although an eastward-facing wall receives the same radiation as one facing west, the westward-facing wall is irradiated during the hottest part of the day, making it the warmest wall of the dwelling. When this is anticipated, some care can be taken to avoid placing a kitchen behind that wall or to devise shielding to check heat absorption. Because of the low angle of the sun in winter, the south wall is heated more in winter than in summer; on the other hand, the north wall receives most of its sunshine and the roof twice the radiation of any of the walls in the summertime.

But this constantly changing heat load on a house can be predicted, and steps can be taken to cope with it by the basic orientation of the house, design features, and landscaping. Since zoning often confines both the architect and builder by allowing only a certain orientation a defined number of feet back from the street, the architect often finds it impossible to arrange the rooms to accommodate both summer shade and winter sun. Well-designed plantings may help the situation. In brick or stone houses, the warm west wall can be shielded from the late afternoon sun by the strategic placement of trees or a deciduous vine such as Boston ivy. When the ivy sheds its leaves in the fall, the wall is favorably warmed by the winter sun. In the summer the ivy leaves form an insulating layer of air shielding the masonry from the heat of the sun.

The force of cold winter winds can be broken by planting hedges or belts of evergreens. Windbreaks have been built throughout much of the Midwest in the Dakotas, Iowa, Nebraska, and Kansas with excellent results; thirty-foot trees significantly reduce wind velocity 400 feet to the leeward. Dwelling design plus strategic placement of trees, shrubs, and vines can warm a house as much as 18°F in winter and cool it by 16°F in summer. This is equivalent to moving from New Jersey to upper New York state for the summer and down to Virginia for the winter. Although these expedients incorporate neither new materials nor techniques and indeed are only extensions of common sense, often they are not utilized. This is perhaps a result of ignorance but is more likely due to the wide availability of air conditioners and other technology.

Naturally, with modern heating systems and air conditioning, a standard tract house could probably be built atop Pike's Peak or in Death Valley and be made livable if not comfortable, but only at great expense. The whole point of adaptation of a dwelling to its setting by design, orientation, and planting is to gain maximum benefit from the natural heating and cooling afforded by the environment. But so far we have been complacently running roughshod over the natural landscape, using our technology to make every lot and every dwelling suit the builder's convenience.

The Internal Environment

Whatever its relationship to the outside environment, a dwelling functions as a shelter from the elements. Since man is most comfortable at a temperature of about 72°F, we assume man's need for shelter developed when migration from tropical climates made necessary artificial means of recapturing this optimum zone of comfort. The diversity of houses that we have examined illustrates attempts to recreate this comfort range by whatever means were available.

Considering a shelter or house as an enclosed environment, what factors are important to our health and comfort? Certainly temperature, but also light, humidity, ventilation, and perhaps that intangible aesthetic that transforms a "house" into a "home."

Temperature. Man can tolerate, with protection, the extreme natural range of minus 105° to 145°F. Since the outside air temperature is only occasionally within man's ideal comfort range, 70° to 75°F, one prime environmental function of a house is to add or remove heat.

Heat is intentionally supplied to houses by a great variety of sources, combustion of wood, charcoal, coal, oil, natural gas, or by electricity, directly or indirectly by heat pumps or heat exchangers. Since all of these sources are derived ultimately from the combustion process, air pollution is a direct consequence of heating houses. Of all the heat sources, electricity provides a much greater opportunity to control pollution. The emissions from a few power plant stacks are easier to control than pollution from millions of combustions from individual heat sources (Chapter 10). While it would be nice to have individual power plants, nuclear or solar, that would provide all our power needs, heat as well as electricity, such units are well in the future; the problem of pollution is not.

Artificial heat is not, of course, the only supply of heat in a house. As we saw in the previous section, conduction and radiation from roofs and walls heated by the sun are important sources of heat, which have been taken advantage of in regions with a fair amount of winter sunshine without severe cold. By manipulating the size, shape, and orientation of the windows, a well-designed house can be comfortably heated much of the winter using solar heat, which need be reinforced by conventional sources of heat only during long periods of overcast weather or extreme cold.

Other sources of heat, of minor significance in private dwellings but of some magnitude in a crowded store, apartment, or office building, are body heat and lighting appliance heat. Measured in British thermal units (1 Btu is the amount of heat required to raise the temperature of one pound of water by 1°F), a 100-watt bulb gives off about 500 Btus an hour and a human body about half that. When thousands of people and lighting fixtures of much greater power than a single bulb are involved, the resulting heat can be of major significance. Several buildings have already been constructed to make use of this source of heat, a large post office in Houston, a sixty-story tower in Chicago, and other buildings in Pittsburgh and Portland, Oregon. The system is basically simple; air leaving the building and warmed by the activities inside is passed over a system of water-filled coils. The heated water is compressed, increasing its temperature still further. Fresh air entering the building is warmed as it passes over the hot water coils. Some of the buildings have water tanks to store heat when the building is unoccupied. This heat recovery system works until the outside temperature falls to 10°F, then supplementary heating becomes necessary. This is a particularly obvious example of the economic and ecological good sense implicit in the use of what would otherwise be considered waste.

Heat removal is usually achieved by air conditioning, either central or room units. If both the roof and walls of a dwelling are well insulated, reasonable control over summer temperature can be achieved simply by opening doors and windows at night and closing them early in the morning. Since it is more important and difficult to prevent entrance of excess heat than to prevent its loss, the location and overall design of a house becomes important in controlling internal temperature.

Humidity. Another obvious function of a dwelling is as a shelter from rain, but almost any type of structure will afford this protection. More important is the control of moisture in the internal environment of a house. There are two ways to determine the moisture content of air: by measuring the vapor pressure or the relative humidity. Vapor pressure is a measure of the actual amount of moisture in the air at any given temperature; relative humidity is an expression of how much moisture the air is holding compared with how much it *could* hold at that temperature. A relative humidity of 80 percent at 80°F means that the air is 80 percent saturated with moisture. This is not equivalent to a relative humidity of 80 percent at 50°F because as the temperature of air falls its capacity to hold moisture decreases. For this reason 80 percent relative humidity at 90°F may be unbearable, while 80 percent humidity at 60°F may be quite comfortable, because warm air can hold so much more moisture than cold air.

In the winter we may have the apparent paradox of a house at 15 percent relative humidity and 75°F losing moisture to an outside environment of 50 percent relative humidity at 20°F. Even at 15 percent relative humidity a house at 75°F contains more moisture than half-saturated air at 20°F, so moisture diffuses from a point of higher concentration inside to a point of lower concentration outside.

The chief problem in dwellings over much of the United States in summer is not the heat but the humidity. Hence, any air conditioning system must dehumidify as well as cool the air passing through it.

In the winter, the problem is just the opposite. The small amount of vapor released from our breath, washing, and bathing tends to diffuse out through the walls and roof of the house, thereby reducing the relative humidity of a heated dwelling to as low as 10 percent—comparable with many desert regions of the world. This combination of high temperature and low humidity has a destructive effect upon furniture, causing wood to shrink and the glue to dry out. The effects on people are analogous; the nasal and throat membranes dry out, especially at night, creating conditions favorable to many respiratory viruses and bacteria. It is no coincidence that winter is the season we contract the most colds. Vaporizers can often alleviate the dryness of a small room, but it is more effective and convenient to do this centrally with a humidifier on a furnace. This can be easily accommodated in a hot-air-type of heating system but is more difficult to install with hot water, steam, or electric heat.

Light. Previously, illumination by candle or lamplight made the location, number, and size of windows or skylights to let in natural light of critical importance to household activities. Electric lighting assures plentiful artificial light making the location of windows in a building much less important today than it once was. Windowless buildings can, and are being built and, if anything, seem to be more effectively illuminated than windowed structures with their glare and uneven lighting. There is more involved, however, than good or bad illumination, for windows not only let in light but allow people to see out. There is a psychological satisfaction in being able to relate yourself to time and place by glancing out of a window and, if the view is at all pleasant, there is a mental respite from the pressure or boredom of routine.

A controversy rages as to whether schools should have windows or not. There seems to be no good evidence of an increase in student claustrophobia in a windowless school although it does afford more space. While one suspects that arguments for windows are based more on tradition and sentiment than on demonstrable undesirable effects, students already suffer from too great an isolation from the environment. Perhaps the constant awareness of the real environment viewed through a window might in a small, even insidious way, keep both student and professor attuned to the relevance of life in a natural as well as a synthetic environment. Beyond the mere presence or absence of natural light, sunlight plays an important psychological role in determining mood. Sunlight pouring into a room on a cold winter day is relaxing and cheering and raises the spirits of those confined indoors.

Glass transmits 82 to 90 percent of the visible portion of sunlight but almost no ultraviolet rays, so it is impossible to get a winter tan by lying in sunlight under glass. However, special glass has been developed which will allow up to 60 percent of the ultraviolet to be transmitted. If it were simply a case of getting a tan, a sunlamp would do as well, but ultraviolet light, besides its tanning effect and importance to vitamin D synthesis, is an important germicide; most germs are killed by exposure for little over an hour to the ultraviolet in sunlight. Exposure of the interior of a house to the natural disinfecting power of unfiltered sunlight might be sufficiently important for normal sanitation and cleanliness to be considered in the design, orientation, and glass specifications of a modern house.

Windows also generate the greenhouse effect described in Chapter 10. The shortwave radiation of sunlight easily passes through the window glass, strikes an object or surface in the room and is converted into longwave radiation or heat that cannot pass back through the glass and so accumulates in the room. This effect is quite noticeable in a room with many windows admitting sunlight; for this reason houses with large areas of windows or glass walls must be very carefully designed and situated so that this effect can be exploited in winter and carefully controlled in summer.

Ventilation. Beyond the perception of strong kitchen or bathroom odors or perhaps a vague notion of the air being "stuffy," most people are unaware of the need for ventilation in their homes. Starting with the usual mix of nitrogen, oxygen, carbon dioxide, water vapor, and dust, people give off twenty-three liters per person of carbon dioxide per day, plus variable amounts of water vapor, hydrogen sulfide, ammonia, propionic acid, and various organic compounds. Large quantities of grease aerosols are spread about by frying foods, carbon particles are added from burning foods and tobacco, and dust is constantly generated from bedding, rugs, and clothing. Occasionally malfunctioning furnaces or space heaters may release carbon monoxide in such quantities as to cause death by asphyxiation. More common is the combustion of cigarettes in poorly ventilated rooms where carbon monoxide buildup may lead to dizziness or impaired judgment.

Both burned grease and tobacco smoke may contain carcinogens. While there are no statistics on how many people have died of lung cancer by inhaling burned fat aerosols or someone else's tobacco smoke, there is no reason to believe that these materials are harmless simply because our relationship to them is passive. In general these man-made materials decrease our sensitivity to the stimuli that are integral to our functioning as alert and creative individuals.

Ventilation is, of course, particularly important in the winter when doors and windows of most northern homes are covered with storm windows to conserve heat. But it is necessary all year round where tall apartment houses abut each other. Most cities require air shafts in such buildings to assure interior apartments of at least some ventilation, but these shafts are rarely effective. Air conditioners often serve the multiple function in a city apartment of cooling and circulating the air, as well as removing some of the pollutants. The filter of an air conditioner removes only particulate matter, however, so the urbanite is still exposed to sulfur dioxide and smog, air conditioner or no.

It is ironic that technology, with its potential for offering us a more comfortable personal environment, so often seems to be misused. Instead of providing innovative solutions to basic human needs, technology is often used in lieu of good design, and the basic environmental factors that define a house, within and without, are ignored. If a house is drafty, we push up the thermostat, if it is stuffy, we turn on the air conditioner, and if it is smelly, we squirt a perfumed aerosol about. Each of these problems could be solved through careful study of man's basic life needs, and by applying both the best in design and the most relevant technology to the satisfaction of those needs. But this rational approach will be indefinitely deferred if we continue to be satisfied with the mediocre in housing, which, by its very acceptance, denies the possibility of anything better.

There is much that the individual can do. With existing housing he can select a heating scheme which contributes the least amount of pollutants to the outside environment; he can insulate to control heat entrance and loss; he can humidify or dehumidify the air; and he can landscape to ameliorate temperature and wind velocity extremes. In planning or selecting new housing he can be aware of the importance of choosing a site for temperature control of a house interior and for potential flooding or drainage problems. He can examine floor plans carefully to control summer and winter natural light values. Where large tracts are still in the planning and development stage, individuals can determine if local zoning regulations function merely to keep people or structures out, or if they inhibit the freedom necessary to properly orient a house on its lot. At that point they can determine what provisions are to be made for power supply (wires above or below ground), sewage disposal, water supply, potential air pollution, schools to accommodate more children. Special hearings are often required by planning commissions before a subdivision can be developed, and these hearings allow public opinion to be voiced on the desirability of the project and its impact on the outer environment of surrounding property owners.

In public housing in large cities the quality of the home environment can be improved and maintained by adoption of strictly enforced building codes requiring minimum standards in all dwelling units. Pressure can be brought to bear on public officials to consider the outer and inner environments of public housing much more carefully than has been done in the past.

The key, then, to providing a comfortable and healthy home environment lies in recognizing our environmental needs, making sure they are included in our housing plans, then taking appropriate political action to assure that these plans are realized. When this is done, and we have the requisite knowledge and technology, no human need suffer the indignity of enduring a personal living space that would be regarded as unhealthy for a pet tropical fish, an orchid, or a computer (Figure 18.11).

Figure 18.11 While this urban slum or favela in Brazil seems built into the air, the houses are actually clinging to the face of a very steep hill. (Planned Parenthood—World Population)

FURTHER READING

Aronin, J. W., 1953. *Climate and architecture.* Reinhold, New York. Views the relation of houses to the climate of their site.

Fitch, J. M. and D. P. Branch, 1960. "Primitive architecture and climate." *Scientific American,* **203**(6), p. 134. Informative discussion of representative types of housing adapted to cold, hot, and wet climates.

Geiger, R., 1965. *The climate near the ground.* Harvard University Press, Cambridge, Massachusetts. A fascinating text that looks at micro-climates and man's relationship to them.

Kira, A., 1966. *The bathroom, criteria for design.* Center for Housing and Env. Studies, Cornell University, Ithaca, New York. Critique of the most neglected room in any house.

Olgyay, V. G., 1951. "The temperate house." *Arch. Forum,* **94**(3), pp. 174–194. Many tables, charts and graphs relating the house to the pattern of the changing seasons.

Ordish, G., 1960. *The living house.* J. B. Lippincott, Philadelphia, Pennsylvania. A charming study of the forms of life that have inhabited an old English cottage, reflecting changing styles.

CHAPTER **19**

THE URBAN AND SUBURBAN ENVIRONMENT

Regardless of its suitability as an environment, the house functions for most people as an extension of their personal space, enhancing a sense of personal satisfaction and well-being. Unfortunately when houses are packed into dense concentrations or when apartments are stacked in huge impersonal towers, some personal space is sacrificed for common space; hallways, lobbies, streets, and parks become the living room of the urbanite.

Because of this concentration, the city has changed the natural environment in which it was situated, providing a new environment with unique demands. All environmental problems seem exacerbated by cities—air and water pollution, noise. Even the suburbs share an increasing burden of environmental problems. Yet well over half of the population of the United States lives in cities or towns that occupy only 1 percent of the total land area. While many of these urbanites would prefer to live in the suburbs and more than a few suburbanites would like to move to the country, a surprising percentage of people living in both suburbs and cities live there by choice and are quite happy to limit their rural experiences to occasional vacations. For, despite monumental problems, cities are exciting places to live, offering a quickened tempo both for cultural and intellectual opportunity. But these are subjective values. Viewed as the one man-made environment, urban areas are sharply differentiated from the countryside in many ways.

THE LOCATION AND FORM OF CITIES

Gertrude Stein once described a rather nondescript American city by saying "when you get there, there's no there, there." Most cities that do have a sense of identity are strongly related to topographical forms. London and Paris are located on major rivers, Istanbul on a strategic strait, Rio de Janeiro and San Francisco on superb harbors. More recent cities have been shaped by railroads and major highway intersections, Indianapolis, for example. But transportation routes themselves are strongly influenced by the dictates of topography (see Chapter 20).

Although the location of cities was originally related to the surface features of the countryside, once located, a city is subtly affected by the less obvious subsurface features. Venice was built on the marshy Po River delta in Italy both for protection from attack by land and for easy access to the sea, reflecting the extensive mercantile interests of the city. For centuries Venice prospered but in recent times, the lowering water table beneath the city has begun to cause the slow sinkage of its build- ings. As the piles on which the buildings rest are exposed to the air, they decay and their erosion is quickened by the wakes of the motor boats, which have largely replaced the gondolas and rowboats. The city is now endangered.

Mexico City, seemingly so far removed from the plight of a city like Venice, has a similar problem. Built on a dry lake bottom at an altitude of 8000 ft., Mexico City has experienced a gradual lowering of the water table causing many buildings to sink; most spectacularly, the Palace of Fine Arts has sunk many feet below the street level. Removal of oil from an oil field under Long Beach, California, caused parts of

Figure 19.1   When viewed from the air, the binodal distribution of high-rise buildings in New York City is evident. (Photo by Allen Green, New York Convention and Visitors Bureau)

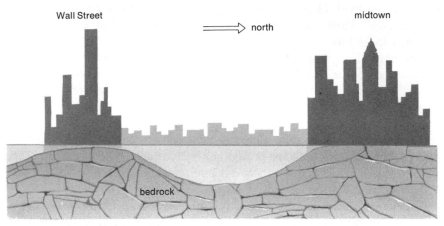

Figure 19.2 The binodal development of Manhattan shows the relationship of tall buildings to bedrock.

that city to subside as much as twenty-nine feet, until water was pumped into the field to stabilize the ground level. While subsurface features can alter the future of a city once it is already there, they can also determine the shape a city may take.

Approaching New York City by car or plane, one can see the rather uneven distribution of tall buildings on Manhattan Island (Figure 19.1). There is one group of skyscrapers toward the southern end, the Wall Street area, and a second group in the midtown area from 34th to 60th Streets. In between, skyscrapers are conspicuously absent. While this is due partly to historic and economic factors, it is largely caused by the accessibility of stable bedrock. The taller the building, the firmer must be its footing in bedrock. Manhattan is fortunate to have a stable platform of hard rock under much of the city, but at different depths. To the north this formation comes to the surface in the midtown area, and is easily seen in Inwood, Morningside, and Central Parks. But as one goes south the bedrock begins to dip below ground. South of 30th Street it is deeply buried, until it lies hundreds of feet below the surface (Figure 19.2). Further south again, in the Wall Street area, it comes up within forty feet of the surface, and then, with the exception of the outcrop of Governor's Island in New York Harbor, it plunges far below the surface again. The expense of excavating beyond a depth of a hundred feet to reach bedrock has so far prevented the erection of very tall buildings in the section of Manhattan between 30th and Wall Streets.

Flying over the Los Angeles Basin you may have looked in vain for "downtown" Los Angeles. If you know exactly where to look you can make out a few multistory buildings in a sea of one- and two-floored structures (Figure 19.3). Los Angeles is located near the San Andreas

Figure 19.3 Because of the ample building space available, as well as the danger of earthquakes, Los Angeles until quite recently built out rather than up. (Los Angeles Chamber of Commerce)

Figure 19.4 The verticality of New York City is emphasized by this fish-eye lens view of midtown. The East River is at the bottom. (Planned Parenthood—World Population)

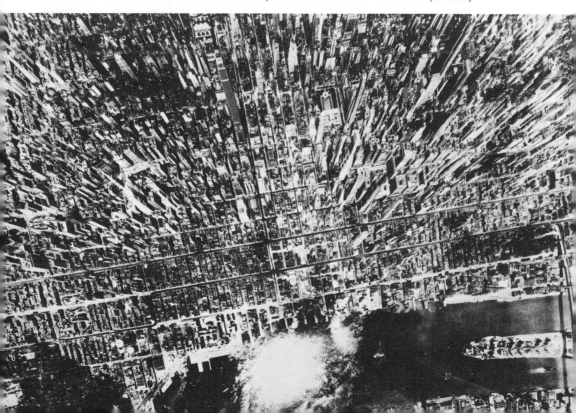

fault on the earthquake-prone rim of the Pacific, and has so far avoided the risk of high-rise structures. Modern advances in engineering and construction techniques have partially overcome, if not the destructive potential of earthquakes, then the inhibitions of the builders. Therefore, this pattern may change. In the Bunker Hill section of Los Angeles, a 100-story building flanked by sixty- and eighty-story companion towers is expected to be completed by the late 1970s.

There is another reason why Manhattan prickles with tall buildings (Figure 19.4), and Los Angeles still has so few. New York City has developed on a small island area. When Manhattan was developed in the early twentieth century, the value of land rose sharply and the available land was quickly bought up; the city could only expand upwards. Fortunately for the landowners this coincided with the development of the elevator. In contrast, in the Los Angeles Basin there was always open land just beyond any congested area, and this encouraged occupation of new land rather than more intensive development of the old.

THE MICROCLIMATE OF CITIES

As large cities grow in size, so does their impact on the environment. The overall climate of a city reflects the region where the city is located. But large cities can modify some of the climatological factors in their immediate vicinity, resulting in a relatively small-scale but important variation in climate, which is called a *microclimate*. When you think of the ways that a city differs from an equal area of forest or farmland, the reasons for the existence of microclimates become apparent.

Temperature

With increase in growth and three-dimensional complexity, the mean temperature in the city tends to rise. This results from the loss of evaporative cooling normally provided by vegetation and exposed soil, the gain of reradiated heat from pavement and building surfaces, and heat produced directly by factories and buildings. Walk down a city street on a hot summer's night and feel the hot breath of air conditioners from apartments or the heat shimmering up from the pavement. In wintertime, notice the heat rising from subway grills in the sidewalk, or the clouds of steam pouring out of manholes or storm sewer gratings. These heat sources have a cumulative effect. The rough profile of the city deflects the wind which might otherwise sweep some of this heat away.

The most striking temperature differences between the city and the country are seen in the daily high and low temperatures which tend to be attenuated in cities. This is emphasized when one listens to a city weather report. Expected lows are always lower in the suburban areas than the city center and the rural areas beyond are still lower. Conversely, although the city center gets quite warm on a summer day, the highs for the area may be in the suburbs or even in the country. The difference lies in daytime shading and nighttime reradiation.

Despite the radiative mass of the city, shading by tall buildings or narrow streets prevents sunlight from striking many potential radiation surfaces, thus lowering temperature maxima. On the other hand, the minimal shading in suburbs allows maximum heating of paved surfaces, roofs, and the walls of low buildings. Long after the country has become chilly on a summer's night and the suburbs comfortably cool, the city is still sweltering, perhaps only a few degrees lower at midnight than at sundown. By sunrise when the city is beginning to cool, the tall buildings are already being warmed. The mass of the city, then, acts as a buffer, damping the temperature extremes experienced in the suburbs or the countryside beyond.

While these microclimates do not produce dramatic changes in climate, over the years statistically valid changes have been noted. The city of Kyoto, Japan, experienced an increase in mean temperature of 1.8°C between 1880 and 1935, a period of rapid industrialization and growth. Now 1.8°C may seem insignificant, but remember, this is a mean. You should remember too that, as we saw in Chapter 9, an elevation of just 5°C in the mean temperature of the earth would be sufficient to melt all remaining ice caps, and to raise the ocean level by about 500 feet, flooding the world's coastal cities (Figure 9.2).

Precipitation

Particulate matter provides nuclei for the condensation of atmospheric moisture into precipitable drops or rainfall. The city's effects on rainfall are difficult to quantify, but there are some data which suggest that as cities grow in size the rainfall on the city proper increases. As the English city of Rochdale became industrialized and grew, its rainfall measured over ten-year periods increased: 1898–1907, 42.81 inches; 1908–1917, 45.83 inches; and 1918–1927, 48.65 inches. Interestingly enough, there was a statistically significant drop of 0.37 inches on Sundays over the time period that measurements were taken, when particulate matter in the atmosphere was lower than on the six working days of the week.

Particulate matter may affect surrounding towns as well as the urban area. Gary, Indiana, is covered much of the time by an enormous pall of smoke from its steel mills. As the cloud drifts eastward, it apparently triggers rain, for the towns over which it drifts have recorded noticeably larger amounts of rainfall than areas to the north or south.

Cities may also produce an orographic effect, that is, their domes of warmer air force the moisture-laden clouds upward into colder air which triggers rain further on. At times when the temperature of Buffalo, New York, is 20° higher than its environs, precipitation plumes can be traced many miles inland—a dramatic demonstration of the climatological impact of cities.

Although rainfall seems to be greater over cities than their surroundings, at least over the long-term period, there is distinctly less snowfall in cities than the surrounding countryside. In temperate areas, especially along the coast, most heavy snow falls when the temperature is near the freezing mark. Often the city temperature is just high enough to change the snow into rain as it falls through the warm air enveloping the urban area. On such occasions the city may remain bare while the surrounding suburbs receive several inches of snow. When snow does fall on a city, the city-generated heat may cause the snow to melt as fast as it falls; even when it accumulates, it melts more rapidly than suburban snow.

Wind

While certain streets or areas of most cities can be quite windy because of local updrafts between buildings or the channel effect of towering buildings lining a narrow street, an equal number of places in a city are quite protected from wind. Because of the rough surfaces they present to the wind, cities tend to experience reduced *average* wind velocity in direct proportion to their size and density. Between 1909 and 1930 the mean wind velocity in Detroit, even when measured 259 feet above the street, decreased from 14.8 to 8.7 feet per second.

Solar Radiation

Regardless of its velocity, the air of cities, as we saw in Chapter 10, is especially rich in fly ash, dust, and particulate matter of all sorts. These materials have a climatological effect as well as an aesthetic and public health impact. Particulate matter can reduce the natural radiation received by a city by 15 percent, a reduction that is uneven across the radiation spectrum and tends to be greatest in the ultraviolet range. This portion of the solar spectrum is extremely important in the production of vitamin D in the skin (see Chapter 1).

URBAN SPACE

For anyone who regularly commutes between a city and its suburbs, the microclimate of the city is obvious enough. Even more obvious as one enters a city from its suburbs is the almost complete loss of contact with natural vegetation. With the exception of a few scattered parks and occasional tree-lined streets, the urban scene is too often one of stark, dirty buildings. Unfortunately, suburbs, rather than acting as green wedges pushing into cities to relieve their starkness, act instead as gray fingers extending from the city and extinguishing the interplay of man-made and natural features that soften and humanize man's environment. The problem is not purely one of a lack of open space, for parking lots and plazas can provide that, but of space softened by trees, greenery, water, and different textured surfaces that mitigate harshness and muffle city noise.

A lack of these amenities contributes to the poor mental health of many urbanites. A study made about twenty years ago by the Cornell Medical Center found that of a sample population of 1700 non-Puerto Rican, white adults between twenty and fifty-nine years of age, living in a midtown neighborhood of New York City, four out of five had symptoms of psychiatric disorders and one out of four was neurotic to the extent that his daily life was disrupted. This area was not considered a slum but a cross-section running from luxury apartments to working-class tenements. By contrast, another study in a rural Hutterite colony (a religious sect living in a tightly knit community) found only three out of 100 people with recognizable mental problems.

The reasons for the poor mental state of urbanites are complex—the many stresses of urban life, the tendency of people with personality problems to migrate to cities where somehow they find a niche unavailable in small towns—but any problem is aggravated by the constant grating of a harsh environment. Jackhammers in the street, wailing police cars and fire trucks, subway trains that alternately shriek and roar, dirt and grime—all contribute to the impression that the environment is "out to get you." Apparently it often does.

Space assumes a far more important role in ameliorating the city environment than merely that of an aesthetic afterthought of the developer. To appreciate the importance of properly designed or sensibly retained urban space and to evaluate its role, we must first take a brief look at the relation between cities and their suburbs and the natural environments that once surrounded them.

Older cities realize more clearly than newer suburbs the value of open space. Many cities have tried to pace their growth with either internal park systems or peripheral green belts to insulate them from the surrounding suburbs. Cleveland and Washington, D.C. have attempted to preserve some open space by creating parks from the flood plains of many small streams that drain the region. These parks form necklace-like skeins running through the city, which help break up the monotony of heavily urbanized areas and provide open space for nearby residents.

Both London and Vienna have pioneered in the attempt to separate the city from its suburbs by a broad belt of undeveloped land, called a greenbelt which encircles the city. Other cities have been able to set aside large blocks of land that today are irreplaceable; New York's Central and Prospect Parks, Philadelphia's Fairmount Park, and San Francisco's Golden Gate Park, for example. Many more opportunities exist to utilize open space which at present lies unused. Most waterfront cities on rivers, harbors, or lakes have turned away from their natural waterfront orientation and, by expressway development or redevelopment schemes, have isolated the public from the most obvious focus for their recreation. Ian McHarg, a landscape architect, estimates that, of the twenty-two miles of Delaware River frontage available to Philadelphia, eight would more than suffice for port facilities; the remainder could be reclaimed for parkland, refocusing Philadelphia on the river. Much the same can be said of New York or Boston where miles of rotting piers testify to past use but present inactivity. In New York, after great pressure by citizen's groups, people were allowed to use a city-owned pier (the Morton Street Pier) giving them the only access to the Hudson River between 72nd Street and Battery Park, a distance of more than five miles, while dozens of other unused piers rotted into the Hudson. More recently the city government has risen to the challenge by using five more strategically located old piers as unique recreational areas.

One hopeful sign which indicates increasing awareness of the importance of recreational open space has been the development of vest-pocket parks and playgrounds in many large cities. Parks need not be large, with acres of grass and groves of trees. The simple juxtaposition of different textures of stone or small fountains or pools can be just as aesthetically rewarding (Figure 19.5). Ideally every block should have such a respite from the trauma of city life. But such space must be carefully utilized for a strip of carefully tended grass is just not effective.

The conflict of natural and man-made environments is not restricted to larger cities and their suburbs. Growth in small towns, starting with the building up of a few main streets, is accompanied by much protest as lawns are paved and old trees removed. The smaller the town the more important the destruction of these symbols of past provinciality seems to be. Trees and unbroken concrete are incompatible. For a tree to remain healthy it must have space for its feeder roots to obtain moisture and nutrients and as lawns disappear and the pavement encroaches, the feeder root system is progressively curtailed until the limbs that exist in

Figure 19.5 While there is little greenery in this playground, and no grass, the imaginative use of textures, forms, and water makes a far more exciting playground for children than the standard grass and swings. The same concept can be applied to parklets scattered through a city. (Pomerance and Breines, Architects)

equilibrium with the feeder roots begin to. die. Old trees are doomed under these conditions and young trees planted hopefully as replacements are severely limited in the size they can attain by the minimal space available.

Urban Renewal

While vest-pocket parks can be superb additions to the city, large-scale urban renewal projects often mar the city's fabric. The American genius for euphemism was evident when the term "slum clearance" was replaced by "urban renewal." Slum clearance suggests removal of slums without implied solutions of what to put in their place or where to put the slum dwellers. Urban renewal, on the other hand, connotes a rekindling or recreation of the essence of urban living that has often been regarded as snuffed out in slum areas. In practice, urban renewal has often meant the inept replacement of an old slum with a new one or the displacement of low-income families into another slum so that the cleared land can make a profitable return, perhaps in the form of luxury apartments, high-rise office space, or cultural centers and sports facilities.

People in the worst slums interact and relate to one another in the only space they have, the streets. Such streets thronged with people almost around the clock are often far safer than elegant tree-lined streets in well-to-do suburbs.

When a slum is "renewed," buildings, streets, and whole blocks are swept away and frequently replaced by vertical towers separated by strips of green lawn. By day the renewed areas may be bright and cheerful but at night they may be deserted and dangerous. Sometimes even the building interiors are not safe; blind corridors, elevators, and stairwells provide ample room for antisocial activity.

Why the antisocial activity? In a street-oriented slum the activity and interest on the street seem to pervade the hallways and corridors of the tenements lining the street. If you were to wander in off the street you would find many doors open, children playing in the halls, men and women chatting from their doorways or leaning out of windows. The street and its extensions into the houses form a social outlet for the people, all the more important when more formal open spaces are unavailable.

In a typical renewal development, the commons of the street and halls is gone, lost in the elegant but impersonal bulk of buildings. Once the interaction and activity of the street have disappeared, the people turn inward. There is no reason to look out of the windows; there is no fire escape to sit on, no front stoop, no sense of intimacy in the hallways. Wander into one of these buildings; doors are closed and halls are empty. Deprived of the opportunity to interact, people are withdrawn and resentful, often without knowing why. Antisocial acts become more frequent and less noticed. This phenomenon is by no means limited to low-income renewal housing. There have been repeated instances of violence in middle- and upper-class areas in New York, Chicago, and other large cities.

City planners sometimes seem unaware of the social importance of streets as environmental space, for although renewal projects provide open space, most of it is apparently unsuitable for social activity. More often a city planner or architect *does* understand the spatial needs of people (Figure 19.6) but is thwarted by archaic bureaucrats and hard-nosed bankers who will not fund imaginative solutions, myopic labor leaders imprisoned by rules no longer needed to protect their men, and manufacturers with a vested interest in yesterday's technology. If mere space were the requisite, parking lots would be filled after business hours with happy people enjoying themselves.

Scale is also important, for the characteristic high-rise redevelopment project is constructed on an impersonal scale. Often much is made of its self-containment—entertainment areas, a supermarket, open space,

Figure 19.6 This creative use of space in the form of an amphitheater and playground facilities does much to relieve the drabness of the public housing surrounding It (Rlls House Plaza, New York City). (Pomerance and Breines, Architects)

and so on. But from the point of view of many tenants, the project is cut off from the city and remote from the quality of their old neighborhood and the rich life it supported.

This is not to say that slums are desirable. Far from it. Poorly heated in the winter, stifling in the summer, often overrun with cockroaches and rats, with trash in the hallways, the smell of urine on the stairs, peeling paint, cracking plaster, broken plumbing—a slum is a rotten place to try to live. But in their ardor to improve on this, city planners and model-happy architects often throw out the baby with the bathwater.

One logical approach to renewal of a slum neighborhood is to rehabilitate as much housing as possible. Many slum areas are made up of solidly constructed houses that, if renovated, could be made more comfortable than most new construction and at less cost. Some excellent examples of rehabilitation can be seen around the country, Wooster Square in New Haven, parts of the West Village and upper West Side in New York, Society Hill in Philadelphia, to name just a few. In most of these instances however, the poor were simply shunted to another slum when the middle class discovered another "charming" neighborhood to "save." But the principle remains—*housing can be rehabilitated* (Figure 19.7).

A pilot experiment in New York pointed the way by cutting a hole in the roof of an otherwise sound but shabby tenement and lowering prefabricated bathrooms and kitchens into the replumbed and rewired building. The operation could be completed in a couple of days with minimum disruption of the occupants' lives by boarding them in a hotel until the operation was completed. Unfortunately, the experiment was abandoned when costs came to over $20,000 per family (enough to buy them a home in the suburbs). But with application of the vaunted American know-how, a mass production approach should bring costs down to the level where whole blocks could be upgraded, preserving the open-space value of the streets without unduly disrupting the social life of the community and maybe even improving on things by knocking out a house here and there for intensive use parklets.

If, however, the quality of housing is low and beyond rehabilitation the most logical approach for a city planner to take is to find a city neighborhood that "works" like Boston's North End, the old French Quarter of New Orleans, parts of New York's Greenwich Village or Chelsea and find out *why* it functions smoothly, has a low crime rate, safe streets, and relatively happy people who interact peacefully with each other. What are the environmental factors at work there and how do they affect the people? Then the planner-architects should do their utmost to build in those features that seem to make people happy. Maybe the world's cities have had enough of idealistic models that work for no one and in some ways are more demoralizing than the slums they replace.

THE RISE OF SUBURBIA

Historically, cities have been compact and densely populated; yet because of their relatively small population, ten thousand to 100 thousand

Figure 19.7 Two townhouses built about the same time. The one on the left has been restored; the one on the right shows the wear and tear of conversion into tiny apartments. (Photo by Dan Jacobs)

until a few hundred years ago, they were able to coexist with the natural environment that surrounded them. Until the use of gunpowder became widespread in the thirteenth century and put an end to city walls, the medieval urbanite could walk through a gate into fields, forests or onto the shore, which were all in intimate contact with the city (Figure 19.8).

Cities remained small because of the attrition of war and disease; the notorious Black Death or bubonic plague reduced the population of some European cities by 80 percent. But the self-contained medieval cities were most effectively breached by changing times; people spilled into the countryside and the cities grew beyond their ancient walls. The industrial revolution drew many people to new jobs, and the resulting factories and housing pushed even further into the surrounding countryside. New cities and towns also sprawled over vast areas where raw materials needed for industry were available, as in the Ruhr Valley in Germany, or Detroit and Chicago in the United States. The final impetus to growth of cities in both the United States and Europe was due to the striking advances in public health services and the advent of public transport systems, first the railroad, then the tram and trolley, finally the subway and automobile. This pattern of city growth is now being seen in other countries undergoing rapid industrialization, Japan and China, for example.

We think of cities in the United States today as being densely packed with people, but urban density has been falling for many years while the suburbs have grown. In 1910 New York City contained sixty-four thousand people per square mile; by 1960 the density had fallen to thirteen thousand people per square mile. Rapid transit systems played

Figure 19.8 Mont St. Michel, both literally and figuratively an island, exemplifies the medieval walled city turned in upon itself but in intimate contact with its environment. (French Government Tourist Office)

an important role in this dispersion, as have changes in the standard of living. The average New York City household sheltered 4.5 people in 1910 but only 3.1 people in 1960. This means that one and a half times as much housing was required in 1960 as 1910 for a comparable number of people. These houses now surround the cities rather than constitute them. Of course, such statistics apply only to those who have come to enjoy both a higher standard of living and the ensuing greater mobility. Those who have neither remain packed in tenements.

Growth of a Megalopolis

The flight from the city began after World War I, continued through the 1920s and 1930s, and exploded after World War II. War-induced recovery from the depression together with the return to civilian life of millions of servicemen caused a tremendous upsurge in the rate of home ownership. Veterans' Administration and Federal Housing Administration loans and mortgages as well as Federal tax deductions on mortgage interest made it, for the first time, cheaper to buy than rent; home ownership suddenly was within reach of greater numbers of people. Of course, some people moved from the country and small towns into suburbs, but the main impetus behind the growth of suburbs was the disenchantment and socioeconomic mobility of former urbanists.

By 1960, suburbs had coalesced into sprawling tracts, smothering the landscape and severing the cities' last link with open space (Figure 19.9). A city dweller no longer could take a bus or trolley to the end of the line and be in the country. To get to the country from a city's downtown became a major expedition, involving hours of driving in very heavy traffic.

By 1961 the large cities of the eastern seaboard of the United States were beginning to bleed together into one large increasingly urbanized mass that Jean Gottmann called a *megalopolis*. The largest megalopolis in the United States runs 500 miles from Portsmouth, Virginia, to Portsmouth, New Hampshire, and includes Washington, Baltimore, Philadelphia, New York and Boston; others are scattered over the rest of the United States and the world.

Will the United States soon be completely covered by merging megalopolises? By 1950 urbanization had occupied only eighteen million of the 1904 million acres of the continental states. Even if urbanization continues at a rate of one million acres a year, and if we assume a half-acre per house and a proportion of 40 percent houses to 60 percent services, streets, factories and other urban uses, by 2010 a population of 360 million would still occupy only seventy-five million acres of the land area of the United States. This might engulf 18 percent of current cropland, but with a doubling of productivity it might be possible to produce more food on less lands as is now done in some crowded European countries. Does this mean then that we can accept urbanization with the assurance that there will be plenty of land in the foreseeable future? It does not.

Loss of Farmlands

The best farmland is not randomly distributed: often it is associated with cities. Much fertile and valuable farmland has already disappeared beneath the suburbs, for example, in Nassau County in New York, Orange County in the Los Angeles Basin, and the Santa Clara Valley southeast of San Francisco. In the latter instance, 70 percent of the farmland in Santa Clara County was classified as prime agricultural land in 1949. But as the city of San Mateo, itself a suburb of San Francisco, grew, the rich valley in which it is situated became desirable to developers. The same characteristics that make land desirable to a farmer (flatness, depth of soil, accessibility, lack of large rocks or ledges), also make it attractive to the developer, who must keep in mind the expense of land preparation before construction is possible.

Many farmers in the Santa Clara Valley were unable to resist a price of $4000 an acre for their land, and soon bulldozers removed the orchards and subdivision housing began to spring up. It appeared that in a few years the entire valley would be completely developed and its superb soil buried forever. In an effort to save some of the valley a group of farmers persuaded the county to declare an exclusively agricultural zone to preserve the remaining farmland. This measure has held the line for a while, but farmers are constantly harassed by their new neighbors, who object to spraying operations and the noise of machine operations in early morning. As the value of this land increases to $15,000 an acre the process of urbanization may resume, making retention of the agricultural zone extremely difficult. In this way, 15 to 20 percent of California's best agricultural land has already been lost. On a national scale only seventy-two million of the total 465 million acres of arable land is considered prime, and one-half of this cropland has already been urbanized.

The problem, then, is not in the total number of acres either in open space or in urban use but the quality and distribution of the land that is being converted from one use to the other. The thousands of unused square miles that are still available in Wyoming or Nevada offer little for either the farmer or the urbanite. While it is not likely that the urban sprawl will soon be stopped, it can be regulated, with provision for more balanced land use.

Control of Land Use

The need for ecological control over land use has been demonstrated again and again. But assuring that an ecologist plays a significant part in land use planning does not necessarily lead to ecologically sound land usage, for there are strong local political and economic pressures and financial control always seems to be in someone else's hands. Only public opinion will sway those who control land development and persuade them that ecologically sound solutions are ultimately the best solutions.

Imagine, for example, a narrow rural valley with a stream flowing through it year round. The slopes are covered with trees and perhaps an occasional house built to take advantage of the rustic qualities of the environment. One day a bulldozer appears on the hillside above the valley and intensive development begins. During the construction period many tons of soil wash into the stream, followed by runoff from roofs and paved streets of the completed development. This rainwater no longer seeps into the ground and recharges the water table; instead it pours from storm sewers into the stream. Deprived of its source of recharging, the water table is lowered, requiring deeper and deeper wells for local water supply. The excess water in the stream causes flooding downstream where basements, previously dry, must be periodically drained or waterproofed. During periods of high water, septic tanks are flooded, the stream is polluted, and sewers become necessary to avoid public health problems. Because of previous silting the stream is dredged to handle the recurring flood waters, but this seems to make matters worse. Then the Corps of Engineers is only too happy to build an expensive dam upstream, which in turn permanently floods much valuable bottom land behind the dam. Wells are running dry now, so a municipal water system must be installed. What is left of the stream alternates between feast and famine, flooding one season, dry the next. Finally it is agreed to confine it to a culvert to avoid future problems. This chain of consequences is not at all unlikely, but is it inevitable? By no means.

If the area including the wooded valley had available a planning commission, or an ecologist who is trained to recognize potentially troublesome sites, the whole destructive process might have been avoided. Quite clearly there are some places where houses *should not* be built or at least built only with extreme care and foreknowledge of the dangers involved. The function of a planning commission is to remove the trial and error approach and to suggest on a rational basis which land is most suitable for development and which land is better left undisturbed. The developer sees a swamp as reclaimable land capable of being filled in and covered with houses: an environmentally informed planning commission might view that swamp as a sponge capable of absorbing 300 thousand gallons of water per acre for every foot of rainfall added to it and wonder where all that water will go if the sponge is destroyed.

Another problem that can be avoided by comprehensive planning is leapfrogging (Figure 19.9). As land close to the city rises in price the developer looks for cheaper land farther out on which he can still make a high profit. He leapfrogs the undeveloped but expensive land nearer the city and builds further out. This results in a chaotic mosaic of developed and undeveloped land that is extremely difficult and expensive to service with rapid transit, sewers, water, fire and police protection, and garbage collection. On a smaller scale, a farmer pressed for cash may slowly sell lots fronting along a road, thereby isolating his remaining land from the road, both decreasing its value and its accessibility to future development. Even more important, this practice puts great pressure on a rural road system incapable of handling the traffic thrust upon it.

A more traditional and piecemeal approach to the control of land use is zoning, which seeks to control the use of land by regulating the size of plots and the shape of buildings and their relationship to each other. Zoning is often used to prevent nuisance construction, such as backyard chicken coops, or to keep population density low and "undesirables" out by establishing minimum lot size. But such planning has

Figure 19.9 After World War II, the increased demand for housing resulted in an unprecedented sprawl of houses which today cover square mile after square mile of fertile land near large cities. (Thomas Airviews)

often backfired. Instead of controlling the pace of development, it may accelerate it to the point where no open land is left.

Suburban open space. Monotony and crowding from ineffective zoning practices are being overcome by the growing popularity of cluster or green-space housing. This plan allows houses to be built closer together in a portion of a tract, and frees the remainder of the land from development. The concept embodied in cluster zoning is quite old. Farm houses in parts of French-speaking Quebec, for example, are grouped in small villages allowing all of the surrounding land to be intensively farmed rather than tying up several acres per family with houses and outbuildings.

The open space saved from housing can be used for recreation and to preserve some of those values which people seek when they move from the city center to the suburbs in the first place—a semblance of a natural environment, free from parking lots, houses, and utility poles. For cluster housing to work, continuing provisions must be made to preserve the open land. Occasionally the whole concept is vitiated by later owners who decide to sell the commons to free themselves of maintenance costs.

Cluster zoning (Figure 19.10) is only one way to reserve open land in urban areas. Easements, which involve the purchase of development rights rather than the purchase of the land itself, can be used to preserve wooded or scenic land. Claiming more land than required for a building project and then placing the excess into an open space program is another approach. Gifts from large landholders or exchanges of land parcels are still other ways to obtain open space.

The question that often arises in suburban or rural areas is whether to develop land suitable for housing or for other uses. The town of Closter, New Jersey, when faced with a controversy over its attempt to acquire eighty acres for an open space program, made some calculations and drew the following conclusions: eighty acres would allow the construction of 160 houses, which would bring at least 200 new children into the local school system. Estimating that it would cost the town $720 per year to educate each child ($144 thousand), $4000 to collect the additional garbage, $6000 for police protection, and $2000 for extra fire hydrants and miscellaneous services, the total expense to the community if this land were developed would be $156 thousand. Calculated land tax revenues would amount to $100 thousand a year. Thus the development of the eighty acres would have *cost* the town $56 thousand a year. In the enthusiasm for growth of urban housing, so often encouraged by merchant's associations eager for new business, the ultimate costs of "growth and progress" are rarely spelled out. If they were, the desire for development might be tempered and effective plans made to control it.

Figure 19.10 By clustering the housing and using forms sympathetic to the environment, much of the natural beauty of this seacoast at Sea Ranch, California is preserved. (Aero Photographers)

NEW TOWNS

As cities have grown beyond manageable size and often have become remote from any possible natural environment, there has been an attempt to overcome the mistakes of old cities by starting new ones. The reasoning usually runs: a near-perfect environment would be within our grasp if we chose just the right site and carefully planned buildings and streets set in parklike grounds. But this planner's utopia continues to elude us.

Much has been said about the impossibility of successfully governing a city with over a million population. The ideal size submitted by planners varies from thirty thousand to a couple of hundred thousand. Clearly a city of ten million has exceeded manageable proportions. However, if the existing population of, let us say, France were evenly distributed into cities of a few hundred thousand people scattered throughout that country, some sixty new cities would have to be built, which would simply urbanize large areas of presently rural France.

The English pioneered the establishment of new towns in an effort to relieve pressure on London and other urban regions. Following the ideas of Ebenezer Howard, who developed the concept of the garden city, several new towns were planned in every detail and then established, but reactions from their inhabitants have been mixed. One town constructed on the basis of twelve houses to the acre provided gardens front and back which the designer included to please working class families from big city slums who had never had more than a flower pot. But young couples resettled in these units felt isolated from old friends and family in the cities they left behind; young wives felt that the gardens insulated them from meaningful contact with neighbors and prevented the making of new friends to replace those left behind (apparently another example of the misunderstood role of space in the human environment); the husbands found there was no neighborhood pub to get away to and no recreation anywhere comparable to that available in the city. Then the need to commute daily to work took time and money, neither abundant to a working man. Many newcomers to these new towns, like the young couples mentioned, leave after a while, preferring the overcrowded cities. After a few years have passed, there is a mellowing. This can be seen in the English new town of Letchworth, yet there is a cloying sameness that one must expect from something conceived in its totality at one point in time. Diversity of building styles is, after all, a strong point of old towns. Letchworth is quite successful, however, compared with Cumbernauld near Glasgow, Scotland. This new town is built around a rambling building called the city center placed at the highest point of the city and apparently designed as a piece of conspicuous architectural sculpture. Unfortunately, environmental considerations seem not to have been well made and the center acts as a wind tunnel, presenting little environmental improvement over conventional cities that just happened. In addition, this fixed center provides the only focus, indeed relief, from the rather drab housing units which effectively insulate the city from the lovely Scottish countryside. Still, these considerations are trivial when Cumbernauld is compared with the Glasgow slums it was intended to relieve.

Another type of new town is an intentional suburb, a satellite town built without the presumption that it need stand alone. Two new Swedish satellite towns, Vällingby and Farsta, have shopping and cultural areas of their own but are connected by a high-speed rapid transit system (see Chapter 20) to downtown Stockholm, which offsets possible shortcomings of the new town. Tapiola, near Helsinki, Finland is another successful satellite town.

Some new towns, constructed on a grand scale, are the capital cities, Brasilia and Canberra. Brasilia has had a cold response from the public servants who are obliged to live there. After dark and on weekends the monuments, grand and imposing, are deserted, as are the recreational areas of many of the apartment blocks. The action is in the shanty town

of temporary buildings constructed to house the workers who built and
are still building Brasilia. Considering that when location is politically
rather than environmentally determined and that civic activities are
sharply limited, the boredom of a new capital is not too surprising. Indeed officials are often given a hardship allowance to redress their isolation from the natural center of a country's culture.

We might be better off if housing and, collectively, cities could be
flexible enough so that people could do their own thing, like children
with building blocks. Somewhat short of this technological miracle but
interesting, nevertheless, is modular housing which is being tried in many
cities. By erecting a superstructure to which they are attached, modular
units can be flexibly and creatively arranged to fit the whole structure into
any existing neighborhood (Figure 19.11). But it makes as little sense for
a planner to try to design a new town without real knowledge of its intended occupants as it would for an ecologist to try to recreate a salt
marsh without any knowledge of the animals that he might expect to live
there. However, Howard was quite probably on the right track, with his
notion of city units separated by green belts, for the largest and most
densely populated city inevitably breaks up into blocks and neighborhoods, as thoroughly isolated from the city as a whole as suburbs miles
away. If one could add the amenity of comfortable housing and convenient open space for a spectrum of recreational outlets, Howard's garden
city might be considered to be a form of neighborhood. Perhaps the
successful neighborhood can provide a prototype for the long sought utopia of planners.

Figure 19.11   The superstructure of this modular construction can be clearly seen
in the vertical pillars. Modular units are then fitted into the three-dimensional grid,
giving unusual flexibility. (Warner, Burns, Toan and Lunde, Architects)

The Future of the City

Cities and their suburbs are likely to continue to sprawl across the United States. All people seem to have an insatiable need for some bit of territory over which they have absolute control. It thus becomes the responsibility of architects and planners to build this privacy and individual control into urban areas. But the chaotic sprawling suburbs with which we are familiar need not be chaotic or sprawling. Preservation of some open space through parkland, agricultural reserves, scenic easements, and cluster zoning can provide breathing space and some sense of relationship between man-made and natural forms. Long-range planning is imperative on a regional basis before any amelioration of the grimmer aspects of urban life can be hoped for. With some idea of which areas should grow, when, and how, transport problems can be anticipated and included in long-range plans.

The present environmental problems of urban areas are not unsolvable, however. Indeed there is every reason to expect that there will be successful cities in man's future if we take the long overdue step of infusing a few billion dollars into a full-scale study of the total environment of cities. We need to know how transportation systems can be designed to meet the needs of *people;* what makes one neighborhood exciting to live in and another boring; how we can make the best use of space in our urban designs; what environmental factors are involved in chronic mental problems and how they can be controlled; what human needs are not met in present housing, both old and new?

These are questions that can be approached logically and answered rationally; the solutions can then be considered in any future plans for redesigning old cities or building new ones. There will soon be too many of us to continue the cruel idiocy of trial-and-error experimentation with people's lives, especially at a time when more people are moving from the country to urban areas than ever before. By the year 2000, it is estimated that over three-fourths of the American people will live in towns and cities with populations of over 5000. If a start is made now toward solving the age-old environmental problems of cities there is a chance that these cities may not only be habitable but enjoyable.

FURTHER READING

Eldredge, H. W. (ed.), 1967. *Taming megalopolis.* Vols. I and II. Praeger, New York. An encyclopedic collection of papers dealing with a variety of urban problems.

Gottmann, J., 1964. *Megalopolis; the urbanized northeastern seaboard of the United States.* M.I.T. Press, Cambridge, Massachusetts. Gottmann's book, though a tome, is a classic work, the first to recognize the implications of regional cities.

Hall, E. T., 1966. *The hidden dimension.* Doubleday, Garden City, New York. Incisive examination of man's private and public use of space.

Herber, L., 1965. *Crisis in our cities.* Prentice-Hall, Englewood Cliffs, New Jersey. Popular but literate account of urban problems.

Jacobs, J., 1965. *The death and life of great American cities.* Random House, New York. A common-sense point of view that cuts through much of the persiflage surrounding city planning.

Mumford, L., 1961. *The city in history.* Harcourt, Brace & World, New York. Leisurely and thorough, Mumford blends scholarship and style.

Rudofsky, B., 1969. *Streets for people; a primer for Americans.* Doubleday, New York. Well-illustrated book pointing out the use of streets as urban space.

Srole, L. et al., 1962. *Mental health in the metropolis, the midtown Manhattan study.* Vol. I. McGraw-Hill, New York. A pioneering, thought-provoking study of the effects of urban life on mental health.

Tunnard, C. and B. Pushkarev, 1963. *Man-made America: chaos or control?* Yale University Press, New Haven, Connecticut. If you read any book on this list, read this one. Beautifully thought out and illustrated, it gives penetrating insight into man's attack upon the environment in the guise of urbanization.

U.S. Department of Health, Education, and Welfare, 1969. *The climate of cities: a survey of recent literature.*

U.S. Department of Housing and Urban Development, 1970. *New communities, a bibliography.*

U.S. National Commission on Urban Problems, 1969. *Building the American city: report.* Praeger, New York. Much information about ghetto areas, zoning problems, urban renewal.

CHAPTER 20

TRANSPORT: LIFELINE AND NOOSE

Cities have become so vast that any vestige of self-sufficiency they may once have claimed has long since disappeared. In order to function they rely exclusively on complex transport systems that move their managers, workers, raw materials, goods, and services. But these transport systems are also responsible for many of the urban ills which threaten the system they support. The dual nature of transport as a lifeline and a noose is a problem rooted in the past.

ENVIRONMENT AND THE EVOLUTION OF TRANSPORTATION

The history of transportation is a curious one. For untold millennia man walked, or rode an animal that could drag his belongings or tools. But it was not until the invention of the wheel that transport began to develop in earnest. As obvious as the wheel is to us today, it was unknown in the Western Hemisphere until its introduction by Asians or Europeans. Transport systems remained crude, if functional, for thousands of years. Then in progression came the carriage, the steamboat, the railroad train, the automobile, the airplane, and the rocket, each succeeding its prede-cessors in successively shorter time spans.

A major stimulus behind this rapid evolution of transportation was, of course, the industrial revolution, both in Europe and the United States, demanding more rapid and efficient transport of raw materials and fin-ished products. In Europe distances are comparatively short; however, the rugged environment and enormous distances which characterize North America threatened to delay development of the New World. Hence the very challenge of the landscape acted as a spur to the application of technology to transport problems.

Although each of the successive modes of transport has had its impact upon the environment, the environment in turn has had a definite effect in shaping and directing both the type of transport and its lifespan. To see this in its historical perspective let us examine the interaction of transportation and environment since 1800.

Wagon roads. One of the most important early innovations in the United States was the development of a sturdy, commodious wagon by German settlers in the Conestoga Valley near Lancaster, Pennsylvania. Freely adapted from the crude wagon of their homeland, the Conestoga wagon quickly came into wide use in the early nineteenth century to transport freight between the rapidly growing cities of the eastern seaboard. Settlement followed the easy wagon routes, north along the Connecticut and Hudson Rivers, southwest following the Great Valley, and west along the James and Rappahannock Rivers into western Virginia and Maryland. This developing country was not trackless however; buffalo, elk, and other large game had long followed the easiest grades over mountain passes and along river valleys. These trails were used by the Indians and finally by the white men with their wagons. Other trails were cut and developed into wagon roads by army campaigns—Braddock's and Forbes' roads to the forks of the Ohio and later Daniel Boone's wilderness road that led over the Cumberland Gap into the frontier country of Kentucky beyond.

Post roads. As the wilderness trails extended westward, the people remaining behind needed more than single-file tracks or crude wagon roads; all-weather roads able to handle heavy freight traffic all year round were imperative. Using a road building technique developed by a Scotsman by the name of McAdam, a paved road was laid between Philadelphia and Lancaster in the late 1700s, soon followed by the National Post Road leading west from Washington roughly parallel to the present Interstate 70. Macadam pavement in the early nineteenth century did not mean asphalt or concrete, of course, just a graded layering of crushed stone. But anything that replaced mud, dust, and frozen ruts was a welcome improvement. Because the federal government was loath to subsidize this construction, such improved roads were opened as toll roads. The resulting high rates for freight movement spurred the development of another transportational innovation of the early nineteenth century—canals.

Canals and riverboats. Following with great interest the enthusiastic development of canals throughout Europe, especially France and England, speculators were ready to compete with toll roads in the growing business of transportation. Politicians were also eager to open new areas to settlement, an attitude due in part to a sharing of the general enthusiasm for growth and progress, but also from a desire to profit handsomely from speculation in land made more valuable by increased access.

In 1825 the Erie Canal finally connected the Great Lakes to New York City. This was the first of many steps that ultimately made New York City the largest and most important city in the Western Hemisphere. The succeeding two decades saw an unprecedented wave of canal construction; over 1700 miles of canal were constructed in Pennsylvania alone.

In 1811, Robert Fulton, inventor of the steamboat, formed the Mississippi Steamboat Company to carry shipping between Pittsburgh and New Orleans. Soon steamboats were on the Ohio, Mississippi, and Missouri Rivers and shipping grew rapidly, greatly increasing the importance of New Orleans and the new towns of Memphis, St. Louis, Nebraska City, Council Bluffs, Leavenworth, and St. Joseph. But by 1860 the heyday of the steamboat as a major form of transportation in the United States had come and was quickly to pass, doomed by the rapidly expanding railroads.

The railroad.   The same year the Erie Canal opened, 1825, Colonel John Stevens had built a steam locomotive. Within four years the Baltimore and Ohio Railroad connecting the Atlantic Seaboard and the Ohio Valley opened for business (Figure 20.1). The options of post road, canal, riverboat, or rail that were successively open to the East were, for many

Figure 20.1   One of the first passenger trains in service in America. The locomotive was built by a watchmaker in York, Pennsylvania in 1832, the handmade coaches in Philadelphia. The locomotive ran for sixty years and is still capable of running under its own steam. (The Warder Collection, New York)

years, closed to the trans-Mississippi West. Until the railroads could span
this almost unsettled vastness, animal power was the only practical form
of transportation. The Conestoga wagons because of their high sides, size,
and canvas-covered tops were still the first choice of settlers pushing
West from the river towns of Independence, St. Joseph, and Leavenworth; later, similar wagons made by Murphy, Carson, and the Stude-
baker brothers joined in opening the West to trade and settlement. The
Oregon, Bozeman, and Santa Fe Trails and the Central Overland Road,
developed in part by Army surveyors to establish and service outlying
forts, were the most heavily used.

Soon, however, the railroads had pushed west to the Missouri River,
and having eclipsed all but the Erie Canal in the East, began to compete with the steamboat trade on the western rivers. New cities such as
Chicago, St. Paul, and Omaha, with rail connections to the east and serv-
ing as jump-off points for travel farther west, rapidly began to outshine
the older river towns of Nebraska City, St. Joseph, and Leavenworth.
Steamboats were slow and, because they were obliged to follow the
rivers, indirect. Like the earlier toll roads in the East, their rates were
high enough to be easily undercut by the railroads.

The impact of the railroad on the West was immediate and long-
lasting. The destruction of the vast buffalo herds was quickly accomplished from the base provided by the railroads, which made possible
and profitable the transportation of hides, meat, and bones to the East.
Railroads had not only the power of life and death over any town along
their prospective routes in the more settled East, but they could and did
found towns en route further west, using land granted by the government. By 1900 the railroads reigned supreme, but the next actor was
waiting impatiently in the wings. Henry Adams once said that his generation of Americans was mortgaged to the railroad; ours is mortgaged to
the automobile.

HIGHWAYS AND THE LANDSCAPE

As the railroads took over trade once carried by wagons and stages, less
attention than ever was paid to roads, but around the turn of the century
the rise of bicycling as a sport of fad proportions again focused attention
on the pitiful state of road maintenance and construction. The point was
heroically made by the epic journey of Thomas Stevens, who bicycled
from San Francisco to the east coast in 1884, 3700 miles in three months.
Today high school boys make coast to coast trips on their bicycles every
summer but they have the benefits of roads which Mr. Stevens lacked. The
League of American Wheelmen lobbied actively for construction of roads,
ostensibly for bicycling, but they were cognizant too of the importance

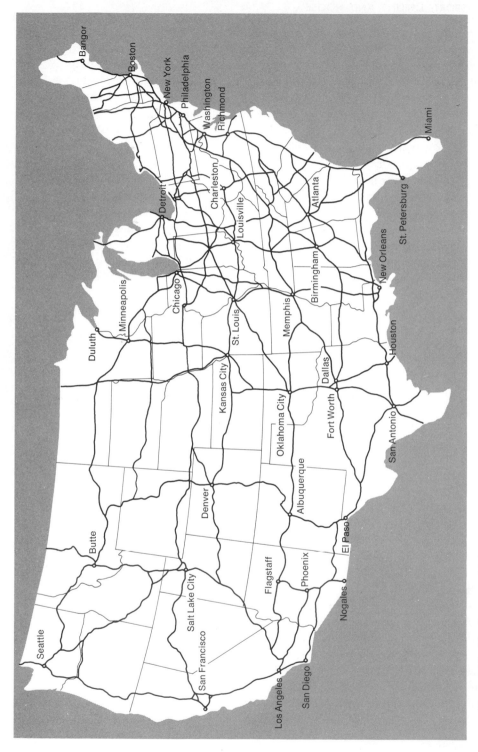

Figure 20.2   The interstate highway system connects all major cities with over forty-one thousand miles of high-speed expressways.

to the local farm economy of farm-to-market roads connecting outlying farms to railroads. By the time the automobile made its debut, the public had been aroused to the need for improved roads, and with the added stimulus of the car, good roads were soon forthcoming.

The Growth of Highways

Initially towns were connected center to center to allow maximum accessibility. As long as there were few cars, there were no problems: traffic and parking were easily accommodated by two-lane interurban highways. These town-to-town roads were gradually expanded into a network. The influence of local topography and land usage on their pattern is obvious to the road map reader. Even today you can easily identify a section of the country by the pattern of its highways. For example, as you can see from a road map, the roads in central Pennsylvania tend to arc southwest to northeast following the curving ridges and valleys of the Appalachian Mountains; in Kansas, although the land is flat and roads could theoretically go in any direction, the land is all farmed, and since no one wanted a road to cross his field, the roads follow section lines with many right-angled turns, giving the appearance of a giant checkerboard; in West Virginia the roads twist and turn as they move across a plateau heavily dissected by streams and rivers.

As the towns and cities continued to grow, often along new or improved highways, and the congestion of their sprawl was added to the increasing number of cars, it became more difficult to get from one town to another. As more and more people became mobile, both through their ability to own automobiles and their widening spheres of interest, they were no longer satisfied to drive from town A to town E, through B, C, and D, but wished to go from town A directly to town E. As truck transport developed and began to give the same competition to the railroads as they in turn had given to the rivers and canals, still more pressure developed for through traffic routes that bypassed the congested towns and cities. Their construction was begun in the 1930s in the more densely populated areas and continued after World War II.

But population growth and commercial development began to attack these bypasses almost as soon as the concrete was dry; clearly what was needed was a road that separated opposing traffic, controlled access, and whose intersections were separated. Divided lanes were available long before the automobile, Commonwealth Avenue in Boston, for example; but the Bronx River Parkway in New York, the Merritt Parkway in Connecticut, and the Pennsylvania Turnpike in Pennsylvania were the first major roads in the United States to combine all three characteristics and become expressways. About the same time, Germany began her system of autobahns, planned by teams of engineers and architects who were sensitive enough to environmental considerations to include nesting places for birds in their bridges. When World War II broke out road building came to a halt for six years. But afterwards, there was a rash of toll expressways or turnpike building, mostly in the northeastern United States. Then in 1956 the federal government began an enormous program of interstate expressways designed to connect virtually all cities with fifty thousand population or more (Figure 20.2). Over forty-one thousand miles of these limited access divided highways were to be provided. The original cost has since grown to $60 billion, and the completion date of 1972 has been extended to at least 1974. As a result of the interstate system, we have gone the full cycle in many cities, from roads built from city center to city center, to bypasses, and back to downtown expressways.

Highway design.   Modern expressways are planned with great care and the roadways fitted to the landscape according to very specific rules. Unfortunately, until quite recently, these rules had nothing to do with the particular aesthetic values of the countryside over which the expressways were rolling. The route was, and sadly still is in many places, determined by the expense of a particular route modified by its projected use. If this directed that an expressway should go through an historic site, a grove of redwoods, a state park, a marsh, or pond, all were sacrificed to the supposed greater good of the transportation corridor that replaced them.

Expressways and standard two-lane roads are blueprinted with as much care as a building, for every drain, curve, grade, ramp, overpass or underpass must be carefully specified. This can result in a well-engineered route with no sharp curves or steep grades—the choice between an attractive road in harmony with the landscape and one that seems to have been carelessly laid down lies with the designing engineer. A properly designed road must be articulated horizontally, vertically, and laterally with the countryside it traverses.

Only rarely can roads be set arrow-straight onto a landscape; even when they can, as on a coastal plain or flat plateau, this is usually undesirable because of the danger of monotony to the driver. Most roads are designed on the basis of tangents (straightaways) and curves, much like the tracks of model trains which come in straight and curved pieces. Engineers, by fitting these tangents to the curves, carry the road along the proper route, climbing or avoiding hills, approaching towns or rivers. When this is done with long tangents and short curves the road looks awkward and the curves appear too suddenly (Figure 20.3); the driver is alternately lulled, then forced to react. One alternative is to shorten the tangents and lengthen the curves; this improves the driver's reaction but it is as difficult to fit in short pieces of tangents as it is short curves.

Perhaps the best alternative, at least in hilly country, is a continuous curvilinear configuration in which the road constantly curves gently to

Figure 20.3 By fitting long straightaways and short curves together a road appears to lurch drunkenly across the landscape.

make the best fit with the landscape (Figure 20.4). Such roads are no more difficult to build, although they may be slightly more expensive to design. But a road with long tangents and short curves may ultimately cost more because of a greater accident rate and because of the potentially greater expense of fitting the road to the landscape by cutting and filling —a process that can be exceedingly expensive. Aesthetically, in the proper topographical setting, a continuous curvilinear design creates a far more attractive harmony between man-made and natural environments than any alternative.

Interstate expressways, by law, and most other federal and state routes, by practice, have grades limited to 5 percent, that is, a five-foot rise in 100 running feet. This vertical fitting of road to landscape can also be achieved by tangents, curves, and curvilinear design. Often a road using the long tangent and short curve approach in its horizontal alignment does the same thing in the vertical plane. The result is a jumble of awkward transitions: downhill runs that seem to bend sharply at the bottom, uphill segments that break into nowhere at the crest of a hill (Figure 20.5). With a curvilinear or at least a long curve and short tangent design, these awkward transitions can be avoided; once more the gain is both in aesthetic appearance and safety.

Figure 20.4 Continuous curvilinear roads present a much more graceful fitting of a highway into the landscape. (Thomas Airviews)

Figure 20.5 Highways must be carefully aligned in the vertical as well as the horizontal plane. The transition at the bottom of this hill is too abrupt.

Figure 20.6 Highway cuts should be beveled and sloped to allow plantings to stabilize the cut. This poorly designed cut will erode and drop soil and stones on the highway indefinitely.

The first expressways followed natural contours, and curved in harmony with the landscape, but these roads, although their windings and curves might have been pleasing on foot or in a Model T, were exceedingly dangerous when packed with fast cars driven by people in a hurry. Testing the power of earthmoving machines, road engineers replaced these parkways with beeline routes cutting through hills, rolling on filled beds over valleys with no attempt to repair, much less avoid, the damage to the landscape. Steep cuts were not only ugly scars that defied revegetation (Figure 20.6) but, prone to erosion, they showered the roadway with rock and mud with every rain, and required constant maintenance. Today enough care is taken in the construction of state and even local roads to avoid excessive cutting and filling, even if the road is longer as a result. Necessary cuts are beveled and planted, improving their appearance and stabilizing them against erosion.

Roadside planting. Most interstate expressways in suburban areas, at intersections, and occasionally along lengthy segments have plantings, well-intentioned and expensive, but sometimes very poorly executed. The whole point of expressway landscaping is to fit the road into the landscape with as little disturbance as possible to the natural environment and at the same time, by ensuring good visibility, to allow maximum speed and safety.

Two areas in which many roadside plantings fail are scale and appropriateness. While it is a fine gesture for a local garden club to help beautify our expressways by planting pansies in the median strip, the effect is utterly lost at sixty-five miles per hour. Many landscaped plantings also fall into this category. Trees or shrubs are strung out in lines or isolated groups that even when mature will have a scattered, chaotic appearance; mass planting on a bold scale can overcome this problem. Ironically this has been done most impressively in the semidesert of Southern California, where nothing would grow along the freeway without extensive irrigation. But mass planting of lush ivies, bayberry, oleander, iceplant, and other ornamentals makes some of the Los Angeles and San Francisco freeways among the most beautifully landscaped in the world (Figure 20.7).

This brings up the other planting problem—appropriateness. If the planting is in a dry region where native shrubs are few or lacking in proper scale but where irrigation is available, exotic ornamentals might very well be appropriate. In a well-watered region with an abundance of

Figure 20.7 Despite the dry climate, or perhaps because of it, California freeways are among the most beautifully landscaped in the world. Note the effective use of mass planting. (California Division of Highways)

native trees and shrubs, it makes little sense aesthetically, ecologically, or economically, to plant expensive exotics. The Long Island Expressway, for example, runs for much of its length through an oak-pine scrub forest. Instead of planting species native to this local forest, scarlet oak, bear oak, and pitch pine (Figure 20.8a), to coordinate the highway with its environment, the "beautifiers" have planted little clumps of Austrian pine and tulip trees (Figure 20.8b), which effectively divorce the road from its natural environment. So planted, the road could as well be in Maine or Georgia.

The great Brazilian landscape designer, Burle-Marx, trained in Europe like most of his contemporaries, returned to Brazil and rebelled against the slavish use of European ornamentals in Brazilian parks and gardens, whose landscapes ignored the incredible richness of the native Brazilian flora. Overcoming provincial prejudice, Burle-Marx made use of native materials with brilliant success. The same principle can be followed on American highways.

Another approach would be to allow natural revegetation to take place wherever possible. This would sharply reduce maintenance since plantings of ornamentals often require much tending. An example of this natural approach to highway planting can be seen on another Long Island highway, the William Floyd Parkway. Rather than removing the oak-pine woods evenly along the right-of-way, the designer made the grassy shoulder of variable width, sometimes stretching to the edge of the right-of-way, sometimes allowing trees to approach the road, giving a flexible, scalloped appearance that ties the road beautifully to the wooded land through which it passes.

An unplanned result of planting grass as a stabilizer along highway margins is that the lush pasture formed attracts deer, especially in heavily wooded country. The wild deer are easily frightened by headlights, and the casualties both of deer and injured motorists are high. Construction of fences has been helpful but is quite costly. An ecological solution to the problem would be to find an aesthetically pleasing roadside stabilizer or a planting that would be unpalatable or unattractive to large, grazing animals.

A final consideration in the planting of roadside areas where snow and ice are a problem is the effect of salt on plants. Although salt is effective in melting ice and snow, when splashed onto the soil in which trees and shrubs are planted, the salt, usually calcium or sodium chloride, kills plant roots and the plants slowly die. The maples which once lined country roads all over New England are dying because of this increased use of salt in winter snow-clearing operations. If tree-lined rural roads are to be preserved, new methods of dealing with snow and ice covered highways must be developed.

Figure 20.8 (a) These beautiful pitch pines overlook the Long Island Expressway right-of-way. (Photo by Dan Jacobs) (b) The right-of-way planting, however, ignores those species well adapted to the dry sandy soil of central Long Island and features instead this lonely tulip tree, which is more at home in a rich, well-watered forest. (Photo by Dan Jacobs)

EXPRESSWAYS IN THE CITY

Building roads in rural areas is relatively uncomplicated when compared with the problems that arise from the planning and construction of expressways through urban areas. No one, rich or poor, wants to sacrifice his property, house, or business to highway construction. Road designers know that city parks are the least vigorously defended parcels of land in most cities; consequently interstate expressways are invariably routed through them. Probably the most absurd example of the lengths to which this practice is carried is the North Expressway in San Antonio, Texas:

> For more than five years a bitter controversy has continued over the location of an expressway route through San Antonio's famed Brackenridge Park and related open spaces. This park . . . includes not only undeveloped land and a major flood control facility and parkland but recreation and sports areas, picnic grounds, zoo, college campuses, a renowned landscaped Sunken Garden and outdoor theatre, a city school stadium, a municipal golf course, and stretches of the natural water course.
>
> The proposed expressway curves and winds through this open space system, crossing an Audubon bird sanctuary and Olmos Creek, a tributary waterway in its natural state. It moves along a picnic ground and recreation area obliterating a Girl Scout camp and nature trail. It plunges across the Olmos Flood Basin and rises to enormous heights to go over the Olmos Dam. It severs the campus of the College of the Incarnate Word and stretches across the lands of the San Antonio Zoo. It blocks off the half-built public school gymnasium, slides along the rim of the Sunken Garden, hovers over the edge of the outdoor theatre—squeezing between that and the public school stadium—blocking a major entrance. It slashes through residential areas and slices along the golf course and a wooded portion of the natural watercourse of the San Antonio River.
>
> How many irreplaceable trees of magnificent size and age, how much spoil of adjacent area, how much auxiliary space for interchanges, drainage, and other highway structures will be needed have not, as yet, been calculated fully.[1]

After park upon park had been compromised, if not ruined, people began to protest, first on their own, and then with the help of their lawyers. The North Expressway has still not been built.

The Highway in the Ghetto

Another stumbling block in the routing of urban expressways has been slum districts. The poor, it was considered, could always find another

[1] Zisman, S. B., 1967. "Open spaces in urban growth." *Taming Megalopolis* Vol. I, Eldredge, H. W. (ed.), Praeger, New York.

slum or be relocated in some form of public housing. The land left over from expressway construction could be redeveloped with luxury apartment or high-rise office buildings and the whole package called urban renewal. But the poor are beginning to find their tongues too and as a result urban expressway construction in city ghettos is also being blocked. The interstate highway system may be nearly complete in 1974, its new target date, but in rural areas, not in the cities.

Citizen protest over leveling of parks and slums has brought about some rethinking of the role of highways in the urban environment. Expressway construction is a big business backed by a formidable lobby of construction firms, labor unions, steel and concrete producers, car manufacturers, gasoline and tire dealers, and organizations of car owners. The more roads that are built, the more workers are employed, the more profits are made. If cities are congested, the reasoning runs, we need more expressways to handle the traffic. When these are filled we add more lanes, then double decks, then parallel roads. Ignoring the limited capacity of city streets, more expressways bring more congestion leading logically to the situation described by Victor Gruen, who calculated the effects if everyone coming into Manhattan today by various forms of mass transportation was required to use a car:

> If 1,000,000 transit passengers were to drive in to work or to shop, they would occupy approximately 750,000 automobiles. If, having reached Manhattan, they were satisfied merely to stand still, bumper to bumper, they would cover 15,000,000 square feet of road surface. To make space for them within the main business area of Manhattan, we not only would have to eliminate all sidewalks, but would have to demolish every last structure, and then double-deck a part of Manhattan Island now covered by office buildings, hotels, theatres, stores, etc. Inasmuch as standing still, bumper to bumper, would be a highly undesirable and fruitless activity, and as people obviously will want to move around, we will have to provide three times that space; that means we will have to build six layers of transportation area covering the entire business core. If we also desire space for taxicabs, trucks, service vehicles; if we consider that there might be occasional accidents and stalled cars; and that it will be necessary to build some ramps, stairs, elevators and escalators, nine levels will be required. On top of the ninth level, we could then start from scratch and construct new buildings to house those facilities which we had to demolish in order to make space for motorization.[2]

Cars and Cities

Gradually the limited capacity of cities to absorb automobiles is being recognized and beltways are being constructed to divert cars away from

[2] Gruen, V., 1964. *The heart of our cities; the urban crisis: diagnosis and cure.* Simon and Schuster, New York.

the city's center rather than force them through the city's business district. Relieved of the burden of through traffic, a city can then have the flexibility to innovate. Some cities have separated cars from people, constructing underground plazas and shopping malls. Ironically it is always the people who are put underground, leaving the vehicular traffic to enjoy the light and air of the surface. Another possibility followed by Fresno, California has been to ban all vehicular traffic from certain main shopping streets, converting them into pedestrian malls with sidewalk cafés, parks, paved areas, planters—an exciting inducement to shop in the adjoining stores (Figure 20.9). Adequate peripheral parking is, of course, essential to the success of any such plan.

Any alternative to expressways through cities must involve use of mass transit. Buses, subways, and suburban trains have long been belittled by the highway lobby for obvious reasons. One result has been the neglect of mass transit, which has discouraged its use to the point of reinforcing the opinion that mass transport is outmoded and inadequate, and that the automobile is the only mass transit of the future. Gruen's calculation should disabuse anyone of that fantasy. To be sure, use of mass transit *has* fallen off; in 1929 the average American used public transport 115 times per year; in 1958, fifty-four times. The reason is simple—mass transportation has not kept pace with the automobile in its technology, convenience, or comfort. When an effort is made to provide convenient, clean, and comfortable *rapid* transit, including buses, trains, and subway cars, many commuters are happy to leave their cars at home.

Figure 20.9 Once typically filled with traffic and noise, this street in the business district of Fresno, California has been transformed into a pedestrian mall. (Fresno County and City Chamber of Commerce)

Man and machine. The problem is deeper than simple merchandising however: not surprisingly, environmental considerations and their psychological ramifications enter the picture. Very often the only privacy a typical white-collar worker has is the daily commuting trip in his car. The road may be jammed bumper to bumper but the personal space of his car is *his*. It represents a glimmer of independence in an otherwise thoroughly routine life. For this reason there will always be those who drive their cars as close to their work area as possible. But even they can be accommodated at least in part by some form of mass transport. Large parking lots or garages on the edges of a city would allow suburbanites to drive part way, since mass transport is often ineffective in low density suburbs, and take a train, bus, or subway to the central business district. Comparing capacity in persons per lane per hour, cars on a city street can carry 1400 people, buses 7500. On an expressway cars can handle 3500 people, buses 15,000, rapid transit trains 60,000.

Future Forms of Transport

In the future there is promise of more than refurbished trains or air-conditioned buses, for there is no reason why technology cannot be applied to the problem of moving people rapidly on the surface of the earth as well as above or beyond it. Popular science magazines are filled with fascinating schemes for magnetogasdynamic propulsion, gravity tubes, and various suspension systems—air cushion vehicles, magnetic suspension systems—all of which are based on perfectly plausible physical principles capable of working, if technology is assisted with investment dollars. More immediately useful are highspeed trains capable of moving people from one end of megalopolis to the other at 150 miles an hour. The Tokaido Express which runs between Tokyo and Osaka in Japan is probably the best example of the application of present technology to a standard form of transport (Figure 20.10). But to achieve even 150 miles an hour in a train requires complete rebuilding of the roadbed. With some modifications of the existing track the metroliner service between Washington and Boston has become not only possible but profitable, weaning commuters away from their cars. Speeds rivaling those in Japan must await further roadbed improvements, though. Surely some such system, metroliner or monorail, can provide an interim form of mass transit suitable for our sprawling megalopolises until the new technology becomes available. But even interim systems require vast sums of money.

One thing is certain, however, there is no *single* answer to the transport problem in urban areas. To preserve the fabric of cities and restore some degree of amenity to their environment the ultimate solution will

Figure 20.10 The Tokaido line connecting Tokyo and Osaka carries commuters at speeds up to 125 mph, making the 320-mile trip in three hours and ten minutes. (Japanese National Railways)

probably involve mass transport systems designed to move people rather than vehicles within the city and from selected collecting points beyond the downtown area. It will require beltways that allow traffic to bypass the limited capacity grid of city streets. It must offer some system of pedestrian and vehicular separation, either by converting districts or neighborhoods into pedestrian islands, limiting the type of traffic or the timing of its access on other streets, or putting vehicular traffic underground, giving pedestrians their proper place in the sun.

MAN TAKES TO THE AIR

For many cities the epitome of growth and progress is the size of their airport, the number of planes leaving and taking off, the number of people who funnel through their transportation centers to other parts of the nation or world. These activities *do* generate money, which accrues to the city, but the problems generated by airport complexes are seldom faced realistically. Air and noise pollution have already been discussed in Chapter 10. The problems that concern us here are the congestion caused by overuse of ground facilities and the huge space demands of expanding international airports.

No city in the world has more congested airports than New York. Kennedy, LaGuardia, and Newark airports, all bursting at the seams, must handle such dense air traffic that holding patterns often have dozens of jets endlessly circling the metropolitan area waiting for the opportunity to land. Something is clearly wrong when you can fly from Miami to New York in two hours, then be obliged to circle the airport for one hour waiting to land.

There is no reason why more people cannot fly directly to Europe or South America from any number of cities throughout the United States. Many cities have large airports which are capable of handling much of the traffic that passes through the ever-narrowing funnel of New York's Kennedy Airport. Why, if you live in Altoona, shouldn't you be able to fly from nearby Pittsburgh to London or Rome? Couldn't customs facilities be set up at these airports as well as at Kennedy, Boston, and Washington? If this feeder traffic were to be diverted, present overcrowded facilities would be adequate until more basic technological changes could be worked out.

But the standard answer to overcrowding at airports is to expand them or build new ones; consequently the search goes on for a fourth huge jetport for the New York area. No matter where a fourth jetport is located in the New York metropolitan area, an enormous chunk of the natural environment will be lost to any other use and the environment of miles of surrounding countryside will degenerate through the resulting development, noise, and air pollution. But the sky can accommodate just so many planes in holding patterns before the whole system breaks down into chaos.

Since each new generation of planes produces larger jets, requiring longer runways, and ever-larger terminals to accommodate increasing loads of passengers, economic alternatives to the present system must be found. But if huge airports are to remain the standard in air transport, efficient mass transit systems capable of moving goods and passengers quickly to and from the airport, regardless of its distance from the city, become imperative. At the present, only two American cities, Cleveland and Boston, have this kind of link with their airports. The time for developing solutions to the problems of congestion and space is growing short. While New York casts about for a site for a huge new jetport, other cities are beginning to eye natural environments that cannot be replaced; Chicago, Lake Michigan; San Francisco, its bay; and Miami, the Everglades. With truly *rapid* transit systems regional airports could be constructed at some distance from congested cities, pooling the resources of a region and at the same time sparing those natural environs of large cities that do still exist.

RIGHTS-OF-WAY

The movement of goods and people requires corridors or rights-of-way which traditionally have been devoted to a single use. No one would think of taking a casual stroll down an expressway or having a picnic on a railroad track. But there are other rights-of-way whose uses are not so clear-cut. The development of huge regional power plants has led to miles of power transmission lines. Pipelines, originally built to transport oil, now move natural gas, chemicals, and even coal, crushed and mixed with water.

Management of Rights-Of-Way

Regardless of the commodity being transported, rights-of-way must be managed so that the service they perform is not impeded by the environment. The basic reason for highway plantings, you will recall, is to stabilize the right-of-way by preventing rock falls or mud slides. Powerline and pipeline rights-of-way, which are often roadless, are usually covered with vegetation of some sort that must be managed to prevent damage to the facility. When land in well-watered areas is cleared, the disturbed ground is soon covered by weeds, then grasses, shrubs, and finally trees in succession. The goal of right-of-way management is to maintain vegetation at some particular stage in this succession, usually the grass or shrub stage.

According to Frank Egler, a distinguished plant ecologist who has studied plant succession and its manipulation for many years, the direction of the succession is greatly influenced by the types of plants already present at the time of abandonment. If these can be manipulated, the end result can be both predicted and controlled.

The standard approach of power companies, pipeline operators, and highway departments has been to spray their rights-of-way with herbicides to kill trees and shrubs selectively, leaving only grass. This works well up to a point; shrubs and trees *are* temporarily eliminated. However, the sprayed trees and shrubs turn brown and become an unsightly fire hazard. Because the spray does not kill the roots of the unwanted woody plants they continue to sprout and spraying must be repeated every year or two. Even if these existing woody plants could be eliminated from the unnatural grasslands, seedlings would constantly be a problem.

The blanket spraying operation, presently favored by chemical companies and by many right-of-way operators has proven to be a mistake. Egler has demonstrated in trial plots and field observations that shrubs encouraged by the selective killing of competing trees form a much more stable ground cover than grass, and one which requires far less maintenance. In addition, shrubs provide a greater variety of browse and food for wildlife than grass. While highway rights-of-way are no place for grazing animals, other types of rights-of-way offer no such conflict of interest. Managed according to Egler's technique, a powerline or pipeline right-of-way would have a strip of low perennial plants—goldenrods and bracken ferns—which can keep out tree seedlings with a little selective spraying. Beyond this there might be another strip of low shrubs—viburnum, shrub dogwoods, blueberries, and low junipers—and beyond that on the forest edge there might be alders, shrub willows, dogwood, and redbud. The particular species would, of course, vary from place to place as with highway shoulders and medians, but the idea of using a variety of plants from the natural flora would be the guiding principle everywhere.

Instead of the expense of spraying large amounts of herbicides over the entire roadside, shrubs could be maintained indefinitely by one man walking the right-of-way once every few years with a knapsack sprayer. Once the shrubs are established, maintenance may even be unnecessary, for seedlings of potentially competitive trees would be shaded out. One section of a powerline right-of-way in New York state covered with shrubs called sweet fern has had no maintenance for over twenty years. Since Egler estimates that close to ten million acres in the eastern United States alone are being used as rights-of-way, any program that would increase their use by wildlife or even people would certainly be worth looking into in a progressively park-hungry world. The reason for the delay in applying ecological principles to rights-of-way management ultimately lies in ignorance and apathy among the utility companies who, encouraged by chemical companies, seem to view every shrub as a potential tree.

Powerline Aesthetics

Power lines, unlike pipelines or roadsides, have one major aesthetic liability: their ugly towers, resembling overgrown erector sets, may completely dominate the countryside they traverse (Figure 20.11). The obvious solution would be to put them underground, like the utility lead-in wires of a house. But high-voltage lines are expensive to bury using present technology. The power loss from buried alternating current transmission lines is so great that without expensive compensating equipment, 345 thousand volts would be dissipated for every twenty-six miles of underground cable. If direct current transmission, which has much lower line loss, were used, it would have to be inverted back to alternating current for distribution, requiring expensive equipment. These problems are being researched and eventually most power transmission lines will be placed underground.

Figure 20.11 A row of powerline towers marching across the countryside cannot be ignored. Much more can be done to bring these towers into scale with the landscape.

Until that time the Hudson River Valley Commission, charged with preserving scenic and aesthetic values in the Hudson River valley, recommends the following guidelines for preserving scenery that might be marred by power lines:

1. Avoid, where possible, prominent scenic features.
2. Follow lower slopes or valleys between hills.
3. Do not cross the contours of hills at right angles and avoid the steep slopes that expose the power line to view.
4. Jog alignments when crossing major roads to reduce the view of the line from the road.
5. Bend the lines to follow the edges of different types of land use rather than plowing through the complex.
6. Route lines along existing pipeline and railroad rights-of-way wherever possible.

The unsightly stringing of power lines across rivers can often be avoided by using existing bridges and either running the wires underneath in conduits or stringing them alongside on outriggers. Finally, the towers now fulfilling engineering requirements but ignoring aesthetic ones could also be designed to harmonize better with the countryside.

As power needs continue to double every ten years, we shall have still more power lines. If they are constructed with some of these considerations in mind with regard both to the appearance of their hardware and to the environment through which they pass, then many of the lines to be built in the next thirty years can at least be made as inconspicuous as possible, until technology can give them a decent burial.

   Out of sight, out of mind.   Although power lines, because of the high voltage which they carry, resist underground placement for the present, the same is not true of the 220-volt lines that enter our houses or the lines strung up and down our streets (Figure 20.12). Fortunately large cities

Figure 20.12 There is no reason why urban eyesores such as this cannot be put underground.

Figure 20.13 Large cities were forced to eliminate overhead wires many years ago, as this 1881 cartoon suggests.

Figure 20.14 (a) In older housing developments utility wires were strung along on poles as suited the convenience of the developer or utility. (b) Today most new subdivisions put their utility lines underground, which allows the planting of trees that soon soften the aspect of the ugliest neighborhood. (Photos by AT&T)

were driven at an early date to place their wires underground (Figure 20.13), simply because of the vast numbers of lines. Many new communities around the country are being constructed using underground utility lines from the start, as a selling point. The cost is either being shared by contractor and utility or is being passed on to the buyer in the purchase price. One factor that prevents older neighborhoods from switching from overhead to underground lines is that the cost would have to be borne by the householder directly, a burden that many would view as unnecessary, particularly when they are also faced with the replacement of septic tanks by expensive community sewage systems, in a nationwide effort to combat water pollution.

Unfortunately, we have grown accustomed to seeing utility poles running along streets and highways, much in the same way as we have grown up with the sound of airplanes in the sky overhead. It is probable that if both were to disappear, most people would never know the difference, particularly when the landscape is ugly to begin with (Figure 20.14a,b).

We have long grown accustomed to thinking that the ugliness and destruction that result from transportation are inevitable, the price we must pay for progress. This is part of the old struggle-against-nature syndrome, which seeks to overwhelm the environment by the brute force of technology and then to make that technology a convenient scapegoat for the environmental chaos that has resulted. Whenever the interaction of environmental factors is understood, technology can smooth the way and lead to a resolution of potential man-environment conflicts. Each of the transportation modes discussed previously can be fitted into the natural environment with minimum adverse effects if we can develop an ecological sensitivity that allows the environment some basic rights. If these rights are protected and if technology gives us the ample means of doing this, we can coexist with nature despite all our roads, airports, and transmission lines. But if we continue to ignore the modest demands that the environment makes upon our use of it, then all the technology that we can devise will not suffice to put right the endless problems that will be generated.

FURTHER READING

Egler, F. E., 1953. *Our disregarded rights-of-way—ten million unused wildlife acres.* Trans. Eighteenth New York Wildlife Conference, pp. 147–158. Recommends multiple use of these strips through intelligent vegetation management.

Gruen, V., 1967. *The heart of our cities.* Simon & Schuster, New York. Absorbing account of the potential still left "downtown," by one of America's foremost city planners.

Hellman, H., 1968. *Transportation in the world of the future.* J. B. Lippincott, Philadelphia, Pennsylvania. While predictions are quite chancy, this book goes beyond cars, trains, and airplanes.

Howlett, B. and F. J. Elniger, 1968. *Powerlines and scenic values in the Hudson River Valley.* The Hudson River Valley Commission of New York, Tarrytown, New York. This well-illustrated little booklet contains much sound advice on the placing of powerlines.

Mowbray, A. Q., 1969. *Road to ruin.* J. B. Lippincott, Philadelphia, Pennsylvania. Exposé of the Interstate Highway System.

Owen, W., 1966. *The metropolitan transportation problem.* Rev. ed. The Brookings Institute, Washington, D.C. Critique of the automobile as the solution to all transportation problems.

Rapuano, M. et al., 1968. *The freeway in the city.* U.S. Department of Transportation Report, Sup. Doc., Washington, D.C. With planning, highways need not destroy cities.

Siddall, W. R., 1967. *Transportation geography, a bibliography.* Rev. ed. Kansas State University Library, Manhattan, Kansas. Fine bibliography on all phases of transportation.

Winther, O. O., 1964. *The transportation frontier: trans-Mississippi west 1865–1890.* Holt, Rinehart and Winston, New York. A history of transportation modes in the region and time mentioned in the title; interesting reading.

CHAPTER 21

SOLID WASTE DISPOSAL: MIDDENHEAP INTO MOUNTAIN

As if our transportation-induced problems were not enough, we must also cope with the staggering quantity of solid waste materials generated by an increasingly affluent society that habitually attempts to throw away the unthrowawayable. Man's first communal living areas were probably caves, well sheltered from the weather and potential enemies, man or beast. Solid wastes—excrement, bones, old tools, charcoal, and ashes—were quite literally thrown away. When the resulting piles or middenheaps got a bit lumpy underfoot, or became too offensive, it was a simple matter to cover up the mess and start all over. After a few hundred years a small cave would fill up, forcing the occupants to find new quarters, but some of the larger caves accommodated thousands of years of continuous occupation and hundreds of feet of wastes.

The cities, though they came much later in man's history, were hardly more subtle in their handling of solid wastes. Discarded materials were freely scattered just outside the city wall or around the dwellings. When destroyed by fire or war, ancient cities were often rebuilt upon their solid wastes and rubble. Over hundreds or thousands of years, the once-level sites of cities slowly became mounds, often abandoned and covered with sand or soil, but occasionally still inhabited by a few hundred people perched above the surrounding countryside on their huge waste heap (Figure 21.1).

An astonishingly complete historical record of daily life in caves and cities can be pieced together by careful archeological digging. But exploration of past cultures by examination of their waste is by no means limited to prehistoric times. Williamsburg, Virginia was restored after very careful archeological scrutiny uncovered the nature of building

Figure 21.1 This mound in Iraq represents 3500 years of human occupation and originally stood seventy feet above the surrounding plain. (The University Museum, University of Pennsylvania)

materials, techniques, and tools, deduced from the scraps and wastes associated with the construction of every building. Excavations in the older parts of eastern seaboard cities often bring to light old bricks, pieces of bottles, and crockery that tell much about the past. New York's Central Park, once a shanty town of squatters' huts, has pieces of glass and crockery scattered throughout, some dating from the 1840s and 1850s.

TODAY'S WASTE

In many ways, however, the solid wastes with which we are faced today are quite different from those of even the recent past. There is, for example, a much greater variety of more stable wastes. A simple glass bottle can at least be broken to take up less space, and a steel can will rust away in a few years; but a plastic bottle or an aluminum beer can lasts indefinitely. Even more serious is the rate at which solid wastes are accumulating. The average person throws away five pounds of wastes per day, or 1800 pounds a year, 121 pounds more per person in 1966 than in 1958, and by 1976 it will be 136 pounds more than in 1966. Of the 73.5 million tons of waste expected to be produced in 1976, only seven million tons will result from growth in population. The next fifteen million tons will represent the wasteful consequences of the packaging revolution.

We have come a long way from a plain brown bag or newspaper-wrapped fish; today and increasingly in the future, items will be packaged "for the convenience of the consumer." Take for example the plastic bubble packs hung on hooks by the thousands in all stores. The merchandise, a few nails or an electric socket, is plainly in view, but the package usually serves as advertising space for the manufacturer; it is less handy to carry, is difficult to open, and its content surely costs the consumer more than the same item bought unpackaged. The so-called convenience foods with their aluminum foil covers and pans, pressurized cans of just about every liquid—all add immeasurably to the trash load to be disposed of and to the cost of the material sold. Those tiny picnic-sized containers of salt, for example, inflate the price of the product to eight times the price of salt bought by the pound.

The full impact of the packaging revolution which began after World War II can be appreciated only when the cost to the public is calculated. In 1960 it was estimated that the average American family spent $500 a year just for packaging. In 1966 the disposal of wastes cost close to $375 million a year; seventy-five percent of this was spent for picking up and transporting household wastes to their disposal site. If you include the costs of air and water pollution from present disposal practices and the loss of the space that present landfill techniques are rapidly usurping, the final cost of packaging wastes is even greater than it appears.

The Kinds of Wastes

Of course, discarded packages are not the sole ingredient of household wastes, any more than household wastes represent all wastes produced by contemporary society. Before we look at disposal techniques we should have some idea of which wastes are causing the biggest problems and why. Since industrial wastes have already been considered in Chap-

Table 1 **Breakdown of household wastes in the United States**[a]

Waste	Weight percent	Waste	Weight percent
Cardboard	7	Leather, rubber,	
Newspaper	14	molded plastic	2
Misc. paper	25	Garbage	12
Plastic film	2	Grass clippings,	
Wood	7	dirt	10
Glass, ceramics,		Textiles	3
stone	10	Metal	8

[a] Subcommittee on Environmental Improvement, 1969. "Cleaning our environment, the chemical basis for action." American Chemical Society, Washington, D.C.

ters 7 and 10, our prime focus here will be consumer, rather than producer, generated wastes.

What is the nature of household solid wastes? One breakdown has been given by an American Chemical Society report (Table 1). Notice that the largest category, miscellaneous paper, accurately reflects a part of the packaging revolution.

Paper. At one time waste paper, particularly newspaper, was carefully saved by most households to be sold to the rag man for small but welcome change, later it was given away to neighborhood charity collection drives, and now it is thrown away with the trash because no one will cart it away. What has happened? The demand for waste paper by the paper industry has certainly declined: in 1945, 35 percent of the waste paper generated was reused; by 1966 this had fallen to 21 percent, and by 1980 less than 18 percent is likely to be reused.

The slackening demand has probably been due to the increased efficiency and lower cost of harvesting virgin pulp; these factors have been matched by a sharp increase in the cost of handling the collection, transport, and processing of waste paper. The packaging revolution has contributed mightily to the pileup of waste paper because of the huge increase in volume and the contamination of paper with aluminum and plastic coats. Furthermore, special inks and treatments, because of their variety, can be extremely expensive and difficult to separate from the useful paper pulp. Some success has been achieved in efficient and inexpensive de-inking of waste newsprint, but fancy paper wrappings and containers frequently are not worth the trouble. To make matters worse, special coatings, however attractive to the consumer, retard the natural decomposition of waste paper in landfill dumps and lead to delays in reuse of the land.

Cans. The tin can, which has been a mainstay of the food industry for over 100 years, is not made out of tin but steel, covered with as thin a layer of tin as the manufacturer can economically apply. Tin does not react with the contents of the can and has a pleasing shiny surface which, unlike the underlying steel, does not quickly rust. Although tin cans have been around for many years, their life in the natural environment is relatively short. The thin tin coat weathers away in the open, exposing the underlying steel to fairly rapid rusting. In humid climates most tin cans will rust away in a few decades—sometimes much less time (Figure 21.2a). Campers have long made it a habit to throw their tin cans into the campfire before burying them because this helps remove the tinned layer and expose the cans to rusting more rapidly.

Ironically, tin coating, which allows rustable steel to be used and so predisposes the can to environmental decay, made the cans virtually worthless as iron scrap because of the contamination with tin, which is a

Figure 21.2 (a) As objectionable as they are, tinned steel cans *do* rust in a few years and disappear. (b) Aluminum cans, however, persist indefinitely.

problem, especially with the newer basic oxygen steel-making furnace. For this reason cans are thrown away rather than salvaged.

But the tin can now seems to be on the way out because chrome and resin-coated steel has become cheaper than tinned steel and is more acceptable as scrap to the steel mills. Aluminum has also made great strides in the container market. As with the new steel cans, aluminum cans can be reused as scrap but they are so readily discarded that their collection is unlikely. Once thrown away piecemeal, their retrieval expense is as much as thirty cents a can. While the new steel cans will corrode if the tough chromium or resin coats are broken, the aluminum can is practically indestructible when dumped into the environment (Figure 21.2b).

Bottles. Whatever the change in metal cans, glass as a packaging material will endure. Since 75 percent of glass is silica sand, the most abundant substance in the earth's crust, cost of raw material is low and likely to remain so. At the present rate of bottle production, sand reserves are calculated to last around three billion years. However, some glass companies have expressed concern that, should bottle production increase significantly in the future, the sand supply might become dangerously depleted a little over one billion years from now.

Glass bottles are reasonably destructible. Despite the danger from broken glass, breaking at least reduces the bulk of a glass bottle or jar. But volume is not so much a problem as number. Once, most glass bottles were stoutly built with the intention of their being used many times. It was later realized, however, that bottle return is a nuisance to the consumer, the middleman, and the bottler, who must at times clean some incredibly dirty bottles.

A few years ago some merchandising genius discovered that for every returnable bottle, used twenty times on the average, twenty non-returnable bottles could be sold; this discovery resulted in the "disposable" bottle. In 1959, 2.7 percent of the bottles produced were non-returnables; in 1967 their slice of the bottle market had grown to 13.5 percent. Yet the disposable bottle is just beginning to be used; in 1966, sixty-five billion bottles were filled but only twenty-eight billion bottles were produced. In the next few years far greater quantities of bottles will be added to the disposal problem.

Fortunately the glass industry is not oblivious to the problem it is generating. Several kinds of self-destructing bottles are being tested by industry-sponsored research. One is a bottle that can withstand normal droppage but when struck hard fractures, like auto safety glass, into harmless granules which are much more easily disposed of than the more bulky bottle. Broken glass of any sort is a good landfill material because it is dense, less subject to settling, and is insoluble. A second approach to disposable bottle-making is the production of a water-soluble glass, covered inside and out with a waterproof layer to last out the shelf life. When the layer is broken the glass would quickly dissolve. Perhaps we shall end up flushing soda bottles down the sink; if so, the problem will not really have been solved. Instead of getting rid of a pound of solid glass bottle, we would ask a sewage plant to get rid of a pound of synthetic material dissolved in the sewage effluent.

Automobiles. Throwing away tin cans or bottles is one thing; throwing away a junked car is quite another (Figure 21.3). At one time

Figure 21.3 Abandoned cars accumulate conspicuously.

the problem was unknown for as recently as the 1950s one could get $40 or $50 for an old car. Wreckers could make their profit by stripping off useful parts and valuable scrap, because a typical car contains sixty-six pounds of aluminum, thirty-seven pounds of copper, thirty-five pounds of zinc, and twenty pounds of lead, and then he could squash the remainder into a small bale worth about $40. The steel industry at the time used a 50:50 mixture of scrap steel and iron ore to make new steel and consumed over forty million tons of scrap in the process. But conversion in the steel industry to the use of the more efficient oxygen furnace changed the bright picture for scrap steel. Not only did the new furnace use only one-half the amount of scrap steel as the old one, but the scrap had to be much cleaner, with a much lower content of metals other than iron, especially copper.

By the early 1960s the new urban phenomenon of abandoned cars was causing consternation. The decrease in the value of a ton of scrap to around $10 plus costs of $12 a ton to bale and another $10 a ton to transport made a dying auto worth $5 at most. Rather than pay $25 to haul a defunct car to a junkyard, many owners elected to leave them where they stopped running. In most cities all valuables are stripped in a few days and the car is often burned to a rusty wreck shortly afterwards (Figure 21.4). In 1965, 190 thousand cars were abandoned in this fashion, close to fifty thousand in New York City alone.

The slack market for scrap steel together with the rising labor cost of separating nonferrous from ferrous elements has caused a pileup of

Figure 21.4   Stripped and burned, this car slowly rusts and accumulates trash that is only sporadically removed by the city. (Photo by Dan Jacobs)

Figure 21.5 When crushed flat, these eight cars take up remarkably little space, allowing their economical transport to regional processing centers.

junked automobiles that can no longer be ignored; the junked autos seem to be everywhere, not just in junkyards, but on roadsides, in farmyards, city lots, and back yards.

Technology is finally rising to the challenge, and several impressive car-eating machines, capable of dealing with the backlog, have been developed by large scrap firms. One such machine can chew 1400 cars a day into small pieces which can be efficiently sorted into steel and nonsteel piles by magnets. This scrapyard separation produces a grade of scrap that is much more salable than the customary bales of mixed metals.

Regardless of the handling techniques, two problems remain: the cost of transporting junked cars to modern regional processing yards, and air pollution from burning the cars, which has been thought necessary to eliminate plastic, rubber, and upholstery materials. The first problem might be handled by small hydraulic crushers that can flatten a standard sized car to a thickness of one foot (Figure 21.5) and thus make it possible to ship large numbers of cars by flatbed truck or rail car to regional collection centers. One alternative has been the development of portable car eaters which work their way around a region, junkyard by junkyard, eating as they go. Air pollution from burning out cars has been eliminated by reducing cars to small pieces of scrap in a powerful hammermill, leaving the nonmetals to be recycled, used as landfill, or incinerated. This approach would tend to concentrate this phase of the business in regional centers that, because of their business volume, could afford the necessary antipollution equipment.

Tires. Usually by the time a car is scrapped the tires have been removed for salvage. But most cars in their lifetime run through at least three or four sets of tires and these too create disposal problems: altogether, 100 million tires are discarded every year (Figure 21.6). Because many states have minimum tread requirements enforced by yearly inspection, tires are often thrown away with a worn tread even though over 80 percent of the rubber in the tire remains. Some are recapped, but more and more people who have experienced poorly recapped tires prefer new tires. In contrast to the 10 percent of tires recapped in the United States, 80 percent of all tires are recapped in less affluent foreign countries. In addition, reclaimed rubber processing for various purposes has fallen from 30 percent in 1945 to 10 percent or less today in the United States.

Most old tires end up in landfill, for which they are poorly suited. Occasionally they are burned as a means of protecting citrus orchards from frost or for burning green timber in right-of-way clearance. But burning rubber in the open air causes about the thickest, blackest smoke of any combustible and is outlawed in many places.

Figure 21.6 Most discarded tires end up as just another item in the general debris that is slowly accumulating in all environments.

One promising solution to the pileup of discarded tires is being developed by Firestone Rubber and the United States Bureau of Mines. Shredded tires are placed in a closed vessel and heated without air to 500°C. At this temperature the rubber yields 140 gallons of liquid oils per ton of scrap that are similar to and as useful as those from coal. Also 1500 cubic feet of gas with a heating value equivalent to natural gas is produced. The residue solids have a very high carbon content that might be useful as a basis for other products.

YESTERDAY'S DISPOSAL TECHNIQUES

Despite some progress in the handling of a few of the major waste categories, techniques for disposal of the overwhelming mass of solid wastes have hardly been improved since Roman times. The town dump is still the site that receives all the solid wastes, usually dumped into a gully, canyon, or quarry pit. Such open dumps are infested with rats, mosquitoes, and flies. They often catch fire and smolder, producing foul-smelling smoke, and the trash tends to be blown about by the wind.

Landfill

One step beyond the open dump is the so-called sanitary landfill (Figure 21.7). Of the twelve thousand landfill sites around the country, less than 6 percent qualify for the term sanitary; the rest are little more than open dumps. To carry out sanitation efficiently, it is necessary for a bulldozer to compact each truckload as it is dumped, then for a dragline operator to quickly cover it with earth, and then for another bulldozer to compress and smooth the site to make room for the accommodation of another load of waste.

The advantage of a properly maintained sanitary landfill is the prompt burial of garbage and trash before it has a chance to catch fire, be blown away, or attract vermin. A good site can accommodate a great deal of trash and can be situated on old strip mines, sand and gravel pits, or abandoned quarries which are presently wasted land found in almost every state. When the site is filled to capacity, a two-to-three-foot layer of earth is laid over the last trash layer, and a park or recreational area can be created.

Since only 6 percent of the landfill sites qualify as sanitary, obviously the ideal conditions described are not often achieved. Too often trash is not covered immediately but once a day or every other day. Much paper and light trash get blown about and the site with its surrounding properties frequently becomes an eyesore. Most sites that are selected do

Figure 21.7 In a properly run sanitary landfill, trash is covered with earth every day by a bulldozer. This reduces breeding sites for vermin and eliminates danger of fires.

not need the improvement of sanitary landfill; salt marshes, for example, which are favorite landfill sites along the coasts, are valuable ecologically and aesthetically, and yet they are in danger of complete elimination, simply because they are handy and "empty" (see Chapter 9). Further, the anaerobic decay of the buried trash causes low and erratic settling, thereby making construction of buildings unwise for many years after the site has been filled. When a landfill site is unsuitable, the water table may be contaminated with obnoxious seepage from the buried trash, endangering nearby wells.

However, sanitary landfill can be a useful disposal technique, especially for small towns, if well-sited on true wasteland, and scrupulously tended to—conditions much more easily promised than realized.

Incineration

Another ancient disposal technique for getting rid of waste is to burn it. In its primitive form, incineration swaps one problem, solid waste disposal, for two, air pollution and solid waste disposal. When anything burns, particularly a heterogeneous mass such as trash, smoke is always produced in great quantity; after the combustibles are consumed the noncombustible glass and metals remain to be carted away, usually to a landfill site or out to sea.

Their poor design and operation makes most incinerators, whether backyard, apartment house, or municipal, simply roofed burning dumps.

When the new air pollution standards were set up, cities were forced to close down incinerators, because such equipment could not pass their own standards. Apartment house and office building incinerators, although handy for occupants and welcomed by city sanitation workers, pour large quantities of soot, fly ash, and smoke into the atmosphere.

Another pollutant being generated in increasing quantities by incineration is exceedingly corrosive hydrogen chloride which is released when polyvinyl chloride (PVC), a component of plastic, is burned. Only thirty-six thousand tons of PVC were burned in 1966, but over one million tons were produced in the diverse forms of garden hose, rainwear, shoe soles, floor coverings, and so on. Since 50 percent of the PVC is chlorine, the potential combustion of this plastic in the future makes emission control of incinerator stacks imperative.

Incinerators can, however, when properly designed and operated, cope with large amounts of combustible trash. Many European cities have been able to combine disposal of trash with generation of heat, and have put to use the thermal energy that in United States incinerators goes up the chimney as waste. One reason that American cities have lagged behind Europe in incinerator technology has been the availability of cheap fuel in the United States compared with its shortage in Europe. A power company in New York City can get quality coal for $5 a ton to generate heat for electricity; in Munich, coal may cost $20 a ton which inspires more creative efforts to use alternate sources of heat. But as air pollution regulations become more strict and fuel becomes scarce, more and more attention will be given to designs which both eliminate smoke release and use the heat produced to subsidize the cost of antipollution devices. One useful proposal is to build small incinerators capable of handling the solid wastes of about 150 thousand people, locating these strategically so that transportation costs could be cut. The steam could be used to generate electricity, easing peak period power loads or perhaps in a desalinization scheme if the city were located in a dry climate.

The twenty million tons of fly ash that are potentially precipitable from incinerator chimneys makes extremely fine bricks and concrete, but because building supply firms seem unaware of this raw material, and incinerator operators seem to have little ability to push a resource, most fly ash either goes up the chimney or, if precipitated out, is dumped as fill along with the noncombustible solids.

One tin can or piece of copper wire is worth very little; but the 125 million tons of incinerator ash produced each year contain nine million tons of metal worth $400 million! The metal in some ash piles has been valued at $12 a ton, making it a more valuable source of metal than many ore deposits. But this metal continues to be dumped as trash since its recovery is presently uneconomic; perhaps in a few decades the trash heaps of cities will be eagerly mined for their valuable supplies of metals.

Composting

Another old method used widely in Europe but much less frequently in the United States is composting. Compost is the rich, dark remains of organic garbage which has been aerobically decayed, losing its odor in the process, and thus making a valuable soil conditioner of materials previously buried in landfill projects.

Despite their obvious advantage in making an otherwise wasted resource useful, composting plants have some drawbacks. Often poorly designed, like incinerators, they produce odors which make them unpopular neighbors. To be properly run they need reasonably pure organic garbage, which means that such garbage must be isolated in the homes of the district served and picked up independently of paper and solid trash. Tests show that less than 10 percent of households involved in such a program seem able to keep their garbage uncontaminated. This means that the garbage must be sorted before composting, or screened afterward, which adds considerably to the processing expense. In addition, most people have a misconception of the nature of compost. Like vermiculite and perlite it is a good soil conditioner; it loosens the texture of heavy soil and increases the absorptive capacity of sandy soil. It is, however, rather low in nutrients and cannot be considered as a fertilizer. Often composting plants have difficulty in locating a steady market for their product even if it is dried and bagged. Still many cities —Houston, Chicago, and Johnson City, Tennessee—using mechanical systems, are able to process garbage in less than a week, reducing almost one half of the 75 percent of their organic garbage that is biodegradable into compost; they then manage to find a market for their product.

Responsibilities of Solid Waste Disposal

When we look at the performance of industry in squeezing the last bit of iron from low-grade taconite deposits or the last hydrocarbons from crude oil, we are impressed with the level of technological skill coupled with economy of operation. Until recently this expertise ended with the marketing of the product. Of course some firms have always stood by their products with warrantees, or double-your-money-back guarantees, but in no instance did the concern of the manufacturer extend beyond the reasonable life span of his product. Today, however, population growth together with the packaging revolution has begun to overwhelm our antiquated system of waste collection and disposal. Many manufacturers and industrial organizations are coming to the conclusion that they bear some responsibility for the design of their products beyond the point of successful marketing. But a great part of any changing attitude is due to the realization that if manufacturers do not volunteer changes that will help the nation to cope with the rising tide of waste, they will soon be *required* to make changes. The detergent industry, which switched in 1965 from nonbiodegradable to biodegradable surfactants, did so because of the threat of state laws outlawing the older type of detergent. It is not a coincidence that the glass industry has begun to explore the possibilities of self-destructing bottles at a time when dozens of state legislatures have considered banning, limiting, or taxing no-deposit-no-return bottles. But industry is no more to blame than a federal government which was willing to spend $300 million a year on chemical and biological warfare research, whose need is questionable, while it was spending only $15 million on the study of solid waste disposal systems, which we need desperately.

An Archaic Disposal System

Although the nascent efforts of industry to make more disposable disposables are commendable, the waste handling system is badly flawed at both ends: initial collection and ultimate disposal. As we saw earlier, most of the expense of waste disposal lies in the cumbersome, noisy, and ineffective method of carting by trucks. Many years ago human wastes were deposited in outhouses, which were cleaned out periodically by a man with a horse and wagon. But today in cities, human wastes are deposited in receptacles connected by sewers to central disposal plants. Can you imagine emptying your chamber pot at the curb on the way to work in the morning? Yet as we pause to watch moon exploration feats on TV, our beer and soft drink bottles go into a trash can that is emptied into a truck and hauled away to some dump or landfill.

NEW APPROACHES

Approached from this point of view, our whole system of waste collection is badly out of date. Buying a fleet of shiny new trucks which make less noise and hold more trash is no more an improvement than replacing a horse and wagon with a truck. Plainly what is needed is a solid waste pipeline analogous to the sewage waste disposal system now accepted as a rather basic element of urban life. One pilot system worked out in Sundeberg, a suburb of Stockholm, Sweden, uses a vacuum sealed pipeline that collects the solid wastes from 5000 apartments. Since the pipe openings are twenty to twenty-four inches wide, most household wastes are easily accommodated. The air stream in the pipe pushes materials thrown into the system along at fifty feet per second to an incinerator over a mile and a half away. The incinerator then supplies heat to the apartments in winter. The ability of the system to move heavy materials was graphically demonstrated recently when an automobile battery was found in the incinerator. The great advantage of the Swedish system is the elimination of presizing; virtually everything can go down a two-foot-wide pipe. The drawback is the difficulty of moving solids by vacuum tube over distances much greater than a mile or two.

A research team at the University of Pennsylvania has overcome this distance difficulty by using a water slurry system to carry wastes. Using a much smaller pipe than in the Swedish system, wastes are pulped and shredded before being transported. Solid wastes would be fed into a combination pulper and shredder, which would chew paper, cans, bottles, bones, and whatever, into small fragments which could then be flushed through an independent system to a plant where the waste could either be incinerated or dried and economically shipped by rail to outlying landfill sites.

Recycling Wastes

But even when the transportation problem is solved we are still faced with the problem of what to do with the wastes, for all we have really done is to collect them and to reduce their volume somewhat. The key to the whole problem of solid, liquid, or gaseous waste disposal is not disposal at all, but recycling. We still have the old caveman mentality about wastes; we think that all we have to do when the local environment fills up with wastes is to throw them a little farther. So we build our smokestacks higher, carry our nerve gas canisters farther out to sea, and truck our garbage to the next town. Now the signs are everywhere that our total environment is beginning to fill up. By the end of this century, a slim thirty years from now, there may be no more landfill sites and the sea may be too polluted to receive any more trash. Then what can we do?

The difficulty with recycling has always been that wastes by definition are worthless, certainly not worth spending any money on reclaiming. This is in part the result of our current level of affluence and the apparent inexhaustibility of sources of raw materials. Ultimately, of course, certain raw materials will become scarce and today's waste will be tomorrow's resource. Can the natural environment survive that often postponed date? The evidence now apparent to us all indicates that it cannot. We cannot wait for the valueless to become valuable, as in time it must; we must deliberately *place* value upon the valueless, by whatever means, regardless of the tyranny upon our social system, bearing in mind that the environment has until now borne the brunt of our disregard. Virtue is never easily inculcated, and recycling materials in our urban environments will not come easily either. On the other hand, we have no choice.

One approach might be a special tax on virgin materials, which would heighten interest in raw materials currently regarded as waste. This would both conserve the virgin materials and reduce the waste burden. Another approach might be to set up tax incentives for industries that find new ways to use wastes associated with either the manufacture or ultimate disposition of their products. Recycling is hardly new. Even today 60 percent of the lead and 40 percent of the copper in use is from scrap. As population levels off, as inevitably it must, 90 to 100 percent of most currently used materials will have to be recycled, just as living organisms have recycled the water, oxygen, and carbon dioxide for the past billion years. In following suit with our man-made refuse, we will finally be returning to that steady-state environmental balance known to all other organisms on our planet.

FURTHER READING

———, 1969. "Glass makers launch solid waste program." *Env. Sci. & Tech.*, 3(1), pp. 17–18.

———, 1969. "Scrap tires, materials and energy sources." *Env. Sci. & Tech.*, 3(2), p. 119. These two brief papers give some idea of the directions of solid waste research.

Blake, P., 1964. *God's own junkyard.* Holt, Rinehart and Winston, New York. Fine collection of photographs illustrating our messy environment.

Carr, D. E., 1969. "Only the giant car-eater can save us." *The New York Times Magazine,* 4 May. Good review of the waste car problem.

Darnay, A. J., 1969. "Throwaway packages—a mixed blessing." *Env. Sci. & Tech.*, 3(4), pp. 328–333. The packaging revolution comes home to roost.

Grinstead, R. R., 1967. "No deposit, no return." *Environment,* 11(9), pp. 17–23. What do we do with solid wastes?

Packard, V., 1960. *The waste makers.* McKay, New York. Planned obsolescence and the economics of throw-away.

Subcommittee on Environmental Improvement, 1969. *Cleaning our environment, the chemical basis for action.* Amer. Chem. Soc., Washington, D.C. Sound recommendations by a panel of chemists.

Zandi, I. and J. A. Hayden, 1969. "Are pipelines the answer to the waste collection dilemma?" *Env. Sci. & Tech.*, 3(9), pp. 812–819. Technology to the rescue.

Part VI

THE PEOPLE PROBLEM

CHAPTER 22

SUPPORTING THE EARTH'S POPULATION

In 1798 Thomas Malthus, an Englishman, published *An Essay on the Principle of Population . . .* in which he wrote:

> the power of population is indefinitely greater than the power in the earth to produce subsistence for man.

> Population, when unchecked, increases in a geometrical ratio. Subsistence increases only in an arithmetical ratio. A slight acquaintance with numbers will show the immensity of the first power in comparison with the second.

> This implies a strong and constantly operating check on population from the difficulty of subsistence. This difficulty must fall some where and must necessarily be severely felt by a large portion of mankind.[1]

Unfortunately for the currency of Malthus' theory, disaster did not overtake England; population did increase, but so did subsistence—spectacularly so. Yet Malthus was correct in his basic analysis of the problem.

HUMAN POPULATION GROWTH

Unlike animals, which breed until limited by predation, starvation, or disease, man has the unique ability to limit both his death rate and his birth rate. Even so, it took from the beginning of man's evolution until 1850 for his population to reach one billion.

Early man, in the hunting and gathering stage, required about two square miles per capita to support himself. At this cultural stage, the

[1] Hardin, G. 1969. *Population, evolution, and birth control.* W. H. Freeman, San Francisco.

accessibility of food stabilized man's population, until the agricultural revolution removed that limitation. But the earth's resources are not uniformly distributed, and while agriculture flourished in some areas, other areas were nonproductive. This led to population increases in countries with adequate resources, but limitation by starvation in others. But by the eighteenth century, when Malthus made his observations, land and food shortages in many European countries had led to some desperate social conditions.

But by several quite fortuitous developments Europe was saved and Malthus' theory was temporarily discredited. First, the discovery of North and South America, then Australia and New Zealand, provided an outlet for surplus population. These new empires generated capital that permitted the industrial revolution to sweep Europe more quickly than it might otherwise have done. In addition, these colonies provided a market for goods produced by the mother country. Then efficient agriculture, stimulated by the industrial revolution, allowed not only a population increase but a slow rise in prosperity. Advances in medicine lowered the death rate, but the birth rate fell also, for the newly industrialized population desired to limit its families in order to feed and educate them better.

Convinced of the desirability of constant population growth, governments had always encouraged people to have large families to run the machinery in factories, and when necessary the guns to serve the national interest. While it was certainly useful to governments and industry to have a large labor pool, it was not long before one of the laborers, the Englishman Francis Place, came to the conclusion that it was in the best interest of laboring men to limit their numbers and thereby increase their value and power. From this humble background arose the concept of family planning, which will be discussed in Chapter 23. While the acceptance of family planning increased gradually and steadily through the first half of the twentieth century, the catalyst to its rapid expansion came with the surge in population growth after World War II. It became obvious then that concentrated study and action were necessary to control a world population beginning to strain world resources seriously.

If we review the ebb and flow of the population of western European countries we can see a pattern of demographic transition from one stage to another that is being repeated throughout the world. The first stage was a population with high but fluctuating birth and death rates. No matter how many children were born, few survived the hazards of childhood or disease to become reproducing adults, so the population was either stable or slow in growing. In the second stage, because of improved medicine and health care, the death rate began to fall much more rapidly than the birth rate. The decrease was not regular; in fact, in some modern instances, the drop was precipitous. This was seen when malaria was controlled in Ceylon after World War II; the death rate fell from over twenty per thousand in 1946 to ten per thousand in 1954, although the birth rate remained high. In the third stage, both birth and death rates were low and the population was again slow-growing or stable.

England passed through just such a demographic transition between 1650 and 1900; at the beginning of this period, both birth and death rates hovered around thirty-five per thousand per annum. By 1750 the death rate had fallen to twenty-one per thousand while the birth rate remained at thirty-four per thousand. By 1880 the birth rate had fallen to sixteen per thousand, nearly matching the death rate of twelve per thousand.

How can we explain this demographic transition? When people are poor in an agrarian context and the death rate is high, children are an economic asset in the struggle to wrest a living from the soil and parents are dependent upon them for security in their old age. Since many children die before reaching a productive age, many must be born to close the gap. When the death rate falls rapidly as it has in most countries in recent years, rural women, long used to almost annual pregnancy throughout their childbearing years, quickly find themselves with larger families than they had anticipated. Thus, the burgeoning birth rate and annual growth rate in many underdeveloped countries do not necessarily represent a *desire* to have large families; rather, more children are surviving than previously.

As farms are mechanized, fewer hands are needed, so rural people flock into the cities for greater opportunity than farm life provides. Now numerous children become an economic liability, expensive to feed, clothe, and educate; hence fewer are desired and fewer are born. Since farm mechanization and movement to cities reflect increased prosperity and industrialization, the process of development tends to reduce the birth rate. Indeed, communist dogma states that overpopulation is an illusion of conservative capitalism, and that, if development is rigorously pursued in underdeveloped countries, the birth rate will quickly fall in the rising tide of worker prosperity. This proposition is naïve, however, for the industrial development that took a period of 50–100 years in western countries is not likely to be significantly accelerated in underdeveloped countries. In addition, these countries already have far larger population growth rates than European countries had at a comparable stage and may well be swamped by the population increment over this time span, before they become sufficiently industrialized.

Even when the birth rate has been significantly reduced by war, disease, or starvation—the "natural" limits—it is still subject to unpredictable fluctuations and even to fashion. By the mid-1930s Germany, France, and England became concerned at the close approach of birth to death rates, and governments, fearing underpopulation for the first time since the Great Plague of the mid-fourteenth century, started to subsidize large families directly and indirectly. At that time one or two children were considered fashionable, and anyone with four or five was often hard-pressed to support them and subject to a subtle disapproval by friends and relatives. After World War II the return of social stability led to a jump in the birth rates of most western countries, and families of three or four children again became acceptable. But after a few years the birth rate of most developed countries began to fall again. The United States was the exception for quite a while. As late as 1957 its birth rate was around twenty-five per thousand. With a death rate of nine per thousand, the annual rate of growth[2] was 1.8 percent, greater than Italy, Japan, or India at that time. Had the growth rate continued at this level for only 150 years, the population of the United States would have climbed from 150 million to over three billion. But fortunately in 1958 the birth rate began to fall (no one is quite sure why), dropping to 17.4 per thousand in 1968. Even so, the projected increase in population deserves the alarm and concern it has received. At the present, two-thirds of the world is in stages one and two of the demographic transition.

While the overall growth rate of the world is about 2 percent, there is a great disparity between the developed and relatively stable nations of Europe and the rapidly growing underdeveloped countries of South America, Africa, and Asia (see Table 1). At just 2 percent growth, 180

Table 1   Some population statistics of selected countries and regions (mid-1969)[a]

Area	Population (hundreds of millions)	Rate of growth (percent)	Doubling time (years)
Europe	456	0.8	88
U.S.A.	203	1.0	70
Russia	241	1.0	70
Asia	1990	2.0	35
Africa	344	2.4	28
Latin America	276	2.9	24

[a] Data from the United Nations Population Reference Bureau.

thousand babies are born every day, adding sixty-five million to the world's population every year. Small wonder the doubling time of world population is only thirty-five years and continuing to decrease (Table 2). Projections of the present growth rate beyond a few years are extremely difficult because of the many factors, almost unmeasurable let alone predictable, that affect the number of children born in any given year. Yet, one question arises from this projection; considering the thin line

[2] Annual rate of growth = birth rate − death rate/10

Table 2   **Years required to double the population**[a]

Year	Population (billions)	Number of years to double
1	0.25	1650
1650	0.50	200
1850	1.1	80
1930	2.0	45
1975	4.0	35
2010	8.0[b]	?

[a] Dorn, Harold F., 1968. "World population growth: an international dilemma." *Environments of Man.* Addison-Wesley, Reading, Massachusetts.
[b] Projection of United Nations estimates.

between feast and famine for many of today's three billion people, how many people can the earth support? While natural resources in the form of timber, ores, oil, and such are important considerations in determining the supportive capacity of earth, demands for these commodities could be sharply reduced by intensive recycling of these materials, or even lowering our standard of living. This is not possible with food. Therefore, a broad discussion of natural resources has been intentionally omitted to allow fuller treatment of the food problem.

FOOD RESOURCES AND PEOPLE

As a general rule one acre of fertile land per capita is required for a nation to be self-sufficient in food production. France, Sweden, and Denmark, with about one acre per capita all achieve this self-sufficiency. India and Pakistan, on the other hand, lie very close to the boundary between self-sufficiency and dependency with 0.8 acres per capita. China with 0.5 acres is a chronic food importer, while England with 0.4 acres per capita despite high yields and intensive farming must import significant amounts. In contrast the United States has 2.9, Australia 4.0, and Canada 6.5 acres per capita, enabling all three countries to export large quantities of food. Despite this margin of food production at least two-thirds of the world's people suffer from undernourishment or malnutrition.

There is a distinct difference between undernourishment and malnutrition. Undernourishment is quantitative—an insufficient intake of calories leading to simple hunger; malnutrition is qualitative—the number of calories may quash hunger pains, but there is not enough protein in the diet to avoid chronic debilities or secondary infections and diseases. Children up to the age of six have an especially great need for protein. When this is not available a protein deficiency syndrome or kwashiorkor develops, a disorder characterized by digestive tract disturbances, skin sores particularly on the legs, swelling of some parts of the body, wasting of others, and a reddening of the hair (Figure 22.1). A few years ago during the Nigeria-Biafra civil war magazines were filled with the haunting pictures of Biafran children suffering and dying from malnutrition. But older children and adults suffer from malnutrition too, and the susceptibility to illness, lethargy, and apathy it induces place a steady drain on the resources of developing countries.

Whatever the effects of inadequate diet, only part of the blame can be laid to economic inequities. Other more peripheral problems prevent full realization of a nation's agricultural productivity. Religious attitudes,

Figure 22.1   Protein deficiency or kwashiorkor has resulted in face and body sores on this child from Guatemala City. (WHO/Photo Pierre Pittat)

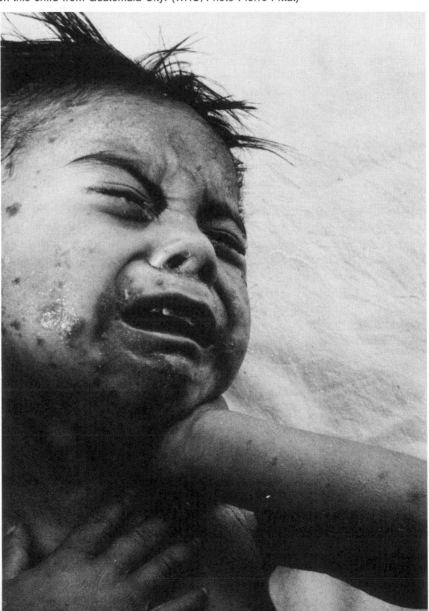

such as those that protect cows and monkeys in India, have been very difficult to resolve. These animals consume huge quantities of food that could contribute directly to agricultural production. Nationalism, understandably important to nations that have only recently emerged from the shadow of colonialism, has led to irrational attitudes and goals that have absorbed much scarce capital that underdeveloped countries could put to much better purposes.

Faced with the problem of widespread hunger in the world today, what must be done, using presently available technology, is to increase productivity of underdeveloped countries to where they either become self-sufficient in food production or sufficiently industrialized so that they have enough foreign exchange to purchase food on the world market.

Present Technology

Only North America has maintained an average annual 2 percent growth rate in agricultural production over a long enough period (thirty years) to assume continuance. Today food production is increasing at 2.3 percent a year in the United States, but at less than half that rate in Latin America and Africa. In the Middle East, rates in many instances are substantially below the rates of population increase. There are several possibilities for improvement.

Plant breeding. Of the 700 thousand species of plants in the world nearly eighty thousand are edible, but only 3000 are normally considered crop plants. Three hundred species are in abundant use but only twelve species or genera provide 90 percent of all edible crops. The leading food plants in order of their importance are rice, wheat, corn (two thirds of the annual crop is eaten by livestock), potatoes, soybeans, sorghum, barley, rye, millet, cassava, sweet potato, coconut, bananas, and plantains. Obviously an incredible number of potential food plants growing in all conceivable environments is unexploited. More will be said about developing new crop plants, but in the short term much can and is being done to improve the major crop plants already in worldwide use.

For example, the International Rice Research Institute at Los Banos in the Philippines developed a variety of rice over a five-year period that has increased production of rice in one instance in the Philippines from 710 to ten thousand pounds per acre where intensively cultivated (Figure 22.2). This strain, IR-8, has been introduced to other Asian countries as well and has increased average rice yields from two to four times. In the few years since its development IR-8 has been planted on twenty million acres in Turkey, India, Pakistan, and the Philippines. With the introduction of IR-8, Pakistan became self-sufficient in rice production for the first time in many years.

Figure 22.2 Although IR-8 rice yields heavily, it does so only when fertilized. The rice on the right has been fertilized; that on the left is unfertilized. (FAO photograph)

In a parallel development Mexico gathered together a "germ bank" of over a thousand different varieties of corn, and was able to develop several high-yielding new varieties adapted to high elevations, low soil moisture, and other special environments. This improvement in corn yield, together with irrigation where feasible, greater use of fertilizer, and better transport facilities, has enabled Mexico to raise its agricultural productivity by 4 percent per year in recent years. Unfortunately, this gain has been mitigated somewhat by a population growth rate of over 3 percent per year.

These developments have been particularly important because, until recently, most of the agricultural expertise was derived from temperate zone climates, soils, and crops, which had limited application to crop cultivation in the tropics.

Figure 22.3 Triticale, a new grain resulting from a cross between wheat (*Triticum*) and rye (*Secale*) is widely adaptable to different climates. The grain on the left was grown in Mexico, the grain in the center in Manitoba, and the grain on the right is ordinary wheat. (University of Manitoba, Department of Plant Science)

Another improvement in a basic cereal crop has been accomplished by crossing wheat and rye to form a new species called *Triticale* (Figure 22.3). Perfected in Canada, this new grain is highly resistant to lodging (the tendency of grain plants to be blown flat by wind, making them unharvestable), has a higher protein content than most other cereals, is a good livestock feed, has a greater yield than either of its parents, and combines vigor with wide adaptability to climate.

Other recent plant breeding developments include a wheat with an increased yield of 30 percent, a leguminous (a member of the pea family) livestock feed well-adapted to the tropics, and a new variety of corn, opaque-2, with a much augmented supply of a very valuable amino acid.

Intensive agriculture. New high-yield plant varieties are not enough, however. The land must be properly prepared by tractors capable of breaking the soil more quickly and effectively than water buffalo or oxen; it must be fertilized with the right formulation applied in the proper concentration at the best time; and the crop must be quickly and efficiently harvested, processed, and stored. Finally, if the increased yield is to be passed on to the people, distribution routes and new transport techniques and systems must be developed, otherwise isolated people may be reduced to starvation simply because there is no effective way to get food to them.

Presenting peasants with improved seed, then, is clearly only the beginning. The orchestration of the necessary accompanying technology requires time and usually outside help and money. Once begun, however, it can generate its own continued momentum. In the meantime, people continue to go hungry. While the diet of undeveloped countries is roughly comparable with developed countries in the caloric intake per day, there is a critical lack of protein in the diet of the former.

Protein supplements.   At least three protein concentrates have been developed, incaparina, laubina, and fish protein concentrate (FPC). Incaparina is a mixture of cottonseed flour, whole corn cooked and ground, yeast, calcium carbonate, and vitamin A. The resulting powder contains 25 percent protein and can be mixed with water and drunk like dried milk or it can be added as a supplement to food. Laubina is made of wheat, chick peas, bone ash, sucrose, and vitamins A and D. Like incaparina, it is compounded of materials whose taste is known and recognized by the people it is intended to serve, and this goes a long way to ensure its acceptance, for people are exceedingly conservative about the food they eat. Rice eating people would often rather starve than eat wheat or barley, which are unknown to them. Americans are no exception. When was the last time *you* ate roasted grasshoppers, lizard, or rattlesnake meat? Many confirmed beef eaters will not touch even the liver, much less heart, kidneys, tripe, or brain.

Fish protein concentrate has the advantage of being produced from just about any kind of fish, especially trash fish not normally eaten. When the fish is ground up and the oil and water removed, the end result is a fine, free-flowing powder. Odorless, tasteless, and containing 80 percent protein, FPC can be produced for about 25 cents a pound. As a powder it can be added to a great variety of baked and cooked foods to supplement the diet with protein. Just about any country with a fishing industry, fresh or salt water, should be able to generate this superb source of animal protein cheaply enough to raise protein consumption to at least minimal levels, buying time until the slower techniques of increasing agriculture production are implemented.

Game ranching.   Another underexploited source of protein is wild game. One of the symbols of progress in an underdeveloped country is the establishment of a livestock industry to supply protein for internal consumption or for export income. Where no suitable stock exists, as was true in New Zealand, domesticated grazing animals may be profitably introduced. But New Zealand is an exception; most other grasslands of the world have numbers of native grazers well-adapted to the environment and the available forage (Figure 22.4). Until recently the value of wild game as a protein source, equal to or surpassing any stock that might be introduced, was ignored.

Then some pioneering work in East Africa by Pearsall and Dasmann disclosed that wild game is more efficient at grazing and even more productive of protein than introduced stock. The difference lies in the food habits, water requirements, and mobility of game. Cows are quite fussy, eating only certain grasses and leaving the rest. If the number of animals is increased, overgrazing takes place, leaving stock and range in very poor condition. Since cows are not very mobile they tend to overgraze and they require sizable water supplies.

Game, on the other hand, is beautifully adapted to the many niches in the environment: elephants eat the bark and roots of trees; giraffes feed on the higher branches of trees; eland, a species of large African antelope, feed on the lower branches, and so on. Because of this diversity, a much broader range of vegetation is utilized by game than domesticated cattle, and thereby becomes potentially available to man. Wild animals are much more mobile than stock, and when the browse becomes thin, they are capable of traveling miles in search of better range; therefore, overgrazing in the natural environment is rare. Finally, most of the wild grazers in East Africa have low water requirements; for example, the gemsbok and oryx take no water at all and the zebra can go from one to three days without water. One of the major drawbacks to the introduction of stock in many parts of Africa is their susceptibility to disease caused by the tsetse fly. Wild animals that evolved with the fly are immune.

Figure 22.4 Wildebeests and zebras are only two of many grazers efficiently harvesting a broad variety of herbs and grasses in Nairobi National Park, Kenya. (Kenya Tourist Office)

Recognizing these advantages, Dasmann made a cost analysis study comparing stock and native grazers on a large ranch in Rhodesia. He found that while a cow was supported by one acre in England and was ready for market in a year, in Rhodesia the comparable figures were thirty acres and four years; while native grazers supplied only a pound more of meat per acre per year, there was a major difference on production cost: the profit margin on the beef was $1416 per year while the profit on the game amounted to $8960 per year. This difference was due to the greater cost of maintaining species introduced into an alien environment. Harvesting of large native game that is abundant and in no danger of extinction is clearly desirable both from an ecological and economic point of view, for any habitat can only support so many animals. Any surplus starves or increases stress on the other species.

Food surpluses.   Another time-buying factor is the continuing surplus of food produced by the United States, Canada, Australia, and New Zealand. Today up to 60 percent of the United States wheat crop is distributed abroad under Public Law 480, the Food for Peace Program. The USDA estimates that by 1984 demands by hungry nations for this surplus will exceed the supply. More pessimistic estimates of this meeting of supply and demand suggest 1975 as the critical year. In any case, if in the remaining five to fifteen years the recipient countries cannot increase their food production to at least the self-sufficiency level or if they cannot control their population growth, there seems no alternative to mass famine as Malthus predicted in 1798.

Future Technology

With great application and perhaps a bit of luck, the world can avoid the famine predicted for it in the next two or three decades. But what of the year 2000 and beyond? What hope does technology provide for the accommodation of two, three, or four times the present population of the world or even a more slowly growing population over a long period of time?

Increase of cultivated land.   Only about 40 percent of the earth's total land area is capable of productive use, at least in the direct sense, of cropland, grazing, or timber production. The other 60 percent is tied up in ice and snow, mountains, and desert. Of the potentially cultivatable land only 25 percent could reasonably be brought under cultivation and only one-third of this land is in Asia, where the demand is most critical: six out of every ten humans live in Asia.

Figure 22.5 Because of its wild profusion of plants, the Amazon basin seems a potentially rich producer of crops. But most of the nutrients are locked up in this vegetation rather than in the soil and are lost when the forest is destroyed. (American Museum of Natural History, New York)

For a better idea of why it is so difficult to bring new land under cultivation despite its apparent suitability, let us look at a commonly cited example. The terra firma soils which occupy most of Amazonia bear a luxuriant rain forest (Figure 22.5). But as we saw in the discussion of soil (Chapter 3), the forest, once established, generates and then recycles most of its mineral nutrients, using the rather infertile soil mostly as a means of absorbing moisture. When the forest is removed, the action of sunlight, temperature extremes, and alternation of wet and dry conditions causes rapid erosion, laterization, or both. If Amazonia were, as naïve dreamers like to claim, the future bread basket of the world, Brazilians would have already discovered the fact and would have cleared and farmed most of it. Today this enormous area produces less than one-half the food eaten by its two million inhabitants. Far from being the economic mainstay of the Brazilian economy, it contributes only 1 percent of the gross national product while requiring 3 percent for its maintenance. It is extremely unlikely that the Amazon basin will ever be used for conventional crop farming.

One other example of attempting to bring new land under cultivation was the great East African groundnut scheme. Much of East Africa is a savannah, scattered shrubs and low trees mixed with grass. In 1947 the British, then administrators of the territory, decided to open over three million acres of the countryside to the cultivation of groundnuts, or peanuts, as they are known in the United States. Research was done to be assured that peanuts would grow satisfactorily in the climate and soil of East Africa. Then thirty thousand acres were bulldozed and plowed, with great difficulty as it turned out, for the roots of trees and shrubs in areas of low rainfall are long and ropey and exceedingly difficult to remove. This accomplished, it was found that the stones in the soil, which were extremely hard and sharp-edged, damaged farm implements, compacted under the tread of tractors, and made the soil almost impervious. In addition, alternation of heavy rain and drought scuttled crop after crop. In 1951, after investing over $100 million, the British admitted failure.

As a general rule, any area of the world not already grazing cattle, growing trees, or producing crops is not capable of doing so using current technology. With much research and development, some of these marginal areas may yet be made productive, perhaps in ways beyond our present imagination. But we cannot tacitly assume that presently unproductive land is simply waiting for the touch of man to be brought into production.

Irrigation. There are large tracts of desert around the world that are neither too sandy, too mountainous, or too cold to be farmed, if water could be supplied. The problem is not water supply per se, for in the next few decades desalinization will probably reduce the cost of fresh "manufactured" water from its present $1.00 per 1000 gallons to the point where economics will allow its use in irrigation (25 cents or less per 1000 gallons). However, once desalinated, water must be moved against gravity, over mountain ranges, and up high plateaus, for few deserts are located at sea level, even if they are adjacent to the sea coast. All of the long distance water supply systems of large cities—New York's from the Delaware River and the Catskills, Los Angeles' from the Sierra and Owens Valley—operate on a gravity flow system. The billion-dollar Feather River scheme, which will bring water from northern California to southern California can compete with local desalinization only because the transport of this water is partially free due to gravity. If some way is developed to overcome the cost of moving desalinated water up from sea level, it may well be possible to bring water to the desert, but that is only a part of the problem.

However sweet, water carries some dissolved salts in solution that are left behind as the water evaporates in the hot desert sun. Unless the soil is periodically flushed and drained to remove these salts, they accumulate, making agriculture progressively more difficult (Figure 22.6). Furthermore, many deserts, otherwise suitable for irrigation, lie in basins having no drainage to the outside. How would the water used to flush ac-

Figure 22.6 Sugar beets grow only sparsely along the salt-encrusted ridges between irrigation furrows in this field. Irrigation waters containing these salts rose to the ridge surfaces through capillary action and there evaporated, leaving their salts behind. The deposits thickened as time passed, and the irrigation waters leached the salts into the ground. (State of California, Dept. of Water Resources)

cumulating salts out of the fields be evacuated? Could we afford the cost of pumping it against gravity, back to the sea? Perhaps some deserts can be made to bloom, but not in the foreseeable future.

New food plants. Assuming marginal lands and deserts could ultimately be cultivated, what crops would be planted? We have seen how few of the edible plants have been exploited; most plants, particularly those which have been intensely bred, are native to and grow best in the temperate zones of the world. But any new land being brought under cultivation is likely to be in the tropics. What is needed is an extensive research program that will screen the thousands of potentially edible tropical plants, isolating those that seem to offer the best possibilities: suitability to soils characteristic of the new lands, high protein content, resistance to drought and pests, and acceptability to the people in need of these foods. These initial efforts must be followed by an intensive selection and breeding program to combine the most useful features into new crop plants. Since the development time is long in plant breeding and selection programs, present efforts must be greatly accelerated if suitable crops are to be ready to take care of any sudden technological advances in marginal land use.

The ocean. What kind of harvest *can* we expect of the world ocean? Altogether, the ocean "fixes" around eighteen billion tons of carbon each year in the process of photosynthesis. If this seems high, remember that 90 percent of the world's vegetation is in the sea. The open sea, about 90 percent of the total, produces fifty grams of carbon per square meter per year. Shallow coastal waters and estuaries produce about twice this, and local upwellings (areas where nutrient-rich deep water comes to the surface, causing a sharp increase in productivity) may produce up to six times that of the open sea.

With our present technology we are harvesting in most instances only the fish at the other end of this production (Figure 22.7). In the open ocean food chains may have five links between the algae producers and the fish that are harvested, coastal waters usually have three links, and the areas of upwelling only two. Because energy is lost between each link,

Figure 22.7 This etching by Breughel embroiders a bit on the "big fish eat little fish" aspect of the aquatic food chain. (Academy Collection, Philadelphia Museum of Art. Photograph by A. J. Wyatt, Staff Photographer)

the shorter the food chain, the greater the amount of energy that can be harvested.

When the yields of the three types of oceanic water are compared, the open ocean appears to be a biological desert, producing only a small fraction of the fish harvested and with little likelihood of any increase. In contrast, the upwellings that form only 1 percent of the ocean surface produce *one half* of the world's harvest of fish.

Annual production of fish is about 240 million tons, but not all of this can be harvested. Some fish must remain to breed and, since man is not the only fish predator, others are taken by sea birds and mammals. Perhaps 100 million tons are available to man on a sustained yield basis. In 1967 the fish catch was sixty million tons. At the current rate of increase, 8 percent per annum, the ocean will be yielding as many fish as it can *within ten years*. Indeed, some parts of the ocean are already overexploited: the area between the Hudson Canyon and the Nova Scotia shelf produced one million tons of fish in the early 1960s but the catch has been falling off ever since.

Quite irrationally, at least one-half of the fish caught commercially are thrown away because of finicky public attitudes about which fish are edible and which are not. Since fish contain all the amino acids necessary for man to manufacture protein, a huge amount of valuable protein is being wasted. An efficient system of trawlers working from a mother-ship base could easily deliver all the fish caught to the factory ship, which could clean and freeze the most desirable fish, then process the trash fish into fish protein concentrate. The Russians are exploring this possibility; indeed it seems too efficient and profitable a possibility to ignore.

Another promising possibility is the raising of some types of seafood under controlled or semi-controlled conditions, opening the way for more convenient and efficient harvesting and processing than is possible today with wild populations of fish and shellfish.

Of all the thousands of edible species of plants and animals in the sea, surprisingly few are eaten and even fewer (50 out of 20,000) are systematically cultivated. A few species of seaweed are grown in Japanese estuaries and surely far more could be grown in other countries. But seaweed is not to everyone's taste, although we all eat considerable quantities as alginate, a stabilizer that is finding its way into more and more processed foods (see Chapter 14). Oysters have been raised for centuries and have proved to be a valuable and nourishing food capable of yielding 6000 pounds per acre (Figure 22.8). But increasing pollution of estuaries is destroying more beds than are currently being developed. Mussels, a favorite shellfish in the Mediterranean countries, are almost totally ignored as a food supply in North America. Considering their superb flavor, far superior to most clams, and their extreme abundance along rocky coasts, mussels are one of many unexploited food sources in North Amer-

Figure 22.8 Oysters require a firm substrate or point of attachment. By providing these wooden frames oysters can be conveniently transported from place to place and ultimately harvested quite efficiently. (Australian Dept. of Information)

ica. Some forms of shrimp, milkfish, and mullet are raised in ponds with considerable success, particularly in Asia. Conch, abalone, lobster, blue crab, and other species of shellfish could be raised in tank farms or salt water ponds.

While the annual production of fish and shellfish may seem enormous, most of the true production of the ocean lies, like the greatest mass of an iceberg, out of sight. When we catch fish, we are harvesting only a tiny portion of the food chain energy represented by that fish. Behind every ounce and a half of tuna that we eat, there is one pound of mackerel, ten pounds of herring, 100 pounds of zooplankton, and 500 pounds of phytoplankton. Phytoplankton consists of the unicellular algae that begin the whole food chain by fixing carbon into food by photosynthesis. Rather than accumulating all the loss of energy involved in long food chains, it would make far more sense to harvest algae directly. On an average the algae in one acre of sea surface could yield about three tons dry weight, compared with the average yield of twenty pounds of fish per acre of ocean.

The problem is that harvesting algae is not easily done: getting those three tons of algae out of the water is one major problem, converting them into palatable protein is another, and persuading people to eat this protein in preference to fish is yet another. It would be more convenient to grow algae in tanks under optimum conditions than to try straining them out of huge quantities of water, but pilot operations of this type have never been as productive as expected. Apparently much more research is necessary before algae can either be harvested or raised directly as a major source of protein.

New techniques in crop production. The potential yield resulting from shortened food chains applies to the land as well as the sea. Beef as a harvester of energy is incredibly inefficient; the yield of protein is forty-three pounds per acre per year. If the protein is in the form of milk, the yield is increased slightly to seventy-seven pounds per acre per year. Soybeans can yield 450 pounds of protein per acre per year and processed alfalfa, 600 pounds. When the demand for protein becomes great enough, eating beef as a major source will become an increasing luxury, and then disappear altogether. Two-thirds of the earth's population has already, by necessity, made the shift.

Even within the spectrum of plants already utilized, we can do much better than the 4.5 million calories per acre per year provided by wheat; potatoes give eight and sugar beets thirteen million calories per acre per year. The difference is due to the small fraction of the wheat plant we actually eat, less than 30 percent. In contrast, 50 to 65 percent of the sugar beet or potato plant is directly edible. When the pinch becomes sharp enough we may have to shift to crops that are more completely edible or we may breed new varieties of grain that have edible stems as well as heads.

Direct synthesis. The short-circuiting of food chains is but one way to eliminate as much waste as possible by going to the primary producer. The next logical step would be direct synthesis from an abundant organic source. Three approaches have already been worked out on a pilot basis. Roughage waste from cereal crops, corn cobs and straw, when treated with a hot acid, forms a molasses-like substance that can be fermented to produce yeast protein. Although the expense would be ten times as great as using molasses from sugar cane or sugar beets, it could be done. Conversion of most of the woody residues from today's crops could lead to a doubling of our present food supply.

Another approach treats soybeans with a mild alkali that aids in the extraction of the soybean protein. This extract can be processed like synthetic fabric fibers; when extruded, the extract coagulates in a suitable

Figure 22.9 (a) Soybean protein can be processed like synthetic fabric fibers. (Ralston Purina Co.) (b) When oriented at right angles to form a mat and artificially flavored and colored, luncheon meats can be synthesized: left, "smoked beef"; center, "corned beef"; and right, "smoked turkey." (Worthington Foods).

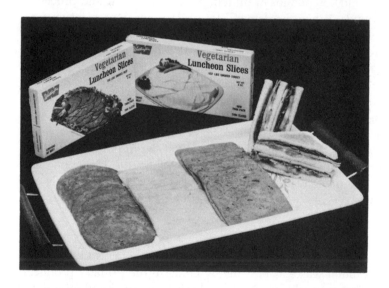

bath into fibers ranging from one- to thirty-thousandths of an inch in diameter. These can be oriented in parallel or at right angles forming a mat (Figure 22.9a). The resulting material is tasteless but can be appropriately flavored and colored to resemble in taste and texture the familiar forms of meat: chicken, beef, sausage, and luncheon meats (Figure 22.9b). Although at present more expensive than the real thing, something true of every pilot project, it probably tastes as good as many meats currently available and could serve the valuable function of weaning the meat-eating public to an acceptable substitute when the real thing becomes too expensive and scarce a source of protein.

The most revolutionary approach to direct synthesis of food involves crude oil. Anyone who has spent time in the tropics realizes that there are microorganisms that will attack anything, no matter how indigestible or poisonous it may seem to us. A few years ago a French research group found some microorganisms that could use petroleum as a food source and thus be harvested as a substantial yield of protein. One enthusiastic estimate suggests that the equivalent of all animal protein needed by the world today could be produced from but 3 percent of the annual crude oil production of the world; however, tooling up from a test tube to the scale necessary to supersede beef as a protein source presents extensive technical problems. In the meantime, the world's oil supply is being rapidly depleted. Direct chemical synthesis of food, though feasible, is still some years and perhaps decades away.

Where does the world stand then when present and future technology is turned to the task of producing food? If the rate of world population growth could be reduced to 2 percent at the same time the rate of food production everywhere could be raised by the same amount, the population of the earth could double by the year 2000 without mass starvation.

Balancing serendipitous technological developments against natural disasters, we are still left with a big if. While many countries have made great strides toward balancing their population growth and food production rates by decreasing the birth rate, by increasing food production, or both, many more have made no progress in either.

As late as it is, we still have before us the choice of voluntarily limiting our population and our consumption of raw materials. Before long, however, limited availability of food and natural resources will remove that choice forever.

FURTHER READING

Borgstrom, G., 1967. *The hungry planet.* Rev. ed. Collier Books, New York. Fairly well-balanced prognosis.

Brown, H., 1954. *The challenge of man's future.* Viking Press, New York. The optimist speaks.

Brown, L. R., 1970. *Seeds of change.* Praeger, New York. A balanced view of the problems as well as potentials of the agricultural revolution set off by new varieties of rice and wheat.

Calhoun, J. B., 1962. "Population density and social pathology." *Scientific American,* 206 (2), p. 139. A classic paper on social pressure in rats, with implications perhaps for man.

Dasmann, R. F., 1964. *African game ranching.* Pergamon-Macmillan, New York. Experimental evidence that native game can outproduce domestic stock and still conserve the landscape.

Iverson, E. S., 1968. *Farming the edge of the sea.* Fishing News (Books) Ltd., London, England. Broad survey of the potential sources of food from the sea.

Paddock, W. and P. Paddock, 1967. *Famine—1975!* Little, Brown and Co., Boston, Massachusetts. The most sensational and alarmist of the dozens of books dealing with the population explosion. Compare with calmer accounts.

Political and Economic Planning, 1955. *World population and resources, a report.* Allen & Unwin, London, England. An important source, though figures are dated. The discussion is quite thorough and balanced.

Rhyther, J. H., 1969. "Photosynthesis and fish production in the sea." *Science,* **166,** pp. 72–76. The most up-to-date survey of the fish resources in the world ocean.

Smith, F. G. W. and H. Chapin, 1954. *The sun, the sea, and tomorrow; potential sources of food, energy, and minerals from the sea.* Chas. Scribner, New York. The title tells all.

Wharton, C., 1969. "The green revolution: cornucopia or Pandora's Box?" *Foreign Affairs,* **47** (3): 464–476. Gives valuable insight into some of the problems associated with greater food production.

CHAPTER 23

POPULATION CONTROL

Of all the problems discussed in this book, overpopulation is the one problem the entire world must share. But we have become so concerned about the quantitative aspects of the billions of people we fatalistically anticipate in the next few decades that we forget the critical importance of the *quality* of life on earth. For hundreds of millions living in under-developed countries, life is so impoverished that quality, in our affluent terms, is unknown.

There are signs posted all over downtown Edinburgh, Scotland, that read "The amenity of our streets is commended to your care." The Scots sign is thought provoking far beyond its literal intent, for the amenity of future life on earth *is* in our care. Amenity is being able to show your children skunk cabbages pushing up through the corn snow of March, going swimming and being able to see your toes, visiting a National Park rather than a national parking lot. It is very easy to pick some point in time past and say "those were the days," regretting the subsequent loss of amenity, for the passage of time never seems to add to the quality of life, only the quantity. Like the realtor who advises buying land now for it will never be cheaper, we must make a stand now, for the quality of life will only lessen in the future.

But an even more important aspect of amenity is personal freedom. The wild west of the frontier days was wild because of the freedom, not simply opportunity, to lead your life as you wished. As populations grew, the freedoms began to wither away. For a while one could always follow a frontier north, south, east, or west; but there are no more frontiers in our country. Backpack into the most remote wilderness and you will surely run into another family doing the same. As the number of people increases, the number of problems arising from interactions increases, more laws are passed, more freedoms are limited. If you feel hemmed in now, sharing North America with 300 million other North Americans,

451

think of the restrictions necessary to allow 600 million or a billion North Americans to live peaceably together.

The only way to preserve what amenity and personal freedom remain is by population control. Despite centuries of chamber of commerce propaganda, growth is not always coupled to prosperity; neither is it necessary or inevitable. Its control is possible with the means at hand, but first we must overcome our superstitions, religious objections, political barriers, and just plain ignorance. Not until *all* the people of the world fully understand the consequences of unrestricted growth can present solutions be applied, and better ones developed.

MOTIVATION AND POPULATION CONTROL

However desirable as a national goal, neither population growth nor population decrease can be imposed. People respond not to external coercion but to internal motivation; the most remote goals can be achieved if people are properly motivated. Conversely, government policies can be effectively undermined if motivation is lacking, as Prohibition conclusively demonstrated in the United States two generations ago. Thus, family planning has no hope of success unless people are motivated by moral, economic, legal, or social means.

Family planning is not synonymous with birth control, for birth control is specifically designed to limit the number of births. Since the purpose of family planning is to plan for the size of a family, help is also given to couples who have problems conceiving children. But far more people are desirous of controlling births than eliciting them, and family planning is commonly used as a euphemism for birth control. For when questioned, most women express the desire to control the size of their families and not simply to give birth to child after child throughout their childbearing years. To help such women, many underdeveloped countries —India, Pakistan, Turkey, Colombia, Morocco, Kenya, and Jamaica, to name just a few—have set up birth control programs and clinics.

Since the family planning movement dispenses birth control information as well as information to help overcome sterility, there has been much opposition to family planning activities in Roman Catholic countries. While the Catholic Church is still unrelenting, in 1961 the National Council of Churches sanctioned the practice of birth control, relieving some of the religious deterrents to a realistic program of population control. Attitudes on birth control are sometimes further complicated by nationalist or political groups who feel that birth control is being applied toward a political or even racist end; some black groups in the United States and Hindus in India have made this charge. But as long as birth control information and devices are made available to everyone who wants them without coercion, such a charge is unwarranted.

Plainly, the social and moral attitudes within our society and throughout the world will have to undergo considerable modification before birth control becomes a way of life for the entire population. At present, young single women are under considerable social pressure to marry and raise a family, despite education or challenging career possibilities. While greater economic and social opportunities have become available to women quite recently, through the activities of such groups as the Women's Liberation Front, much more will have to be done in this area if population growth is to be effectively checked.

Population Policies

Governments have been directly influencing population growth for years. Governments could, however, discourage population growth by the same means. This might be accomplished by rewarding childless individuals with at least equal, if not lower, tax burdens than parents, hence removing the attractive tax deduction reward for having children. Other proposed schemes involve setting up an inverted tax scale, increasing the tax with every child. Delayed marriage, already beginning to take place as a result of increased educational opportunities and requirements, might be further encouraged by government subsidy. The government of India is considering a law that would raise the minimum marriageable age, by which it hopes to shorten the childbearing years of its citizens.

However these are *material* rewards for remaining single, delaying marriage, or having few children. What about the psychological rewards? Social pressure, which usually tends to push people into marriage and then into raising families, may at times have a negative effect on population growth. During the Depression the birth rate fell dramatically in most countries and the one- or two-child family became the norm.

Finally, those parents who can financially and emotionally support more than two children should be encouraged to increase the size of their families not by further natural births, but by adoption. Circumstance and accident will always result in parentless children who need love and protection as much as one's own, for the family remains the only institution capable of fulfilling the needs of *all* of society's children.

In Europe birth control has varying degrees of official recognition. In the Scandinavian countries and Great Britain it is legal; in France, Belgium, and the Netherlands it is considered semi-legal, that is, a couple may practice birth control for medical reasons. However, in Italy, Portugal, Spain, and Ireland, still predominantly Catholic countries, birth control remains illegal. Neither the United States nor Canada has a national policy on family planning, although this situation may change in the United States since in 1969 President Nixon appointed a commission to study population growth as it affects the future of the United States. Sweden and Great Britain officially encourage family planning, but most other Western European countries have no defined population policy. France is an exception; it has an active government sponsored program providing subsidies for larger families, with increasing benefits as the number of children increases. The dissemination of appropriate propaganda also helps the French government to express its approval of larger families. Ironically, France has been plagued by periods of population decline over much of the twentieth century.

Japan presents another type of government policy on family planning. For years, small families have been the socially acceptable practice in Japan. Now, however, with increased industrialization, the government is experiencing a labor shortage. As a result, a subtle change in government policy is taking place. Instead of removing the heavy subsidy on rice farming and retraining the farmers released for industrial jobs, a return to larger families has been suggested as the easiest solution to the problem.

This last example is interesting because it illustrates the major problem with population control today. Countries have no coordination of their political and economic policies. When one country experiences a decline in population it does not draw on the population resources of another country; instead, it chooses to view people as a natural resource and to seek to increase the size of this resource. But people are not a natural resource; rather, people consume natural resources. If we have overpopulation in any one country, that country must tap the resources of the countries which are still underpopulated to meet its consumer needs. If we are to have realistic population control we must change the view that population is related to specific political economies.

BIRTH CONTROL METHODS

Once the economic and social motivations for smaller families are established, the means of achieving the goal of fewer births can be implemented. Ideally, people should have children only when they want them. This is the essence of family planning. Too many families have children who are unwanted, and often resented as well. The unwanted child often matures into an unwanted adult, maladjusted and a burden to society.

Sterilization. The means of avoiding childbirth are many, running the gamut from celibacy to abortion, but the most reliable method is sterilization. The term sterilization has an ominous ring to it because of the implied permanence of the operation. Often confused with castration by the uneducated, sterilization of the male simply prevents the release of sperm. This is accomplished in a minor operation that severs the sperm ducts and can be performed in a few minutes with a local anesthetic. Females may also be sterilized but the operation to tie off the oviducts or fallopian tubes which conduct the egg from the ovary to the uterus requires internal surgery. Sterilization has *no* physiological effect on the sex drive, although if the significance of the operation is not clearly understood, there may indeed be psychological effects on the libido. Studies of Japanese couples where the male was sterilized showed that in only 3.5 percent of the couples was the frequency of sex reduced; in contrast, 28.5 percent reported an increase, probably due to the release from anxiety over possible conception.

The major drawback of sterilization is its implied permanence. Actually, due to improved surgical techniques, over 50 percent of sterilized males and females can be restored to fertility. Still, many doctors in the United States hesitate to sterilize because of the ever present threat of lawsuits from patients who have changed their minds.

India, for example, encourages sterilization by cash incentives, and over four million people, mostly males, have been sterilized in the past ten years. With the possibility of new techniques for temporary sterilization, such as blocking rather than severing the fallopian tubes or sperm ducts, this contraceptive technique may become as accepted in developed countries as it is in some underdeveloped ones.

The rhythm method. Of all the modern contraceptive techniques available, the rhythm method, although widely practiced, is the least effective. This is most unfortunate since it is one of the few contraceptive techniques endorsed by religious bodies otherwise opposed to contraception, notably the Catholic Church.

In theory, if a woman knows when she has ovulated and does not have intercourse until the egg has left her body, conception can be avoided. Unfortunately most women are not aware of when they ovulate, although by careful monitoring of temperature and glucose levels the time can sometimes, but not always, be pinpointed. To make things worse, at least one-sixth of all women in their reproductive years have irregular periods that make calendar watching useless. Then too, the rhythm of a period can be upset by tension, anxiety, and illness. Finally, expecting an uneducated and unsophisticated person to avoid conception by using the rhythm method is a bit ingenuous.

Mechanical methods. There are many mechanical contraceptive devices available, for example, the condom, the diaphragm, and the IUD (intrauterine device). IUDs are made of stainless steel or nonreactive

Figure 23.1 The drawing shows some of the many shapes in which IUDs (intrauterine devices) are manufactured.

plastic and may be just about any shape (Figure 23.1). When inserted into the uterus they serve as a mild irritant which interferes with the implantation of a fertilized egg into the uterine wall that is necessary to begin embryonic development. While this irritation is tolerated by most women, some 30 percent are made uncomfortable or bleed, necessitating removal of the IUD. Another 10 percent eject them spontaneously. A second generation IUD, T-shaped with pure copper attached, has reduced side effects to less than 1 percent. The great value of this device for those women who are receptive to it is that once inserted by a physician it can protect against conception indefinitely. Generally after two years, 55 to 60 percent of the women fitted with IUDs are still successfully wearing the device. The remainder must use alternative methods. The IUD has become quite popular in both developed and underdeveloped countries as well; over two million IUDs are in use in India alone.

Chemical methods.   The first contraceptive pill, marketed in 1959, contained two hormones, progestin and estrogen, which prevented ovulation. However, estrogen mimicked some of the effects of pregnancy: nausea, bleeding, weight gain, and nervousness, and more seriously, the tendency of the blood of pregnant women to form blood clots. The incidence of deaths from this side effect is low—3 per 100,000 users—much lower than the death rate from normal pregnancies.

Unfortunately the possibility of side effects was not thoroughly explored at the time the pill was introduced. So when Senate hearings in late 1969 publicized some of the problems of contraceptive pills, many women gave up the pill in panic, without being made aware of the odds, which as far as side effects are concerned, favor the pill over normal pregnancy. Still, millions of American women are taking the pill with close to 100 percent effectiveness when the sequence is closely followed. The real problem with the pill is its cost, which, unless subsidized, makes it available only to the higher income groups. Another problem interfering with its use in underdeveloped countries is the irregular pattern of intake that is required.

A second generation of chemical contraceptives is being researched and will probably be available in a few years. By enclosing a prescribed dose of progestin in a permeable plastic ring which can be inserted in the vagina much like a diaphragm, a small but steady dose of hormone can be absorbed. Progestin alone seems to have an effect on the sperm, making it unable to fertilize the egg. Further, progestin, unlike estrogen, does not prevent ovulation. The ring lasts for three weeks and is replaced after menstruation if contraception is to be continued. Another approach is to impregnate with progestin a small piece of plastic, perhaps the size of a pencil lead, then insert it just under the skin. Theoretically such an insert could regularly meter out the prescribed dose of progestin over a year or perhaps even the reproductive life span of a woman. If pregnancy were desired, the implant could be easily removed and conception could take place on the next cycle.

While new oral or implanted chemical contraceptives look promising, there is a testing period of eight years required by the FDA to anticipate possible side effects. This is done to protect the public as is the FDA's charge, but every drug, including aspirin, has side effects. Considering that during an eight to ten year development-testing program some underdeveloped countries are well on the way to doubling their population, this is a long time to wait. Many women are far better off taking the pill than running the risk of complications in pregnancy, or worse yet, being unable to sustain the life they have created.

The effectiveness of four of the major contraceptive techniques is listed in Table 1, which will allow you to take your own measure of the risk of pregnancy.

Table 1 **Comparative effectiveness of various contraceptives**[a]

Method	Pregnancies per million users
pill	5,000
IUD	30,000
diaphragm	120,000
rhythm	240,000

[a] Djerassi, C., 1969. "Prognosis for the Development of New Chemical Birth Control Methods." *Science.* 166:468–473.

THE ABORTION QUESTION

For centuries man has practiced abortion, despite censure on religious, moral, and legal grounds. The widespread practice of abortion to terminate unwanted pregnancies has recently been brought more forcefully to the attention of governments, since abortion rates are particularly high in those countries where birth control is illegal. In Italy, for example, there is probably one abortion for every live birth and in Santiago, Chile, 40 percent of all hospital admissions involve abortion attempts. Where contraceptives fail, abortion should be considered as a method of birth control, but its legitimacy or acceptance is a far more complicated question.

Biologically, egg and sperm cells are neither more nor less "human" than any other cell in our bodies. Hence it makes little biological difference whether an egg and sperm are killed by their failure to be united or shortly after their union by removal from the body naturally or artificially. The question arises then, at what point does this fertilized egg become human? Medically and legally, humanity is recognized after 5 months. Should an embryo die before this point it is regarded merely as tissue and disposed of like an excised gall bladder or tonsil. After 5 months the embryo becomes a fetus, and under common law it has a legal identity. Should a natural abortion or stillbirth take place after this point the body is buried and the woman is regarded as having given birth.

Since most induced abortions are performed within the first 3 months of pregnancy, a good two months before the law formally recognizes civil identity, there should be no legal interest in such abortions. This has not been the case, however, and the reason is historical. In the last century, when most anti-abortion statutes were written, abortions performed at any time by anyone were dangerous. This was a time when surgery was quite risky because of the almost total lack of sanitation in the operating room. Abortions were regarded as 15 times more dangerous than natural childbirth. To protect women from this risk, abortions were declared illegal with almost no exceptions.

Today an abortion performed by a physician during the first trimester (pregnancy is usually divided into 3 periods of 3 months) is quite safe, involving only one eighth of the danger of complications associated with live birth, with no risk of subsequent sterility.

In recognition of this change in the medical status of abortion, anti-abortion laws have been under attack in the courts on the basis that the right and choice of abortion should reside with the woman who, after all, must bear the responsibility for her actions. Medically and legally, there is no reason why a woman should not have an abortion in her first trimester of pregnancy if she so desires.

There has been increased approval of abortion in the United States. By 1970 a number of states had passed laws that permitted abortion upon request. While abortion laws are presently in a state of flux in the United States and are quite liberal in England and Scandinavia, most of the countries of western and southern Europe and South America still have restrictive abortion laws. India is considering the legalization of abortion as Japan did just after World War II. Apparently shocked by the huge increase in abortions that followed liberalization, the Japanese government subsequently embarked on a massive program to encourage the use of contraceptives and is even considering reinstating controls on abortion. Similar public reactions are seen in other areas after abortion laws have been relaxed. But much of the increase in abortion simply reflects the shift from furtive, unreported abortions often resulting in injury or death, to hospital- or doctor-supervised abortions that are reported as a routine statistic along with chicken pox or measles. In the United States, for example, there are currently at least one million illegal abortions per year. If abortion were legalized nationally, these presently illegal abortions would doubtless be viewed by outraged moralists as a "rising tide of immorality." Forgotten would be the millions of abortions which have always taken place, legally or illegally.

The chief opponents of legalized abortion through the years have been religious bodies, particularly the Roman Catholic Church. The Church has maintained, not as dogma but as current teaching, that the soul is present from the moment of conception and that any unnatural interference with its progression through to birth is a sin. This concept is called immediate animation. But the Church's view on the matter has not always been so; no less an authority than St. Thomas Aquinas held that the soul developed gradually as the embryo slowly became human, a view congruent with contemporary legal opinion. But after the death of St. Thomas this mediate animation concept fell out of favor, although it has never directly been eliminated as an option. It is quite possible that in future years the Church may again alter its position in view of the seriousness of overpopulation and its threat to *all* of man's institutions.

What then of the role of abortion as a form of birth control? Despite its present ease and increasing legality, no one would seriously recommend abortion in place of one of the contraceptives previously discussed. It is far more desirable to prevent a conception than to terminate one. Yet an unwanted pregnancy is often a traumatic experience even if resolved by abortion. Abortions can perform a valuable service to society by allowing those women who are least fit physically, emotionally, and financially to avoid the birth of an unwanted child. The ultimate social and moral cost to society which tolerates the production of unwanted children is surely far greater than any circumstances surrounding an abortion.

FUTURE POPULATION AND ECONOMIC GROWTH

Birth control measures can be effective in reducing population growth: Japan cut its birth rate drastically by passing a broadly permissive Eugenic Protection Act in 1948 which made abortion available to everyone, while South Korea and Taiwan have also reduced the birth rate by use of the pill and IUDs. Ceylon, Pakistan, and India are, after several years of somewhat disappointing and grossly underfunded efforts, also beginning to make progress.

India plans in the next ten years to reduce its birth rate from 41 to 25 per thousand, which will require that 40 percent of married couples of childbearing age use contraceptives, an incredible task. But if and when it is accomplished, the Indian rate of growth will still be 1.5 percent, compared with 0.78 percent for the United States, 0.29 percent for Sweden, and 0.18 percent for Luxembourg.

Without any effort to control population there will be 7 billion people in the year 2000. With a decreased rate of 2 percent yearly growth for all nations there would still be 5.6 billion people by 2000. Two billion more people can probably be accommodated on earth, but adding four billion more staggers the imagination, for the earth is barely able to support its present population; indeed most men live at a subsistence level far removed from the embarrassing affluence of western Europe and North America. When you consider that Americans, who account for 6 percent of the world's population, are absorbing a disproportionate amount of the world's raw materials to maintain their affluent standard of living, it is not surprising that the underdeveloped nations have thought it hypocritical for us to urge them to control their populations while developed nations continue to grow, albeit more slowly, and use still more of the world's resources. Our insistence upon pollution control is also misinterpreted by nations just beginning to industrialize. Underdeveloped nations should not be denied the benefits of technology simply because of the problems it has generated. Just as developed nations have shared their medical and food production expertise, they can also share the birth and pollution control technology they have gained through the hard experience of overpopulation and badly polluted environments. Underdeveloped nations, then, could be spared the ill effects of development by careful planning while enjoying the benefits of the rising standard of living that development allows.

If the population of the world were frozen at its present level there might be some hope that technology could in the next several decades begin to feed and house the world's population adequately and perhaps even begin to close the yawning gap between the few who have and the many who do not. But any increase in population beyond present numbers, 1 or even 2 percent, so widens the affluence gap as to make it virtually impossible for the average Asian or African *ever* to have the good life which we as Americans now consider our inalienable right. A reduction in the growth rate is not enough. Population growth must cease altogether, not just for India, China, Egypt, or Jamaica, but for the United

States, Russia, New Zealand, and Denmark. For it is not the Indian, Chinese, or Egyptian peasant who is making unprecedented demands on world resources, generating sonic booms, or releasing radioactive wastes; it is the middle class of the world. Caught up in a cycle of more-demanding-more, *they* must bear the responsibility for much of the world's pollution and share the responsibility for the population control that must come.

Since the only rationale for continued growth is its traditional association with progress, the first step in making the transition from an expanding population to a zero growth population is the realization that growth and progress are *not* inseparable. Indeed, if we take the stance that progress has a qualitative as well as a quantitative aspect then growth is clearly *inimical* to progress. For the same population growth which traditionally provides new markets, more profits, and greater consumer affluence at the same time has eroded our personal freedoms and steadily debased the viability of our environment. Once we realize that our definition of progress must be broadened to include the achieving of quality in our future, we can begin to plan our economy to accommodate this change. This will certainly mean a shift from a growth to a steady-state economy.

Since much of the gross profit of industry is reinvested in expansion to meet the demands made by population growth, a steady-state economy should free some of this capital for by-product utilization, recycling of wastes, pollution control facilities, modernization of physical plant and the research and development necessary to supplement diversification. While the details of this shift from a growth to a steady-state economy are best left to economists, the positive impact of such a shift on the environmental problems that we have discussed in this book would be impressive.

With a stable population and a steady-state economy, the world's industry should be fully occupied for decades in satisfying the minimum demands of the world's people for decent housing with electricity, pure water, and sanitary facilities that most Americans now take for granted. This is an enormous market which has barely begun to develop, but when it is satisfied, diversification to meet the increasing leisure needs of the earth's now stable population should assure continuing profits for industry and a better life in both quantitative and qualitative terms for all men.

FURTHER READING

Calderone, M., 1950. *Abortion in the U.S.* Harper–Hoeber, New York. This is the book that made abortion a respectable word; a basic study.

Djerassi, C., 1969. "Prognosis for the development of new chemical birth-control agents." *Science,* **166,** pp. 468–473. Good comparison of effectiveness and relative danger from the use of various contraceptive techniques.

Hall, R. E. (ed.), 1970. *Abortion in a changing world.* Vol. I, Columbia University Press, New York. A fine collection of papers dealing with all aspects of abortion.

Hardin, G. J., 1969. *Population, evolution, and birth control.* 2nd ed. W. H. Freeman, San Francisco, California. Superb collection of basic sources. The section on abortion is especially well done.

Kapp, K. W., 1963. *Social costs of business enterprise.* Asia Publishing House, Bombay, India. Considers the costs which have been externalized by industry and hence borne by the public at large.

Kistner, R. W., 1969. *The pill: fact and fallacies about today's oral contraceptives.* Delacorte Press, New York. A thorough and practical survey of the topic by a gynecologist.

Mishan, E. J., 1970. *Technology and growth; the price we pay.* Praeger, New York. An iconoclastic view of the economic cost of growth.

Mumford, L., 1967. *The myth of the machine; technics and human development.* Harcourt, Brace and World, New York.

Nisbet, R. A., 1969. *Social change and history.* Oxford University Press, New York. A philosophical and historical view of growth and social change.

Shepard, P. and D. McKinley (eds.), 1968. *The subversive science.* Houghton Mifflin, Boston. A superb collection of writings about the ecology of man.

Zimmerman, A., 1961. *Catholic viewpoint on overpopulation.* Hanover House, Garden City, New York. The Church speaks.

EPILOGUE

All animals and plants, whatever their habitat or form, follow one inexorable rule: their populations grow, slowly or rapidly, to some point of equilibrium with their environment, then level off in a change from steady growth to a steady state. This point of equilibrium, known as the carrying capacity of the environment, may be determined by food supply, moisture, nesting sites, competition, or various combinations of these, but it is inevitably reached. If the equilibrium is disturbed, the excess population is lopped off by the environment; directly, as in a patch of seedling ponderosa pine thinned by fire and competition for moisture and nutrients; indirectly, by social stress which in overcrowded animal populations may break down normal reproductive behavior.

For thousands of years, man was little different. While he had tools which certainly differentiated him from other animals and unquestionably gave him a competitive edge, the margin was not marked and his numbers reflected fairly closely the carrying capacity of his environment. But now this has changed, as we saw in the final chapters of this book. Man has manipulated his environment until it has reached a dangerous point of disequilibrium.

Most of the environmental problems we face today result directly from increasing population. Man demands more products that pollute more water and more air, makes more wastes than can be disposed of, takes up more and more suburban space, and puts impossible demands on overexpanded cities. A stabilized population would not solve all these problems overnight, of course, but it would give time to work out more permanent solutions.

Achieving a steady-state population is not of itself enough. A stable population, it is true, will allow a greater allocation of resources to attacking environmental problems, but far too much emphasis has been placed on the *technology* of pollution control. The truism runs, "the technology that has caused the problems can solve them," but as fast as old problems are solved, new ones appear. Clearly technology is not responsible for our current environmental woes any more than a hammer would be for a bruised thumb. The basic cause is man's lack of ecological sensitivity. Until man can become aware of his fellow organisms and the correspondence between their well-being and his, until the environment is regarded as a responsibility rather than an economic opportunity, population control can only partially solve environmental problems.

The first step toward developing an ecological sensitivity is to inform yourself. Just reading this book will give you an overview of environmental problems. Use the bibliographies provided to get more details about areas that interest you, read a good daily or Sunday paper, or a weekly news magazine. Appendix 1 provides a list of periodicals that treat man-environment interactions. Subscribe to one of these or read them occasionally in the library. Use your information in town meetings and public hearings in your community. *No* community, however idyllic, is without environmental problems. Interest yourself in these problems, then interest others; organize and make your feelings and opinions known to public officials and businessmen. Using just this approach the League of Women Voters with only 150 thousand members has had an effect quite out of proportion to this rather modest membership. If working with others, joining organizations, and trying to influence public opinion is not your way, you can still exert a telling effect by a number of purely individual actions.

Some environmental problems seem to defy the individual; suggestions on how to deal with these have been made throughout the book. But there are many opportunities for each of us to make a small effort that will have a cumulative effect. Throwing a beer can out of a car window is a trivial action; 100 million such acts become a national problem. Similarly, the short list of unilateral opportunities for action that follows may seem trivial, but there is no better way to develop an ecological sensitivity than to think about the environmental implications of ordinary daily activities and then exercise the free choice you have in these areas to preserve environmental values.

1. Repair and keep in repair leaky faucets. The steady drip of a faucet can easily waste four gallons of water a day.
2. Keep a bottle of water in the refrigerator for a cold water supply. Running the tap until water is cold may waste gallons for every glass.
3. Be reasonable in your demands on power, particularly during peak periods. Turn off air conditioners and outdoor lights when they are not necessary.

4. Don't burn household trash in the backyard; use waste collection services. Solid waste disposal is easier to control than the acrid smoke from backyard incinerators.
5. Put lawn clippings and fallen leaves into a compost pile or into the garden rather than burning them. This will recycle some of the mineral nutrients and reduce air pollution at the same time.
6. Keep your car properly tuned to reduce unburned hydrocarbon emission. Maintain pollution control devices.
7. Replace faulty, noisy mufflers and use your car horn only when absolutely necessary.
8. When you buy a car consider safety features and pollution control devices as well as luxury features and color schemes.
9. Avoid persistent and broad spectrum biocides around the house and garden; try for some natural control of garden pests by encouraging their natural enemies. Replace some of your lawn with shrubs that will attract birds and give them food, shelter, and nesting sites.
10. Use nonleaded gasoline in your car. It will soon be available from most companies in all areas.
11. Read the labels on the food you buy. While it is impossible and unnecessary to avoid all additives, the fewer in the diet the less risk to your health.
12. Refuse to buy clothing, accessories, or novelty items made from endangered animals; if in doubt, check it out.
13. Be aware of environmental values when you buy or build a house; make your requirements known to builders and realtors.
14. Leave your car whenever possible; walk wherever you can. Take some form of mass transport to work; if it is dirty or inefficient, bring pressure to bear on public officials to improve service. Vote for bond issues designed to upgrade or expand mass transit facilities.
15. Keep a litter bag in your car at all times and use it.
16. Use returnable bottles when you have a choice; avoid "overpackaged" food and goods when possible.
17. If you are single, seriously consider whether marriage is for you. Everyone is not suited for the responsibilities, as the divorce rate suggests, and there are professions which offer rewards for single people commensurate with marriage.
18. If you marry, limit the size of your family to two or three children. If you want more children, adopt as many as you have financial and emotional resources to support.

Time is growing short. Unless population growth is reduced to zero by the year 2000, problems which are just beginning to concern us now —air pollution, water and food supply, waste recycling, the inequitable distribution of affluence, and the lack of ecological sensitivity—will become impossible to solve.

The usual moral drawn from the fable of the tortoise and the hare is that steady efforts to solve problems will ultimately win out. But if we let the hare symbolize man's yearning for a "good life" and the tortoise represent steady population growth, another moral can be drawn: despite the slow pace of the tortoise there is a point beyond which the late-starting hare cannot possibly catch up, no matter how hard and fast he runs. We are probably within a few decades of a point in time where the sheer mass of people on the earth will make continuing degradation of the environment irreversible for the affluent *and* the deprived. While the cost of "catching up" is already staggering, it is not yet beyond man's reach. But if we should falter now, not only will the universal goal of peace and happiness for mankind slip from our grasp, but man's very humanity will become extinguished in a struggle for mere survival.

APPENDIX 1

Keeping informed of the current developments in man-environment inter-actions requires more than reading an occasional newspaper. There are many periodicals which, wholly or in part, are concerned with man and his environment. The following list is by no means complete, rather it includes several points of view. The subscription price is quoted merely as a guide, for many periodicals are intended for library subscription only. Many of these publications can be found in local libraries, most in college or university libraries. The asterisk indicates organizations whose main thrust is toward active participation in preserving the landscape or improving the quality of man's environment.

A. Professional journals, but often containing articles of interest to everyone.

American Forests
  The American Forestry Association
  919 Seventeenth Street, N.W.
  Washington, D.C. 20006
  $6/year (monthly)

Biological Conservation
  Elsevier Publishing Co., Ltd.
  Ripple Road, Barking
  Essex, England
  $15.60/year (quarterly)

BioScience
  American Institute of Biological Sciences
  3900 Wisconsin Ave., N.W.
  Washington, D.C. 20016
  $18/year to members of AIBS only (bimonthly)

Environmental Pollution
  Elsevier Publishing Co., Ltd.
  Ripple Road, Barking
  Essex, England
  $15.60/year (quarterly)

Environmental Science & Technology
  American Chemical Society Publications
  1155 Sixteenth Street, N.W.
  Washington, D.C. 20036
  $5/year to members of ACS; $7/year nonmembers (monthly)

Journal of Range Management
  The American Society of Range Management
  P.O. Box 13302
  Portland, Oregon 97213
  $15/year (bimonthly)

Journal of Soil and Water Conservation
  The Soil Conservation Society of America
  7515 Northeast Ankeny Road
  Ankeny, Iowa 50021
  $10/year (bimonthly)

Journal of Wildlife Management
  The Wildlife Society
  Executive Director, Suite S-176
  3900 Wisconsin Avenue, N.W.
  Washington, D.C. 20016
  $20/year (quarterly)

Nature
  Subscription Department
  Macmillan (Journals) Ltd.
  Brunel Road, Basingstoke
  Hampshire, England
  $48/year (weekly)

Pollution Abstracts
  Dept. Y–5
  P.O. Box 2369
  La Jolla, California 92037
  $70/year (bimonthly)

Science
  American Association for the Advancement of Science
  1515 Massachusetts Avenue, N.W.
  Washington, D.C. 20005
  $12/year (weekly)

B.   Periodicals intended primarily for nonspecialists.

Audubon
  *National Audubon Society
  1130 Fifth Avenue
  New York, New York 10028
  $8.50/year (bimonthly)
  Beautifully illustrated.

The Canadian Field-Naturalist
  The Ottawa Field-Naturalists' Club
  Box 3264, Postal Station "C"
  Ottawa 3, Canada
  $5/year (quarterly)

Environment
  Committee for Environmental Information
  P.O. Box 755
  Bridgeton, Missouri 63044
  $8.50/year (10 issues)
  Probably the broadest coverage of all these magazines.

The Futurist
  World Future Society
  P.O. Box 19285, Twentieth Street Station
  Washington, D.C. 20036
  $7.50/year (bimonthly)
  Exceedingly interesting speculations about the future.

The Living Wilderness
  *The Wilderness Society
  729 Fifteenth Street, N.W.
  Washington, D.C. 20005
  $7.50/year (quarterly)

National Parks Magazine
  *National Parks Association
  1701 Eighteenth Street, N.W.
  Washington, D.C. 20009
  $8/year (monthly)

National Wildlife
  *The National Wildlife Federation
  1412 16th Street, N.W.
  Washington, D.C. 20036
  $5/year to members (bimonthly)

Natural History
  The American Museum of Natural History
  Central Park West at 79th Street
  New York, New York 10024
  $7/year (10 issues)

New Scientist
  New Science Publications
  128 Long Acre
  London, WC2E 9QH
  $16/year (weekly)

Oceans
  Oceans Publishers, Inc.
  7075A Mission Gorge Road
  San Diego, California 92120
  $9/year (monthly)

Saturday Review
  Saturday Review, Inc.
  380 Madison Avenue
  New York, New York 10017
  $10/year (weekly)

Science and Public Affairs
  Bulletin of the Atomic Scientists
  935 E. 60th Street
  Chicago, Illinois 60637
  $8.50/year (10 issues)
  Primarily concerned with atomic energy and disarmament but increasingly
  oriented to broader environmental issues.

Science News
  Science Service, Inc.
  1719 N. Street, N.W.
  Washington, D.C. 20036
  $7.50/year (weekly)

Sierra Club Bulletin
  *Sierra Club
  1050 Mills Tower
  San Francisco, California 94104
  $5/year (monthly)

Whole Earth Catalog
  Portola Institute
  558 Santa Cruz Avenue
  Menlo Park, California 94025
  $8/year (6 issues)
  Devoted to "the power of the individual to conduct his own education, find
  his own inspiration, shape his own environment, and share his adventure
  with whoever is interested."

APPENDIX 2

The following organizations and those marked with an asterisk in Appendix 1 are representative of the many environmentally concerned groups which seek to improve the deteriorating relationship between man and his environment. For details of their programs, write to the organization directly.

American Shore and Beach Preservation Association
P.O. Box 1246
Rockville, Maryland 20850

Conservation Foundation
1250 Connecticut Avenue, N.W.
Washington, D.C. 20036

Defenders of Wildlife
731 Dupont Circle Bldg.
Washington, D.C. 20036
Mostly concerned with humane treatment of wildlife and elimination of unnecessarily cruel methods of trapping and capturing.

Ducks Unlimited
P.O. Box 66300
Chicago, Illinois 60666

Environmental Defense Fund
P.O. Box 740
Stony Brook, New York 11790
An organization of lawyers and scientists which undertakes legal action to bring about amelioration of environmental abuses.

Friends of the Earth
30 E. 42nd Street
New York, New York 10017
Emphasizes political and legislative aspects of environmental problems

Izaak Walton League of America
1326 Waukegan Road
Glenview, Illinois 60025

Nature Conservancy
1522 K Street, N.W.
Washington, D.C. 20006
Main interest is preserving examples of various environments: grasslands, forests and deserts, for educational purposes.

Trout Unlimited
5850 E. Jewell Avenue
Denver, Colorado 80222

World Wildlife Fund
910 17th Street, N.W.
Suite 728
Washington, D.C. 20006
Seeks to preserve endangered species throughout the world.

Zero Population Growth
330 Second Street
Los Altos, California 94022

INDEX

Illustrations are identified by **boldface** numbers, explanations of basic terms by *italicized* numbers; explanations which appear in the course of a discussion of several pages are identified by italicizing the entire discussion; illustrations occurring in the course of a discussion of several pages are not identified, except if they occur on either the first or last page.